METALLURGICAL COKE MAKING IN INDIA (2021)

INDIA'S UNIQUE JOURNEY IN BECOMING #2 IN THE WORLD

A MANIFESTATION BY

TARUN CHATTERJEE

METALLURGICAL COKE MAKING IN INDIA (2021)

INDIA'S UNIQUE JOURNEY IN BECOMING **#2** IN THE WORLD

A MANIFESTATION BY

TARUN CHATTERJEE

Complacency is a killer. From an exporting nation in 2021, India may become the largest importer in the world by 2023 if we fail to take CARE of our coke making plants.

First paperback edition September 2021

Cover design by Final Impressionz
(final.impressionz@gmail.com)

ISBN: 978-93-5530-022-5

DEDICATION

This book is dedicated to those unsung heroes of the oven-top men, the doormen, the machine operators, technicians, welders, fitters, masons and unskilled labour who toiled and unknowingly suffered while working in the most eco-hostile Coke Making Plants till the mid-80s until the new breed of Engineers, Scientists and Environmentalists declared "**ENOUGH IS ENOUGH**" and India started implementing new generation Coke Making Plants meeting the Environmental Laws and Global benchmarks.

CONTENTS

FOREWORD

Prof. Dr Samaresh Bandyopadhyay

MA, LLB, PhD, Gold Medallist (University of Calcutta), FRAS (London)
Former Professor and Head of The Department of Ancient Indian History and Culture, University of Calcutta.
Govt. of India's Representative (Ex.) to the Council and Standing Finance Committee of the Asiatic Society, Kolkata.
Vice-Chairman (Ex.), Numismatic Society of India, Varanasi, Uttar Pradesh.
Honorary Fellow, Ancient Sciences and Archaeological Society of India, Mysore. Karnataka.
Member (Ex.), Academic Council, Indira Gandhi National Tribal University, Amarkantak, Madhya Pradesh.
Chairman (Ex.), Indira Gandhi National Tribal University Centre Establishing Committee.
Recipient of the Highest Award of Government of Karnataka Sponsored Academy of Sanskrit Research, Melkote, Karnataka.
Principal Advisor, North American Institute of Oriental and Classical Studies, USA.

In 2016, I was extremely happy to introduce by writing the 'Foreword' Tarun Chatterjee's pioneering work entitled "***Hundred Years of Metallurgical Coke making in India: 1914 – 2014***" which earned praiseworthy popularity and was updated in 2018. This time Tarun has come up with the concept of collective mindset involving as many as twenty experts in the field and producing this exceedingly interesting work titled "**Metallurgical Coke making in India (2021) - India's Unique Journey in becoming #2 in the World.**" Instead of sitting idle at home, he and his team fruitfully utilised the most pathetic pandemic condition for preparing the work. Such inspiring efforts of him, his team of editors and the invaluable articles incorporated in the work definitely deserve high commendation. There is absolutely no doubt that the work will be of great help not only to the professionals of the Metallurgical Coke Industrial arena but also to the young learners of Engineering and Management.

I extend my hearty congratulations to Tarun Chatterjee and his team and wish them all the very best.

Samaresh Bandyopadhyay.

29.07.2021

(Samaresh Bandyopadhyay)
"NIL-KAMAL", 203 LAKE TOWN, BLOCK-A, KOLKATA-700 089.
PHONE: (033) 2534-9549 / 3918. MOBILE: 094323 33918
EMAIL: felicitationsb@yahoo.co.in

FOREWORD

Rajiv Asthana

BTech, Metallurgical Engineering, IIT Kanpur, 1975

Adjunct Professor. IIT. Bombay

At the time of my graduation, there were just two companies in India with Integrated Steel Plants - Tata Iron & Steel Company, Jamshedpur, and Hindustan Steel with plants in Bhilai, Bokaro, Burnpur, Durgapur and Raurkela. The total annual crude steel production capacity of India at the time was somewhere around 8 MTA. Therefore, mainly because of my passion for Steel Making, I had only two choices, of which I chose Hindustan Steel, and was posted to their Bhilai Steel Plant. After a short period of orientation, I got assigned to the Continuous Casting Department, an up-and-coming unit of the modernisation programme in Hindustan Steel. From then on I went on to commission more than 13 Continuous Casting Machines in my career both in India and abroad.

Unknown to me, Tarun Chatterjee, a Chemical Engineer from Jadavpur University (a university from where so many Steel Plant stalwarts seem to come) was also working with Bhilai Steel Plant since 1969 but being so senior to me I did not really have the pleasure of knowing him then. But in 1981 he moved to Visakhapatnam Steel Plant (VSP), a unit of the Rashtriya Ispat Nigam Limited. At VSP he completed his career as a dedicated Engineer/ Manager. His core expertise, no doubt is then Coke making. It was there that I got to know him well when I myself joined Visakhapatnam Steel Plant in 1989 as head of the Steel Melting Shop. And what a pleasure it was to meet and get to know him. A devout Coke Oven engineer, he also had command of the process intricacies not known to many, and discussions with him always lead to gaining immense knowledge. Therefore, I am confident in my statement that this book titled 'Metallurgical Coke Making in India (2021)' could not have been written in a better way by anyone else but him and him alone.

Coke Ovens (more correctly, Coke Oven batteries) are the BASE on which the whole structure of the Integrated Steel Plant is standing. For personnel working in the Coke Oven Department is very demanding. There are no sheds where one can seek refuge. All are in the open, completely open to the elements for both personnel and equipment, with hot, dusty and humid conditions and with danger lurking at every corner. Standing over the acidic water sumps reminds me of horror films where villains throw people into such tanks.

Control of emissions, the safety of the personnel, the high yield from Coking Coal to Coke, the productivity of the batteries, quality of the product, techno-economic of the whole process is more of a challenge in the Coke Oven Department than in other departments of the Integrated Steel Plant. As compared to the downstream processes, Coke Ones lack automation, have fewer tools for control of product quality, and to top it all, less technical literature is available to help/guide personnel for the betterment of the process. In the Coke Ovens, Process Control is more intuitive, and this is where the expertise of Tarun Chatterjee (and many other generations of Coke Oven Stalwarts) is on display.

There are a few articles available if one does a search on the internet, and there are also some textbooks available on the subject. In this book, Tarun Chatterjee has made the effort of collecting articles from others with expertise in various fields related to Coke Making. This is what has given a unique flavour to this book. Coverage of safety and environmental considerations, technology and Industry status in India as of today, refractories and many other topics are well covered. The future of Coke making and road map will surely generate a lot of discussions in the future.

01.08.2021 **Rajiv Asthana**

PREFACE

Dr B Sengupta

Former Member Secretary
Central Pollution Control Board
Ministry Of Environment, Forest & Climate Change

I visited Visakhapatnam Steel Plant (RINL) as the Senior officer of the Central Pollution Control Board sometime in 1991 to study the Air and water pollution issues of Coke Oven plants of Integrated Iron and Steel plants. My mission was also to study the first Coke Dry Cooling Plant (CDCP or CDQ) of India which is required for reducing toxic air pollutants and GHG emission from Coke oven plants. There I met Mr Tarun Chatterjee for the first time as the then Head of the Department of the largest coke oven and chemical complex. He took me around the plant and explained the technical features of this Environment-friendly waste energy recovery unit. Thereafter, through The Central Pollution Control Board, we had given a project to MECON to examine the feasibility of dry Coke quenching in coke oven plants and standard for other toxic Air Pollutants emission like Benzo(a)pyrene, Polyaromatic hydrocarbon etc. including implementation of this clean technology for the existing units and all future Greenfield coke plants.

As the Member Secretary, CPCB, I had invited all steel plants under SAIL, RINL, TISCO & MECON in a meeting to discuss the new draft standard of Coke oven plants recommended by MECON including CDQ. In that meeting, Tarun Chatterjee represented the upcoming Greenfield plant (KMCL, a joint venture of MMTC and the State of Orissa) at Duburi, Odisha. By and large, all Steel Plants had resisted the proposal of MECON to implement CDCP technology due to the high cost of investment, except Tarun Chatterjee, who stood firm and thoroughly justified the investment. At a later date, on the recommendation of the Peer and Core Committee of CPCB and CPCB Board, MOEF approved the new standard for Coke Oven and Issued the gazette notification under EP ACT 1986. Further MOEF under EIA 2006 while considering new greenfield Coke oven project of Iron and steel plants made CDQ mandatory for all Greenfield coke-making plants.

Tarun Chatterjee has been meeting me whenever he visits Delhi. On every occasion, he would speak passionately for stricter environmental controls and technological up-gradation. In 2016 he presented his book, "100 years of Metallurgical Coke Making in India." He followed up his conviction after that, taking the unique role of a self-appointed educator, travelling to all-steel plants at his own cost to create awareness about the historical developments of Metallurgical Coke Making in India, Quality, Safety and Environmental aspects including the new technologies and improved versions of CDQ.

On 15th June 2021, he spoke to me and explained his intention of making a world-class book on "Metallurgical Coke -Making in India 2021... India's unparallel journey to become 2nd in the world."

This time, He has invited papers on various aspects of Coke Making from specialists in the respective fields, currently working hands-on at different Steel Plants. His honest intentions and the respect he enjoys from the 'Coke Making Community', evoked prompt response from all concerned. In just six weeks, he is ready to go ahead with the book's publication. The book deals extensively in all the areas of present-day Coke making in India both under ISP and Merchant coke-making Plants. He has shared some of the papers on Plant Safety, Process Safety and Environmental Chapters, and I find them of the highest standard.

I congratulate everyone who contributed articles for this book and the editorial team who has worked very diligently under the hard taskmaster Tarun Chatterjee. I am sure, in times to come, this book will become a classic example for sharing knowledge. I do not doubt that such an endeavour will inspire people in other areas like Iron Making, Steel Making, Palletisation, Rolling Mills.

I appreciate the hard work by each author, the editorial team and wish Tarun Chatterjee great success with the book. I appeal to the Government and the Steelmaking Plants for its wider circulation.

02.08.2021

B. Sengupta

(Dr B Sengupta)
161, MEDHA APARTMENT, MAYUR VIHAR PHASE – I
EXTENSION, NEW DELHI
EMAIL: bsg161@gmail.com

PROLOGUE

Dr Sushmita Dasgupta

Retired Joint Chief Economist, Joint Plant Committee, Author of Economics of the Indian Steel Industry, Amitabh- The Making of a Superstar and Amitabh Bachchan – Reflections of An Image

I met Tarun Chatterjee some 12 years ago in a Seminar in a five-star hotel in Delhi. I was in service, and he retired into his Golden Blossom Apartment in Bangalore dovecot with his wife and loads of experience and knowledge of his subject and a vast range of many others. He also lived in his memories, vividly recalling both his work and leisure almost by the minute. I calculated that he must have an infinite volume of memory going into googolplex counts inside his brain cells. When we found out that both of us contributed to a website called the social mantra, we got chatting.

Mr Chatterjee came over to our home the following day for Poila Boisakh for a typical Bengali meal that had pastes of tapioca mixed with mustard sauce. Since then, developed a strong friendship of many shared interests of cinema, music, art but most of all, Coke Oven Technologies. He fascinated me as much as coal and coke did and soon, we had a set of common friends in Sanat Gupta, Kenji Saka, Yamada, Yesuda and others, many of whom continue to be friends even now. The Japanese friends were from Nippon Steel, posted in India and it is from our evening chats that I learnt far more about technology than I ever did at the office desk.

I am an economist, but one cannot achieve the desired results unless one is also in command of the principles of technology. Hence coke oven and the technologies of iron and steel making were very much up my street. In those days he was compiling his tome on 100 years of Metallurgical Coke Making in India, a book that was eventually released by Dr Aruna Sharma, the then Steel Secretary. He also made a presentation of the book to a packed hall of audience in the Committee Room of the Joint Plant Committee in Delhi.

The idea of the present book is to invite experts from a wider circle of India's finest coke oven builders, coal firers, clean up technologists and so on. Coke making is a gamut of work especially as the mineral can vary significantly in its compositions leaving each batch of charge a matter of nail-biting apprehension till the smouldering coke is finally loaded onto the cars. Technology innovations have worried as much with the management of variations in the composition of coal as they have been bothered about the carcinogenic emissions from the charge.

In those days when I met Tarun Chatterjee, he was much in favour of the non-recovery coke making process and hence his friendship with Nippon Steel deepened. Ever so fond of the Old Monk, Tarun-da, for most of the times for us travelled across the coal mines through Chhotanagpur and Chhattisgarh to take stock of properties of the deposits. He also travelled to Durgapur Projects Limited where the by-product ovens had collapsed and together with him, we wrote a paper for the West Bengal Government to reconsider a resuscitation with new generation non-recovery ovens with heat recovery and power generation. Dr B Sengupta ex Member Secretary Central Pollution Control Board also became a good friend of Mr Chatterjee as he compiled voluminous data on the effluents and emissions from coke ovens and the need to control them.

Literally, much water has flown under the bridge of coke oven effluent cleaning technologies but when I tried to bring these together under an umbrella, I realized that such technologies did not even stabilize because the new generation of metallurgists and chemical engineers in the privately-owned integrated steel mills were not so robust with the management of the materials when they mostly operated machines. The machine packed technologies that make materials appear innate and passive are anathema to correct yields in metals and their raw materials. Coke Ovens are about the conversion of coal to coke in the absence of air; that conversion is the most important, machines should follow the basic operations. Unfortunately for us in machine dominated production, we have lost our contact with the real substance, and which is the coal at hand.

The book should be read by technologists, administrators, policy economists and even those working on energy and climate thoroughly for the chapters contained between its covers offer a mine of knowledge upon which to build the next generation of technologies, automation, artificial intelligence, machine learning and even video games!

Dr Sushmita Dasgupta

ACKNOWLEDGEMENT

Tarun Chatterjee

Author of "Coal & Coke Manual 2002", "100 Years of Metallurgical Coke Making in India", 2018 and "Gambling with Life", 2021

I thank all the authors of the articles who at a short notice have contributed their articles sharing their rich experiences despite the difficult pandemic situation.

I am grateful to Dr Samaresh Bandyopadhya an eminent historian and Rajiv Asthana, professor at IIT Bombay for writing the FOREWORD of this book.

I am extremely grateful to Dr B Sengupta, ex-Member Secretary CPCB for writing the PREFACE.

I also thank Dr Sushmita Dasguta for writing the PROLOGUE of this book.

I am thankful to my friend Kenji Saka who inspired me to write a technical book on metallurgical coke making in 2014 and his successor Kumakura of Nippon Steel and Sumitomo Metals, assigning me to make a technical report on Coal Washery in 2018 during my vacation in the USA.

I am grateful to my team comprising of Jayanta Sen and Amrita Kirti and both besides contributing their articles have taken utmost care of very important functions, that is, spell-checks, type-setting, liaising with the press for printing and taking care of other commercial aspects of publication. Editing the technical articles has been exclusively my domain. Suresh Sinha, Susanta Halder and R K Sabat, in addition to contributing their articles have always helped me in providing valuable technical documents and up to date tables at very short notice. And last but not least I am very grateful to four gentlemen, two young ladies and a young student of Engineering College who all have guided and helped me at various stages of writing this book. They are octogenarian Bijoy Chatterjee of California, Dr Ranjan Choudhury of Australia, Pritish K Sen Editor STEEL TECH Kolkata, Anjan Chaudhury of Pune, Monica Bachchan of Metalogics Noida and Ramakrishna Batabyal of Shiv Nadar Institute Delhi.

08.09.2021

Tarun Chatterjee

tchatterjee701@gmail.com/

authortarunchatterjee@gmail.com

A Brief Introduction of Authors and Their Contributions

Chapter I – "**Introduction**" narrates the background of writing this book.

Chapter II – "**Emergence of India as the 2nd largest in Coal and Metallurgical Coke" by** Ms Amrita Kirti from Dhanbad is an article rich with fully updated details of all the Metallurgical Coke Making Plants belonging to Steel Plants, Merchant Coke Making Plants and CDQ Plants of India. The author is an Electronics & Communication Engineer from BIT, Mesra who has worked in Infosys, Mysore, Bangalore and NCS, Singapore. With her background in technical education and a sharp analytical mind, she has been able to assimilate the subject well. Shedding her initial reluctance, under my guidance and inspiration, she has produced a brilliant article.

Chapter III - This chapter has two articles:

- **Technology Transfer** narrates India's journey in Technology Developments. The author of the article is Suresh Sinha, a Gold Medallist in Chemical Engineering. He possesses vast experience of working on MECON and Paul Wurth
- "**India's Self Reliance in Engineering & Manufacture**" is written by Suresh Sinha and Tarun Chatterjee. The authors have done full justice elucidating India's Success Stories in making Metallurgical Coke in India.

Chapter IV - This chapter addresses the philosophy of Coal Blanding and Quality. This subject is of paramount importance for any Coke maker. This chapter contains three articles:

- **Development in Coal Blending for Prediction of Coke Quality** by G S Kishore, A Gold Medallist in Chemical Engineering from Andhra University and the Topper in M. Tech from IIT, Kanpur. The article has been immensely enriched by the author's long experience in SAIL, RINL, ISPAT etc.
- "**Coal Blending in the Age of Transition**" by Dr Robert John Boyd of New Zealand. The author is a powerhouse of knowledge and great experience in Coal Blending Modelling.

- **"Evaluation of Coal for Coke making** - A Perspective as input raw material for seven-meter-tall top charged battery" by Sanjit Kumar Seet, GM, RINL. This article presents an in-depth analysis of the subject.

Chapter V - This chapter handles a subject of prime importance. It is on safety, which concerns everyone. This chapter too consists of three articles:

- **"Process Safety in Coal Chemical Plant"** by Ashish Chakraborty, an M. Tech in Chemical Engineering with a long experience at DPL and TISCO Jamshedpur.
- **"Hazard Identification and Risk Mitigation in Chemical Recovery Coke Oven"** by Susanta Haldar. He is a Chemical Engineering Graduate from IIT Kanpur with long experience in SAIL & TATA-BHUSHAN STEEL Plant.
- **"Safety Protocol in Coke Oven Plants"** by Ramakrishna Batabyal. It is interesting to mention that he is a student of Shiv Nadar Institute, Greater Noida, UP, pursuing Electrical Engineering. He aspires to work in a Steel Plant. His article is very focused on the subject with authentic sources from the notifications issued by the Ministry of Steel.

Chapter VI - The topic of Environment is always a hot topic of discussion the world over. Unfortunately, Coke Ovens and Chemical Plans are environmental hostile plants. Both pollute air and water. This chapter on Environment presents four very important articles:

- **"The Environmental Challenges of Indian Coke Making and Future Technologies"** by K.D. Choudhury, Ex G.M, MECON. It gives the complete guideline and practical solutions for controlling Air and Water Pollution in the best possible ways.
- **"Steps Involved in Environmental Clearance of Industrial Projects"** by Dr Jayanta Moitra, Ex CPCB. He is now a Government Analyst in the Ministry of Environment, Forests and Climate Change. He has written about the most complex problems faced by entrepreneurs to obtain Environmental Clearance. He has written this article in simple and easy to understand the English Language.
- **"Smokeless Charging in 7.6 M Tall Top Charge Battery"** In Bhushan Steel Limited and **"State of the Art Wastewater Treatment Plant through BOD Route."** These two articles have been contributed by Susanta Halder, an ever-dynamic leader is known for leading from the front.

Chapter VII - This chapter is all about the real shop floor experiences of a group of stalwarts with exceptional leadership qualities. It has as many as eight Articles:

- **"Challenges to Coke Makers and Some Non-Negotiable Issues to Achieve Best Results"** by Sujit Haldar. The author had a long and very rich inning at various units of TATA Steel.
- **"My Journey with LAM Coke Making"** by Susanta Halder. He has the exceptional experience behind him of commissioning Asia's tallest (7.6M) Battery.
- **"How Four Jumbo Coke Oven Batteries with a Large Coal Chemical Plant were Saved During the Super Cyclones Hudud in 2014"** by R Nagarajan of Vizag Steel Plant. He is an outstanding Engineer. In his article, he has narrated blow by blow account of the blood-chilling experience of facing the Super Cyclone and his valiant battle against it with a handful of very committed executives.
- **"Energy Conservation Initiatives in Steel Plants** "by S.P Bhargava. He is a veteran expert on Energy Management backed by long years of experience in SAIL, RINL and JSW.
- **"The Journey of Electrical System in CO & CCP of RINL – VSP"** by the humble and modest Electrical Engineer S N Singh. He has the credit of implementing a lot of changes in the electrical systems by scrapping the old Russian Modules bringing reliability and a higher degree of safety standards.
- **"Journey of Coke Making at Jindal Steel & Power Limited and Future Prospects"** by RK Sabat. Sabat is one of the most outstanding Coke Oven Technologists I have come across. He, along with his associates of JSPL, Angul Unit, have added value to the book with their article on the latest Green Field Coke Making Complex.
- **"Commissioning and Running the Coke Oven Complex at in RINL / VSP"** by Venugopal Nabhi, a Gold Medallist in Anthropology from Behrampur University. He is the only Non-Technical Contributor. He has brought out in his article one very important aspect which normally does not come in the limelight. It is the HR perspective covering Human Management and Industrial Relations.
- **"Commissioning and Running of Coke Ovens"** by Deepak Agiwal, the Plant Head of JSSL. It is the first stamp charge Recovery type Coke Plant of Orissa, set up at Duburi by Jindal Stainless Steel.

Chapter VIII - This chapter touches upon another very important aspect of Coke Making i.e. The Refractory Repair and Outsourcing.

- **"The Refractory Maintenance Practice of Coke Oven Plant and Coke Drying Quenching Units"** by Maharaj Singh. The author's career path appears like a legend. He had worked his way up from the level of a mason in TISCO to the GM, MECON and thereafter 30 years at JSW. On the way, he picked up a lot of educational qualifications and gathered considerable experience. I am extremely fortunate to be associated with him.
- **"Repair & Maintenance Practices of Coke Oven Batteries"** by Tilak Choudhury. The author is the EXG.M. (SAIL-DSP). He has long years of work experience in Durgapur Steel Plant, and he is known as a trouble-shooter in handling critical refractory.
- **"Outsourcing in Coke Plants in India"** Many plants under the private sector have chosen to outsource the critical operation, maintenance and repair activities by outsourcing to competent contractors and hence Outsourcing has become the new working order. We have two articles written by JK Choudhury and the team of AC Patnaik with PK Mohanty and I had the privilege to work with them at different periods of time.

Chapter IX - This chapter is dedicated to Merchant Coke Making Plants.

- **Factors Affecting the Cokeries of Gujarat** by Dipen Chanchani. During my maiden visit to Gujrat in 2018, Dipen took me to visit the merchant coke-making plants of the coastal Kuchh region allowing me to learn innovative work done by the Indian designers converting top charging to stamp charging which helped the industry to get a fruitful economic advantage.
- **Development of Non-Recovery Coke Oven for Coke Making in India** by Subhendu Bagchi. The author writes on the design features of indigenously developed Non-Recovery Coke Ovens in his article.

Chapter X - Coal Chemical is a subject next to my heart. For the last few years, I have been contemplating writing a book on Coal Chemicals. Prof. August Kekule has been goading me to do it. But I developed cold feet imagining the magnitude of the subject. Moreover, the complexities of the subject needed huge funds and other resources which were beyond my reach. Therefore, I settled down to write an article instead.

- **"The World of Coal Chemicals and the Ground Realities in India."** Jayanta Sen, with his Chemistry background from Calcutta University, made some research work to assist me to write this article.

All Technologies ultimately lead to Business and therefore, Chapter X also includes two articles on the business aspect of Coking Coal:

- **"Supply Chain Logistics in Coal and Coke Trade"** by Wing Commander Amit Kumar Goswami. This veteran IAF Officer, who has knowledge of importing goods from abroad made month-long research on Importing Coal and came out with this brilliant article.
- **"Risk Management in the International Trade of Coal, Coke and Steel"** by P.P. Moharikar, Ex G.M. (Marketing), RINL. The author is my old friend of over four decades standing. He has contributed this well-articulated article to explain in brief the risk involved in International Trade.

Chapter XI - This chapter exclusively deals with the future need of Coke Oven Doctors in India. And the book has two articles, the first one is by Dr, Mukesh Kumar a very knowledgeable person holding a key portfolio in the Ministry of Steel and has written a very important article titled **"The Future of Coke Making In India."**

"The Concept of Coke Oven Doctors – Why and How to implement in India" Here I have given a new idea. There is no doubt that with the growth of the Steel Sector, the Metallurgical Coke industry will grow and that will lead to a new situation in India with the mixed age group of Batteries. It needs a different strategy to handle that situation towards higher service life more than 2 times what we are at it now.

Complacency is a killer. From an exporting nation in 2021, India could become the largest importer in the world in 2023 if we fail to take CARE of our coke-making plants.

Our Illustrious Authors

Amar Chandra Patnaik
acpattnaik@yahoo.co.in

Amit Kumar Goswami
authorgoswami@gmail.com

Amrita Kirti
kirti.amrita@gmail.com

Ashish Chakraborty
ashisc2221@gmail.com

Bhavendra Kumar Sahu
bhavendra.sahu@angul.jspl.com

Damodar Mittal
dmittal@angul.jspl.com

Dipen Chanchani
dipenchanchani@yahoo.in

Deepak Agiwal
deepak.agiwal@jindalstainless.com

Dr Jayanta K Moitra
emtrcjkm@gmail.com

Dr Mukesh Kumar
drmukeshkumar@gmail.com

Dr Rob Boyd
darci.international@gmail.com

Jayanta Sen
jayantasen1964@gmail.com

JK Choudhary
jkchoudhary@bengalenergy.in

KD Choudhury
kd@greencindia.com

PP Moharikar
ppmoharikar@gmail.com

Maharaj Singh
maharaj_singh@rediffmail.com

PK Mohanty
utkalprojects2009@rediffmail.com

Ramakrishna Batabyal
ramakrishnabatabyal@gmail.com

Rathinasabapathy Nagarajan
nagu3060@gmail.com

RK Sabat
rksabat@gmail.com

Sanjit Kumar Seet
sanjit.seet33@gmail.com

Shailendra Prasad Bhargava
bhargavashailendra@yahoo.com

Subendo Bagchi
souamtechnologies@yahoo.com

Sujit Kumar Haldar
sujithaldar.12@gmail.com

Suresh Kumar Sinha
sinha_sk24@yahoo.com

Surya Kishore Gurram
surya_kishore2002@yahoo.com

Susanta Kumar Halder
durga.halder@yahoo.co.in

SN Singh
snsingh@vizagsteel.com

Tilak Kanta Chaudhuri
tilak.kanta02@gmail.com

Venugopal Rao Nabhi
vnabhi@gmail.com

VM Shisani
vm.shisani@angul.jspl.com

I. INTRODUCTION

Tarun Chatterjee

There was the time up to the nineties when eligible bachelors were afraid of showing their girlfriends or would-be bride's family Coke Oven Department for fear of immediate cancellation of the marriage negotiations. At the end of the shift, almost everyone looked like ghosts. The cost of the proper washing being too high, those boys would generally remain dirty. Things are different these days when the Engineers can directly walk into a party, prim and proper, at the end of his shift.

India's journey in metallurgical coke making reaching the position of #2 in the world is unique in many ways. India makes around 43 million tonnes of coke per year of quality comparable with the best in the world. India is the only country in the whole World that adopts all technologies. For the non-recovery route from our CFRI design to improvised Dasgupta/Kunda Singh Design with Stamp Charging. India uses even the Chinese design, "High capacity horizontal and Vertical ovens with stamp charge along with Heat Recovery Steam Generation" which produces Electric Power as well. As regards Recovery Units, the Top Charge Batteries have undergone big changes. The oven height has been increased from 4 M to 7.6 M, Asia's tallest. The taller ovens have significantly added to improved productivity. From the turn of this century, Indian Steel Plants have preferred Stamp Charging in view of options available to use cheaper coals from larger sources. Technology providers from all over the world are part of this journey which includes MECON, Otto, Koppers, Simon Curves, GIPROKOKS, ACRE, Paul Wurth.

In the past, I wrote two books on the subject of Metallurgical Coal & Coke. The first one, titled "Coal & Coke Manual 2002" was written to inspire the Managers to turn around the perennially sick plant DPL (Durgapur Projects Limited). They did it in less than two years. Kenji Saka the Head of Nippon Steel & Sumitomo Metals in India inspired me to write the second book "100 Years of Metallurgical Coke Making in India 1914-2014." The book was subsequently updated in 2018 and presented to Dr Aruna Sharma, IAS, the then Secretary of Steel. Dr Sharma took a keen interest in the subject and advised me to add a chapter on Coal Chemicals and go ahead for wider circulation. Accordingly, I took upon myself the task. In the next few months, I travelled to almost all steel plants in Eastern India, RDCIS-CET-MTI at Ranchi, IIM Kolkata, Duburi, Visakhapatnam etc and made presentations attended by more than five hundred professionals. I intended to travel to other regions, too, viz, Central and Western India.

But I could not do it. Suddenly, lock-down was imposed on entire India. Ever since India has been struggling with the Pandemic inflicted by Covid 19 virus.

On 15th June 2021, a new idea was born in my mind. I planned to write this book titled "Metallurgical Coke Making in India (2021), India's unique journey in becoming #2 in the World." This time I thought of writing this book with a collective mindset featuring articles from experts in various fields of activities. Accordingly, I requested around 20 professionals to contribute. Thankfully, within a couple of hours, I got a positive response from all of them. Their enthusiasm for the project was palpable, despite some of them had been passing through serious issues including losing the lives of near and dear to ruthless Covid Pandemic. Within the next 45 days, this Book is ready for final edit and publication.

I hope the community of Metallurgical Coking Coal makers will find this book interesting and useful.

II. EMERGENCE OF INDIA AS WORLD #2 IN COAL AND METALLURGICAL COKE

Amrita Kirti & Tarun Chatterjee

1. Executive Summary

INDIA HAS EMERGED AS THE 2ND HIGHEST PRODUCER OF STEEL.

India has crossed the benchmark of 100 Mt and is said to achieve 120 Mt in the current fiscal year 2021-22; 140 Mt in the year 2022-23. (Source: World Steel Report 2019-20).

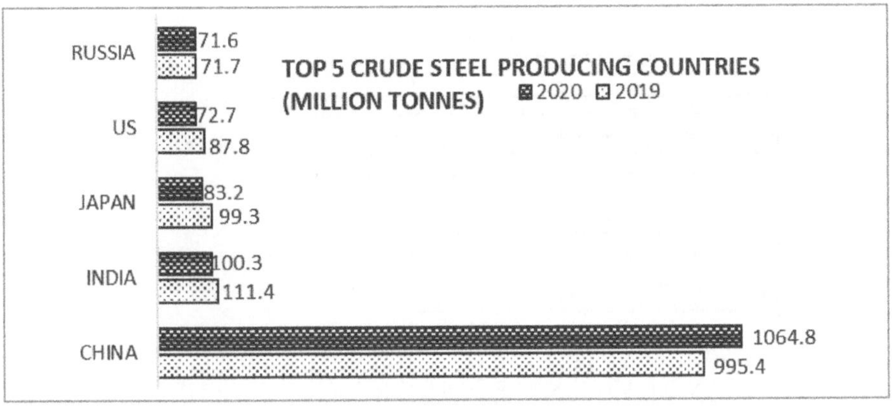

INDIA IS THE 2ND LARGEST PRODUCER OF COAL IN THE WORLD

India is the 2nd largest producer of Coal in the world producing 779 Mt per year. (Source: Global Energy Statistical Yearbook 2021)

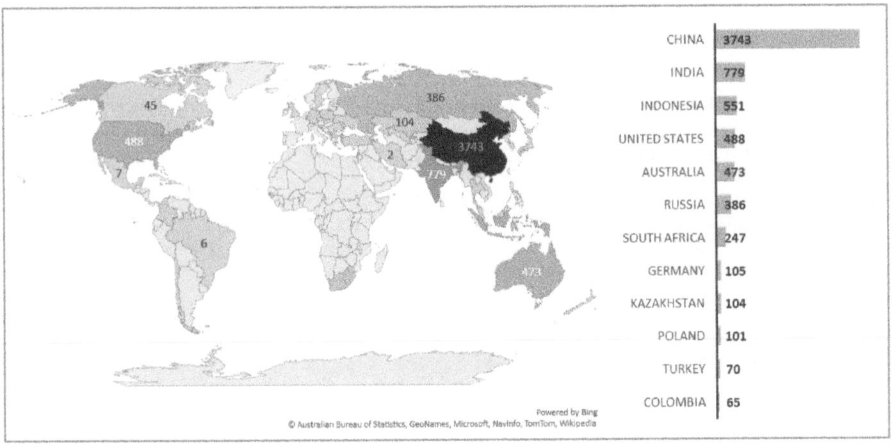

IMPORT OF COKING COAL IN INDIA OVER THE LAST 10 YEARS

Coking coal or Metallurgical Coal is the most expensive input for the production of Steel. India has insufficient coking coal reserves and thus has to import from other countries to meet the demand. In the last 10 years, with the growth of Iron and Steel

Industry, the import has become more than double. Moreover, the use of Stamp Charge Technology in the coke making process has opened up the opportunity to use a wide range of cheaper coals thereby restricting foreign exchange outflow. (Source: Provisional Coal Statistics 2020-21)

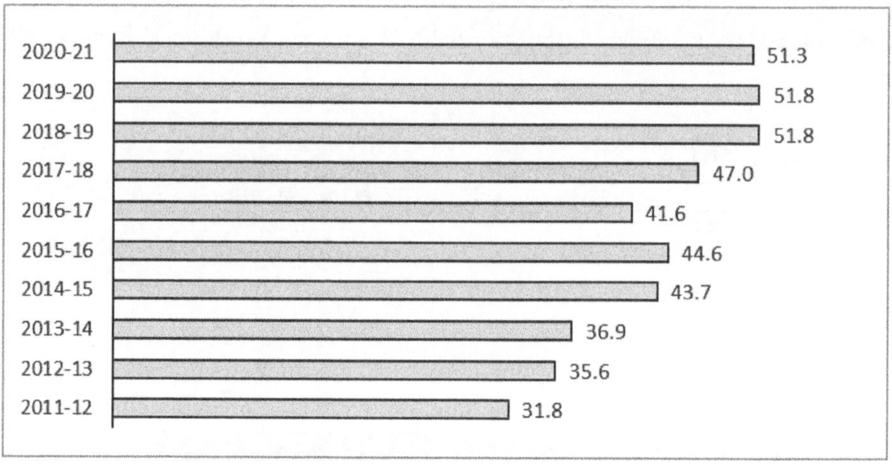

INDIA IS THE 2ND LARGEST PRODUCER OF METALLURGICAL COKE

India is the 2nd largest producer of Metallurgical Coke and the quality of the coke is as per the world benchmark. By 2020-21 India produced around 42.5 million tonnes of Met Coke which is expected to touch 43.5 million tonnes in the current fiscal 2021-22 and 45 million tonnes in 2022-23 because of new production plants and restart of the closed ones.

India which traditionally imported Met Coke has reached this position of self-reliance to meet the domestic demand and during the last three months has even started exporting. (Source: EIA)

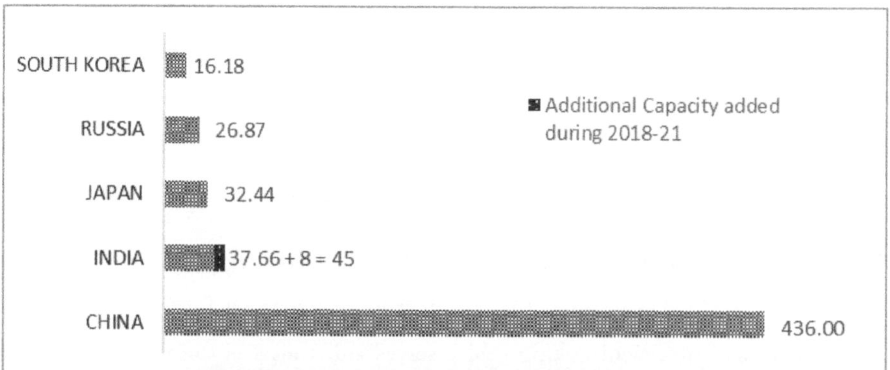

METALLURGICAL COKE PRODUCTION, 2018-19
(MILLION TONNES)

INDIA HAS NOW BECOME SELF-RELIANT TO MEET THE DOMESTIC DEMAND AND ALSO EXPORTS HIGH-QUALITY METALLURGICAL COKE FROM INDIA

With the use of better technology, India now produces high-quality metallurgical coke from coking coal produced within the country. Thus, the average import which was more than 3.5 Mt ten years ago has now dropped below the 3 Mt mark. The country's own production plants meet the domestic demand and also exports to some parts of the world.

During quarter 1 of the fiscal year 2021-22 of India, the Merchant Coke making Plants of Gujrat exported 1.5 Lakh tons of Metallurgical Coke to Southeast Asia, Europe, the Middle East, Pakistan and a few more countries. Despite the huge demand in the domestic market, the production has surpassed the consumption leading to export.

2. Coal Scenario of India

2.1. Coal Reserve of India

As per the World Energy Resources 2013 Survey, India accounts for about 7.0 % of Global Reserves (*source: World Energy Council*). India has a total of 319.02 billion tonnes of Geological Resources of Coal as of April 2018 **(*source: Geological Survey of India*)** out of which a significant number of reserves are of non-coking coal. The reserves of coking coal are very limited.

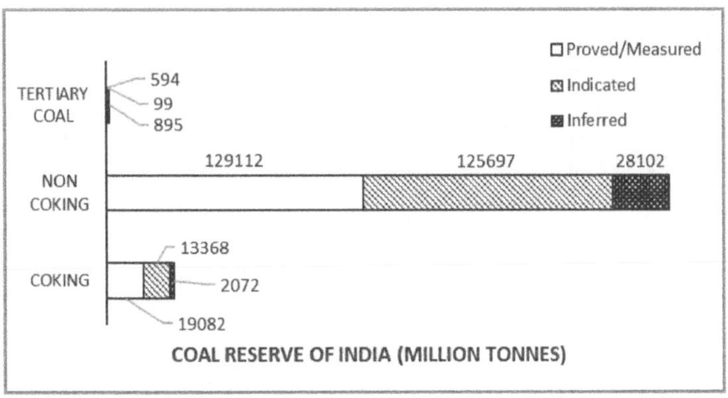

COAL RESERVE OF INDIA (MILLION TONNES)

During quarter 1 of the fiscal year 2021-22 of India, the Merchant Coke making Plants of Gujrat exported 1.5 Lakh tons of Metallurgical Coke to Southeast Asia, Europe, the Middle East, Pakistan and a few more countries. Despite the huge demand in the domestic market, the production has surpassed the consumption leading to export.

2.2. Annual Coal Production in India

In the year 2019-20, the total production of raw coal in India was 730.873 million tonnes whereas it was 728.718 million tonnes in 2018-19, showing a growth of 0.3% over the previous year.

TABLE 2.1 - TREND OF PRODUCTION OF COAL BY TYPE DURING LAST TEN YEARS

P – Production, G – Growth, All figures in million tonnes Source - Provisional Coal Statistics 2020-21								
YEAR	METALLURGICAL COAL		COKING COAL		NON-COKING COAL		RAW COAL	
	P	G (%)	P	G (%)	P	G (%)	P	G (%)
2011-12	16	-8.2	52	4.3	488	1.1	540	1.4
2012-13	15	-10.4	52	-0.2	505	3.4	556	3.1
2013-14	15	3.9	57	10.2	509	0.8	566	1.7
2014-15	14	-8.8	57	1.1	552	8.4	609	7.7
2015-16	14	4.0	61	6.0	578	4.8	639	4.9
2016-17	15	6.4	62	1.3	596	3.1	658	2.9
2017-18	34	122.1	40	-34.9	635	6.6	675	2.7
2018-19	35	3.5	41	2.5	688	8.2	729	7.9
2019-20	35	1.1	53	28.7	678	-1.4	731	0.3
2020-21	32	-9.0	45	-15.4	671	-1.0	716	-2.0

TABLE 2.2 - COMPANY-WISE RAW PRODUCTION OF COAL IN THE YEAR 2020-21

Q – Quantity, S – Share, G – Growth, All figures in million tonnes Source - Provisional Coal Statistics 2020-21						
COMPANIES	Open Cast			Under Ground		
	Q	S (%)	G (%)	Q	S (%)	G (%)
ECL	35.7	5.2	-13.4	9.3	29.8	1.1
BCCL	24.0	3.5	-9.9	0.6	2.0	-41.7
CCL	62.2	9.1	-6.1	0.4	1.4	-39.7
NCL	115.0	16.8	6.5			
WCL	46.9	6.8	-12.4	3.4	10.9	-18.2
SECL	135.2	19.8	1.7	12.2	39.0	-13.6
SECL(GP-IV/2&3)	3.2	0.5	20.4			
SECL(GP-IV/1)	0.0	0.0	-100.0			
MCL	147.5	21.5	5.7	0.5	1.7	-36.0
NEC	0.0	0.0	-93.0			
CIL	569.8	83.2	-0.4	26.5	84.7	-11.9
SCCL	46.2	6.7	-16.6	4.4	14.2	-49.1
JKML				0.0	0.0	-14.3
DVC						
IISCO	0.1	0.0	-66.3	0.1	0.3	-61.7
SAIL						
JSMDCL	0.1	0.0	16.2			
RRVUNL	15.0	2.2	0.0			
NTPC	11.0	1.6	-1.3			
WBPDCL	6.3	0.9	196.3			

COMPANIES	Open Cast			Under Ground		
	Q	S (%)	Q	S (%)	Q	S (%)
CSPGCL	1.5	0.2	194.5			
TSPGCL	2.0	0.3	21.9			
OCPL	2.0	0.3	99.5			
DPL	0.1	0.0	N.A.			
NLCIL	1.0	0.2	N.A.			
Public	655.0	95.6	-0.6	31.0	99.2	-20.5
TSL	5.9	0.9	14.1	0.0	0.0	-100.0
BALCO	0.0	0.0	-99.8			
CESC	2.1	0.3	6.2			
GMR						
HIL	0.3	0.0	-46.3	0.0	0.0	-100.0
JPVL	2.8	0.4	0.0			
SIL		0.0	N.A.	0.1	0.2	-74.8
SPL	18.6	2.7	-0.7			
RCCPL		0.0	N.A.	0.2	0.6	-5.5
TUML	0.1	0.0	-60.8			
OCL	0.0	0.0	-100.0			
AMBUJA	0.2	0.0	-57.8			
Private	29.9	4.4	-3.5	0.2	0.8	-85.8
All India	684.9	100.0	-0.8	31.2	100.0	-23.2

TABLE 2.3 – SECTOR-WISE PRODUCTION OF RAW COAL IN 2019-20

All figures in million tonnes, Source - Provisional Coal Statistics 2020-21			
SECTOR	COKING	NON-COKING	TOTAL COAL
PUBLIC	38.934	647.017	685.951
PRIVATE	5.853	24.280	30.133
ALL INDIA	44.787	671.297	716.084

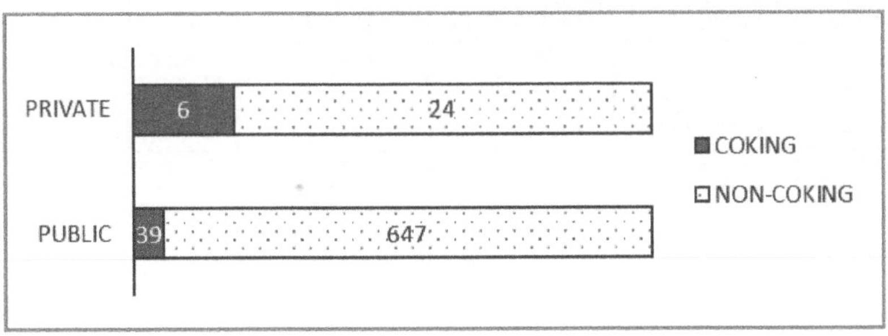

2.3. Coal Import in India

Coking coal accounts for 30-50% of the cost of production for steelmakers that use blast furnace technology. Traditionally the International coking coal price was controlled by Japan as per the JSM index (Japanese Steel Mills Price).

Indian Coal importers for decades were following this practice and also exposed to foreign exchange variations which always moved upwards. This Coal when carbonized in absence of air at series of Coke Ovens called Coke Oven Battery of various technological routes such as Top charge Recovery, Stamp charge Recovery or Stamp charge non-Recovery produces COKE needed for Ironmaking in Blast Furnaces.

TABLE 2.3 - IMPORT OF COAL TO INDIA IN 2020-21

Quantity in million tonnes, Value in million Rs, Source: DGCI & S, Kolkata		
TYPE OF COAL	QUANTITY	VALUE
COKING	52	454354
NON-COKING	164	706017
TOTAL	215	1160371

TABLE 2.4 COUNTRY-WISE IMPORT OF COAL TO INDIA IN 2018-19

Quantity in million tonnes		
COUNTRY	QUANTITY	SHARE
INDONESIA	117	46.94%
AUSTRALIA	47	18.80%
SOUTH AFRICA	42	17.09%

TABLE 2.5 - YEAR-WISE IMPORT OF COAL AND COKE TO INDIA DURING LAST TEN YEARS

Q – Quantity in million tonnes, V –Value in million, Source: DGCI & S, Kolkata										
YEAR	COKING COAL		NON-COKING COAL		TOTAL COAL		COKE & OTHER COAL PRODUCTS		LIGNITE	
	Q	V	Q	V	Q	V	Q	V	Q	V
2011-12	32	424692	71	363683	103	788376	2.4	47585		
2012-13	36	378398	110	490057	146	868455	3.1	56919	0.00065	10
2013-14	37	348319	130	574973	167	923292	4.2	67995	0.00127	24
2014-15	44	337656	174	707411	218	1045066	3.3	43806	0.00064	17
2015-16	45	282519	159	577819	204	860338	3.1	32684	0.00105	15
2016-17	42	412301	149	590013	191	1002314	4.3	54019	0.01912	433
2017-18	47	595226	161	789543	208	1384770	4.6	91525	0.01041	117
2018-19	52	720498	184	988707	235	1709205	4.9	120645	0.01937	403
2019-20	52	612668	197	914652	249	1527321	2.9	60257	0.05425	1074
2020-21	51	454354	164	706017	215	1160371	2.5	44689	0.01886	409

3. Metallurgical Coke for Iron Making

A world-class blast furnace operation demands the highest quality of raw materials, operation and operators. Coke is the most important raw material fed into the blast furnace in terms of its effect on blast furnace operation and hot metal quality. A high-quality coke should be able to support a smooth descent of the blast furnace burden with as little degradation as possible while providing the lowest amount of impurities, highest thermal energy, highest metal reduction, and optimum permeability for the flow of gaseous and molten products. The introduction of high-quality coke to a blast furnace will result in a lower coke rate, higher productivity and lower hot metal cost.

3.1. India Has Adopted the Latest Technologies Like Stamp Charging

In the traditional process of coke making, high-quality coking coal is used to produce coke using Top Charge Recovery technology. The volatile price of prime hard coking coal and its relative availability made the steel industry look for alternatives. Stamp Charge Recovery technology uses cheaper blends of coal. In this process, the coking coal is made into a cake using a metallic box and then using considerable mechanical strength is transferred into a coke oven.

In the last ten years, almost all Coke Making Plants under the private sector have adopted Stamp Charging technology. However, the plants under the public sectors, namely, SAIL, RINL, NMDC with their consultant MECON are yet to adopt the stamp charging technology. Right now, they are using Top Charging Recovery technology for producing high-quality coke. The full and comprehensive list of all Metallurgical Coke Making Plants captive to Integrated Steel Plants under both Public and Private Sectors is presented in Appendix 1 of the chapter.

3.2. Merchant Cokeries Have Also Adopted Stamp Charging

Numerous merchant cokeries have also adopted Stamp Charging technology and thus are making export quality coke. There are more than 100 merchant coke-making plants of small to medium capacities all over the country. The list of some of the prominent merchant coke-making plants is below:

TABLE 2.9 - LISTS OF MERCHANT COKE-MAKING PLANTS IN INDIA 2021

TYPE: SC - STAMP CHARGING, TC - TOP CHARGE, NR - NON-RECOVERY		
PLANT	**TYPE**	**CAPACITY, MTPY.**
GUJARAT NRE COKE, BACHAU & DHARWAD	NR/SC	1.2
TATA STEEL-HMC DIVISION, HALDIA	NR/SC	1.6
BENGAL ENERGY, KHARAGPUR	NR/SC	1.2
ENNORE COKE, HALDIA	NR/SC	0.26
VISA SUNCOKE LTD, KALINGANAGAR, ORISSA	NR/SC	0.4
TIRUPATI FUELS, VISHAKAPATNAM	NR/TC	0.14
FAIR DEAL SUPPLIES LTD, AMRELI, MUNDRA, PIPAVOV	NR	1.05
AUSTRAL COKE, MUMBAI	NR	0.52

PLANT	TYPE	CAPACITY, MTPY.
SAURASHTRA FUELS, MUNDRA	NR	0.5
BLA INDUSTRIES, ARAMBHADA	NR	0.3
GLOBAL COKE, DODAMARG, JAMNAGAR	NR	0.204
GREEN EARTH COKE, LUNVA	NR/TC	0.23
BLA COKE, MITHAPUR	NR/TC	0.16
BCCL, AURA, LODHANA, LOYABAD, SINLABAHL	NR/TC	0.38
CCL, GIRIDH	NR/TC	0.1
SESA GOA	NR/SC	
BRAMAH COKE		
VASUDAH		

3.3. High Capacity CDQ Plants

Coke Dry Quenching or CDQ is a waste energy recovery and environmentally friendly technology used for cooling red-hot coke produced from the coke oven. In this process, gas is circulated in an enclosed system to cool the red-hot coke. Another big advantage is the quality of Coke as it delivers coke with zero moisture and improved hot strength of Coke which have a great impact on the performance of Blast Furnace for Iron Making.

Indian plants under Private Sector have chosen the latest versions of high capacity CDQ that makes zero moisture high-quality coke with heat recovery as mandated by national environmental laws. On the other hand, the Public Sector plants example, SAIL, RINL, NMDC along with their consultant MECON have not yet upgraded. They are using old Russian technology with lower capacity (50 t/hour as against 100-200 t/hour by other technology providers). The comprehensive Status of CDQs in India is as below:

TABLE 2.10 - CDQs AT ISPs IN INDIA (*Source: Suresh Sinha, Sushant Haldar and R K Sabat*)

SL NO.	COKE OVEN PLANT	PLANT CAPACITY (MTPA)	CHARGING TYPE TOP/STAMP	CDQ CAPACITY (TPH)	NO OF UNIT IN PLANT	DESIGNED BY	OPERATION SINCE	PROJECT STAGE
1	RINL, VIZAG	4.2	Top	50 TPH	20	Giprokoks-Mecon	1992	
2	NINL, ODISHA	0.84	Top	50 TPH	3	Giprokoks-Mecon	2004	
3	IISCO, BURNPUR	0.88	Top	50 TPH	4	Giprokoks-Mecon	2013	
4	RSP, ROURKELA	0.768	Top	50 TPH	4	Giprokoks-Mecon	2014	
5	BSP, BHILAI	0.88	Top	50 TPH	4	Giprokoks-Mecon	2014	

SL NO	COKE OVEN PLANT	PLANT CAPACITY (MTPA)	CHARGING TYPE TOP/STAMP	CDQ CAPACITY (TPH)	NO OF UNIT IN PLANT	DESIGNED BY	OPERATION SINCE	PROJECT STAGE
6	JSW, BELLARY	3.4	Stamp	120 TPH	4	Nippon Steel	2013	
		3	Stamp	190 TPH	2	Sinosteel ,China		Under construction
7	JSW, DOLVI	3.5	Stamp	160 TPH	1	ACRE , China		Order Under finalisation
8	JSW, JHARSUGUDA	1.09	Stamp	150 TPH	1			Under discussion
9	TISCO, JAMSHEDPUR		Stamp	135TPH	1	Nippon Steel & Sumikin Engineering Co., Ltd	2011	
		0.7	Stamp	100 TPH	2	Paul Wurth IHI Co., Ltd.	2018	
10	TATA, KPO	1.85	Stamp	200 TPH	2	Nippon Steel & Sumikin Engineering Co., Ltd	2017	
11	TATA BSL ODISHA	1.25	Top	190 TPH	1	Nippon Steel & Sumikin Engineering Co., Ltd	2018	
		0.85	Stamp	120 TPH	1	Nippon Steel & Sumikin Engineering Co., Ltd		under construction
12	JSPL, ANGUL	1.9	Stamp	120 TPH	2	ACRE , China	By Dec 2021	Under construction
13	NMDC	1.68	Top	50 TPH	8	Giprokoks-Mecon		Ready for Commissioning

About the Author

Amrita Kirti, a graduate from BIT, Mesra (2006) in Electronics & Communication Engineering is a writer and editor working under the banner of Final Impressionz. She was once a software professional, developing beautiful .Net codes and worked for Infosys Technologies at Mysore and Bangalore. She then moved to Singapore and worked with NCS Technologies. Later, she jumped into the field of storytelling to pursue her passion. She has a couple of fiction books to her name. Apart from writing, she helps other authors fine-tune their work and provides pre-publishing services like constructional editing, copy editing, typesetting and cover design.

4. Appendix 1

LAM COKE MAKING PLANTS OF INDIA CAPTIVE TO IRON & STEEL PLANTS. 2021

MTPY: million tons per year; Source: Suresh Sinha, Sushant Haldar and RK Sabat)

PUBLIC SECTOR COKE OVEN PLANTS - SAIL

UNIT	COKE OVEN NAME	OPERATION START YEAR	SUPPLIER NAME	REVAMPING YEAR	SUPPLIER REVAMPING	OPERATIONAL STATUS	TYPE (Stamp or top charge recovery or non recovery)	COKE CAPACITY (MPTY)	Number of ovens	Coke weight per oven	Size of oven			Future Expansion plans	
										t/each oven	Height (m)	Length (m)	Width (m)	Capacity : MTPA	By year
BHILAI.	Battery # 1	1985	Giprokoks	-	-	Operation	Top charge, By-product Recovery type, PBCC	0.38	65	-	4.3	14.08	0.407		
	Battery # 2	1982	Giprokoks	-	-	Not in Operation	Top charge, By-product Recovery type, PBCC	0.38	65	-	4.3	14.08	0.407	Cold repair U/progress	
	Battery # 3	1990	Giprokoks	-	-	Operation	Top charge, By-product Recovery type, PBCC	0.38	65	-	4.3	14.08	0.407		
	Battery # 4	1992	Giprokoks	-	-	Operation	Top charge, By-product Recovery type, PBCC	0.38	65	-	4.3	14.08	0.407		
	Battery # 5	1965	Giprokoks	2007	-	Operation	Top charge, By-product Recovery type, PBCC	0.38	65	-	4.3	14.08	0.407		
	Battery # 6	1966	Giprokoks	-	-	Operation	Top charge, By-product Recovery type, PBCC	0.38	65	-	4.3	14.08	0.407		
	Battery # 7	1972	Giprokoks	-	-	Not in Operation	Top charge, By-product Recovery type, PBCC	0.38	65	-	4.3	14.08	0.407	Dismantled for rebuilding of same 4.3 mt	
	Battery # 8	1979	Giprokoks	-	-	Operation	Top charge, By-product Recovery type, PBCC	0.38	65	-	4.3	14.08	0.407	Will be closed shortly for re-building	

	Battery #9	1988	Giprokoks	-	-	Operation	Top charge and Recovery	0.77	67	-	7.0	16	0.407	
	Battery #10	1996	Mecon	2021	-	Operation	Top charge and Recovery	0.77	67	-	7.0	16	0.407	
	Battery #11	2013	Giprokoks	-	-	Under drying	Top charge and Recovery	0.88	67	-	7.0	16	0.45	
	Battery #1	1995	Giprokoks	-	-	Operation	Top charge and Recovery	0.5	69	-	5.0	15.04	0.41	
BOKARO	Battery #2	1976	Giprokoks	-	-	Operation	Top charge and Recovery	0.5	69	-	5.0	15.04	0.41	
	Battery #3	1998	Giprokoks	-	-	Operation	Top charge and Recovery	0.5	69	-	5.0	15.04	0.41	
	Battery #4	1996	Giprokoks	Durgapur, West Bengal	-	Operation	Top charge and Recovery	0.5	69	-	5.0	15.04	0.41	
	Battery #5	down	Giprokoks	-	-	Dismantled	Top charge and Recovery	0.5	69	-	5.0	15.04	0.41	
	Battery #6	1982	Giprokoks	-	-	Not in Operation	Top charge and Recovery	0.5	69	-	5.0	15.04	0.41	Closed for rebuilding
	Battery #7	1985	Giprokoks	-	-	Operation	Top charge and Recovery	0.5	69	-	5.0	15.04	0.41	
	Battery #8	1993	Giprokoks	-	-	Not in Operation	Top charge and Recovery	0.5	69	-	5.0	15.04	0.41	Rebuilding U/progress
	Battery #1	1991	Simon Carves	-	Otto India Pvt. Ltd	Operation	Top charge and Recovery	0.46	78	-	4.45	13.59	0.45	
DURGAPUR	Battery #2	1982	Mecon	-	-	Under rebuilding: will be in operation from June'13	Top charge and Recovery	0.46	78	-	4.45	13.59	0.45	
	Battery #3	1996	Otto India Pvt. Ltd.	-	-	Not in Operation	Top charge and Recovery	0.46	78	-	4.45	13.59	0.45	Rebuilding
	Battery #4	1987	Mecon	-	-	Not in Operation	Top charge and Recovery	0.46	78	-	4.45	13.59	0.45	Dismantled
	Battery #5A	1989	Mecon	-	-	Rebuilding	Top charge and Recovery	0.23	39	-	4.45	13.59	0.45	
	Battery #5B	2015	Mecon	-	-	Under Construction	Top charge and Recovery	0.23	39	-	4.45	13.59	0.45	
	Battery #6	1993	Otto India Pvt. Ltd.	-	-	Operation	Top charge and Recovery	0.46	78	-	4.45	13.59	0.45	
R.S.P.	I-A	1979	Mecon	2007 (rebulding)	-	Operation	Top charge and Recovery	0.2	35	-	4.5	13.59	0.4	
ROURKELA	I-B	1985	Mecon	2007 (rebulding)	-	Operation	Top charge and Recovery	0.2	35	-	4.5	13.59	0.4	
	2A/B	1995	Mecon	-	-	Not in Operation	Top charge and Recovery	0.4	70	-	4.5	13.59	0.4	Dismantled for rebuilding
	3A/B	1992	Mecon	-	-	Operation	Top charge and Recovery	0.4	70	-	4.5	13.59	0.4	
	4A	1987	Otto India Pvt. Ltd.	-	-	Operation	Top charge and Recovery	0.23	40	-	4.5	13.59	0.4	
	4B	1988	Otto India Pvt. Ltd.	-	-	Operation	Top charge and Recovery	0.23	40	-	4.5	13.59	0.4	
	5A	-	Otto India Pvt. Ltd.	2002	-	Rebuilding	Top charge and Recovery	0.8	40	-	4.5	13.59	0.4	Hot repair U/progress
	5B	-	Otto India Pvt. Ltd.	2002	-	Rebuilding	Top charge and Recovery	0.8	40	-	4.5	13.59	0.4	
	6	2013	Giprokoks	-	-	Operation	Top charge and Recovery	0.8	67	-	7.0	16	0.45	
IISCO.	8	1987	Otto India Pvt. Ltd.	-	-	Operation	Top charge and Recovery	0.46	78	-	4.45	13.59	0.45	All 1~9 CLOSED
BURNPUR	9	1991	Mecon	-	-	Operation	Top charge and Recovery	0.46	78	-	4.45	13.59	0.45	
	10	1982	Otto India Pvt. Ltd.	-	-	Operation	Top charge and Recovery	0.46	78	-	4.45	13.59	0.45	
	11	2013	Giprokoks	-	-	Operation	Top charge and Recovery	0.882	74	-	7.0	16	0.45	

NON-SAIL PUBLIC SECTOR COKE MAKING PLANTS

UNIT	COKE OVEN NAME	OPERATION START YEAR	SUPPLIER NAME	REVAMPING YEAR	SUPPLIER REVAMPING	OPERATIONAL STATUS	TYPE (Stamp or top charge recovery or non recovery)	COKE CAPACITY (MPTY)	Number of ovens	Coke weight per oven	Size of oven			Future Expansion plans	
										t/each oven	Height (m)	Length (m)	Width (m)	Capacity : MTPA	By year
RINL	Battery # 1	1989	GPK	-	-	In operation.	Top charge and Recovery	0.8	67	-	7.0	16	0.407	rebuilding shortly	
VIZAG	Battery # 2	1991	GPK	-	-	In operation.	Top charge and Recovery	0.8	67	-	7.0	16	0.407		
	Battery # 3	1992	GPK	-	-	In operation.	Top charge and Recovery	0.8	67	-	7.0	16	0.407		
	Battery # 4	2009	MECON	-	-	In operation.	Top charge and Recovery	0.8	67	-	7.0	16	0.407		
	Battery # 5	2020	-	-	-	In operation.	Top charge and Recovery	0.8	67	-	7.0	16	0.407		
NINL	Battery # 1	2004	Mecon	-	-	Not in Operation	Top charge and Recovery	0.8	67	-	7.0	16	0.407	U/Long SD	
NMDC	Battery # 1	2022	-	-	-	-	-	0.88	67	-	7.0	16	0.407		
	Battery # 2	2022	-	-	-	-	-	0.88	67	-	7.0	16	0.407		
DPL	Battery # 1	-	Otto India Pvt. Ltd.	Operation	-	-	Top charge and Recovery	0.5	30	-	4.5	13.4	0.4	ALL BATTERIES CLOSED	
	Battery # 2	-	Otto India Pvt. Ltd.	Operation	-	-	Top charge and Recovery		30	-	4.5	13.4	0.4		
	Battery # 3	Future Rebuilding	カールスチール	-	-	-	-		-	-	-	-	-		
	Battery # 4	Future Rebuilding	カールスチール	-	-	-	-		-	-	-	-	-		
	Battery # 5	1986	Otto India Pvt. Ltd.	Operation	-	-	Top charge and Recovery		40	-	-	13.43	0.4		

NON-PSU SECTOR COKE MAKING PLANTS															
UNIT	COKE OVEN NAME	OPERATION START YEAR	SUPPLIER NAME	REVAMPING YEAR	SUPPLIER REVAMPING	OPERATIONAL STATUS	TYPE (Stamp or top charge recovery or non recovery)	COKE CAPACITY (MPTY)	Number of ovens	Coke weight per oven	Size of oven			Future Expansion plans	
										t/each oven	Height (m)	Length (m)	Width (m)	Capacity : MTPA	By year
JSW															
Bellary CO-3	COKE OVEN-3	2008	MECC			In Operation	Stamp- recovery	1.5	4 X 56	18.34	4.3	14.08	0.5		
Bellary CO-4	COKE OVEN-4	2010	MECC STAMP HUDR			In Operation	Stamp- recovery	1.9	4 X 72	18.34	4.3	14.08	0.5	3	2022
Dolvi CO-1	COKE OVEN-1	2014	Sinosteel			In Operation	Stamp- recovery	1	2 x 55	27	5.5	13	0.5		
Dolvi CO-2	COKE OVEN-2	2018~2021	ACRE CHINA			In Operation	Stamp- recovery (except one Battery)	3	4 x62	34	6.25	16.1	0.55		
Jharsuguda	COKE OVEN	2017	Paulwurth			In Operation	Stamp - recovery	1.09	102	32	5.5	16.25	0.5		
JSPL															
Angul	COKE OVEN	2017	Sinosteel STAMP HUDE			In operation	Stamp- recovery	1.9	4 X 72	18.34	4.3	14.08	0.5	1.6	2023
Raigarh	Non-Recovery	2006	SSIT			In operation	Stamp non-recovery	0.8	16x11	35	1	13.4	3.5	1.1	2023
JSL															
Jajpur	COKE OVEN	2011	Shougang ,Beijing			In operation	Stamp- recovery	0.43	1 X 64	18.34	4.3	14.08	0.5		
AMNSI	COKE OVEN	Not started				Project	Stamp- recovery							1.35	2023
TBSL															
TBSL CO-1	COKE OVEN-1	2011	Shougang ,Beijing			In operation	Stamp- recovery	0.85	2 X 64	18.34	4.3	14.08	0.5		
TBSL CO-2	COKE OVEN-2	2014	Paulwurth			In operation	Top charge - recovery	1.3	1 X 74	45	7.6	20.08	0.55		
TATA															
KPO	B 1&2	2015	ACRE	-	-	In operation	Stamp Charge	1.5	176	27.5	5	16	0.5	3	
TISCO	B- 10 & 11	2015	ACRE	-	-	In operation	Stamp Charge	1.5	176	27.5	5	16	0.5		
	B-5,6,7	1989~1995	OTTO INDIA		ACRE	In operation	Stamp Charge	1	144	18	4.57	13.76	0.46	Revamping planned	
	B-8,9	1998~2000	OTTO INDIA			In operation	Stamp Charge	1	140	18	4.57	13.76	0.46		

III. TECHNOLOGY TRANSFER & INDIA'S SELF RELIANCE

III.1. Experience Of Coke Making Technology Transfer in India

Suresh Kumar Sinha

1. Introduction

At the time of independence, India had only two integrated steel plants viz. TISCO (now, Tata Steel) works at Jamshedpur and IISCO (now, under the SAIL umbrella) works at Burnpur. Both the plants were promoted by private entrepreneurs and were clearly influenced by available British technology. The Coke Ovens in these plants were relatively dwarf in height, probably between 2.5m to 4m heights, and of much smaller capacities. Notable features of these batteries were relatively high-tech by-product plants associated with these batteries. This is quite understandable because of the high demand for coal chemicals during the two world wars.

During initial industrial progress after independence under the leadership of Shri Jawahar Lal Nehru, all the coke ovens in integrated steel plants of public sectors under Hindustan Steel Limited (HSL-now SAIL) were installed with foreign technologies as a package for the entire plant. While Rourkela Steel Plant was set up by Germans, Durgapur Steel Plant by British and Bhilai and Bokaro Steel Plant were set up by Russians. This was the period when the Indian engineers of CEDB (Central Engineering & Design Bureau of HSL-now MECON) & HSL got the opportunity to learn the technologies from experts of different countries.

2. The MECON Story

The MECON story of the development of indigenous coke making technology is typical of a professional approach with success and failures. To start with, MECON purchased the rights to use the brick-work drawings of foreign proven design and then modified and improved by incorporating the changes suitably. This supports the fact that none of the designers throughout the world has started from scratch and only noticeable and incremental development and change are carried out by the competent and knowledgeable group to gain the desired result. This is almost true for any industrial product and Coke Oven and MECON story is not different. Further, development in any area needs the support of customers in a true sense to accept reasonable failures beyond the commercial terms. MECON has been lucky to have support from all PSU steel plants in this regard.

During this nascent period of technology assimilation, CEDB/MECON engineers got the opportunity to work with foreign experts in the areas of design & engineering, construction, heating-up and commissioning of coke ovens at different plants. This was further enhanced by abundant foreign training available at that time. The first attempt of transfer of technology and know-how was done in the late 1950s, when engineering documentation of coke oven batteries built at Bhilai by Giprokoks, USSR was made available to Indian engineers. On their part, Russian specialists trained and nurtured a new generation of Coke Oven technologists in India.

The first real break thorough of self-reliance and the development of indigenous technology in the Coke Oven area, came in the early 70s when the newly formed MECON (As a separate design and engineering company) was assigned the task of designing the new Coke Oven Battery No. 5A at Durgapur Steel Plant. Under the visionary leadership of the then MD of MECON, Shri KC Mohan, MECON signed an agreement with Durgapur Steel Plant for designing and installing the first indigenous coke oven battery. To start with, MECON acquired the rights to use the brick-work drawings of M/s Coppee Company, Belgium, the designer of other batteries at Durgapur. Under the leadership of Late Shri SN Wazir, a team of dedicated coke oven engineers took the challenging task of designing the first indigenous battery. The entire brickwork was modified and checked by physical modelling together with the hydraulic and thermal regimes of the battery. The laborious effort of the team resulted in calculating and sizing of the equipment, development of manufacturing drawings, development of foundations and structural drawings together with associating piping, electrical, instrumentation designs. During the 1970s also, Computer programmes in the FORTRAN language were developed to cross-check complicated calculations.

Finally, the design of the first indigenous battery emerged, which later resulted in the compilation as a book on "MECON design Coke Oven Battery" and few patents in the area. The book is still referred to for internal reference. The whole design concept was translated into the construction of Battery No. 5A at DSP, Oldtimers can still recall the sincerity, dedication, carefulness and labour put in the design, construction, heating-up and commissioning of battery. The historic date was 07.07.77 when the first MECON design Coke Oven Battery of 4.45m height was commissioned. Though the battery had its problems and could not survive the designated life span, it served as a great experience and learning opportunity for MECON engineers for the future. To date, MECON is continuously and successfully executing many Coke Oven projects and is probably the only company in the country, who can offer an almost complete range of indigenous design, engineering and consultancy services in the Coke Oven field.

3. Russian Support

In August 1980, MECON made a major achievement by signing a license agreement with M/s Tyazhpromexport, Moscow for getting access to Soviet technology for 7m tall Coke Oven Battery and Coke Dry Cooling Plant (CDCP) of Giprokoks. By virtue of the license agreement, MECON got the challenging assignments of designing and installing 7m tall coke oven battery no. 9 at Bhilai Steel Plant under 4.0 Mt Expansion and battery nos. 1, 2 & 3 along with coke dry cooling plants at Visakhapatnam Steel Plant under Rashtriya Ispat Nigam Ltd. This was a major challenge for MECON to completely Indianize the Russian design drawings, complete the detail engineering, supervise construction and finally heat-up and commission 4 nos. 7 m tall batteries and 3 nos. Coke Dry Cooling Plants one after the other. The entire workforce (no computer support in those days) laboured entirely – often on long tours to the site. Opportunity to undergo training in Soviet plants and design institutes along with working together with Soviet experts at the site greatly benefited MECON engineers. Battery No. 9 at BSP, commissioned on 31.03.1988 was the first tall coke oven battery commissioned in India.

The Coke Oven engineers of MECON assimilated the knowledge and experience gained in the execution of both Greenfield and Re-building of Top charged Coke Ovens during the course of time and mastered the art of design, engineering, execution and commissioning of batteries for Indian Steel Plants. Incremental incorporation of various design features mainly related to pollution control and automation of the plant were also integrated with MECON design batteries over the years.

4. OTTO Story

After successfully installing batteries of 1 Mt stage at RSP, Dr C Otto Co., Germany passed on the know-how and technology to its Indian subsidiary Otto India Pvt Ltd. Otto India had developed a good competence and did a good business from the 1960s to 1990s. The company successfully designed and executed quite a few batteries in PSUs as well as TISCO. In the 1970s and 1980s, MECON and Otto India were in good completion for Coke Oven Projects. While, the parent company (Dr C Otto) merged with UHDE, Germany, the Indian subsidiary Otto India perished unexplainably.

5. Scenario Emerging in the 21st Century

The end of the 20th century saw the closure of many steel mills in Europe and North America due to concern for a cleaner environment. China emerged as a major player in steel production. The opportunity was well exploited by Chinese companies, who emerged as technology suppliers for Coke Ovens. The big European designers viz. Carl Still, Didier, Willputte, Koppers, Dr C Otto, Simon Carves etc. had either closed or merged with other companies having Coke making as their secondary business line. Thus, UHDE, Germany having inherited Dr C Otto, Carl Still, Thyssen Kruppe and Kopper's design could not penetrate the Indian market in Coke making so far. Further, the collapse of the USSR weakened and affected the dedicated design institute of Giprokoks, now geographically located in Kharkiv, Ukraine. Many companies marketed and offered the basic Giprokoks designs with few modifications across the globe. These include Vukhin, Russia; Kokso Projekt, Poland; Hutni, Czech Republic and Mecon, India.

The last decade of the 20th century and the beginning of the 21st century saw the change in Government policy with a liberal economic environment and many new private entrepreneurs entered the steel production in the Indian market. Jindal's and Bhushan's entered the scene in a big way and ambitiously planned for big capacity steel plants. The sudden growth of demand for new capacities in coke making in India attracted many Chinese technology suppliers.

6. Stamp Charging as the Favoured Technology

Due to scarce good quality indigenous coking coal, India has become largely dependent on the import of coking coal. To offset this disadvantage, Tata Steel resorted to Stamp Charged Coke Oven in the 1990s with Sarberg stamping technology integrated with OTTO design battery. Stamp charging is a 90-year-old technology, developed for utilizing relatively poor quality of coking coals of Saar Valley and other East European Countries. One stamp charged battery in India was installed in Sindri Fertilizer Co. in the early 50s. However, the renewed interest in stamp charged batteries started in India only with successful operation by Tata Steel due to its advantageously using their captive cheaper coal by the stamping method. Because of the non-availability of indigenous coking coal, all new entrepreneurs were naturally attracted to stamp charging technology. Chinese technology supplier's viz. ACRE, Shougang, SSIT, SEDIN etc. could market the stamp charged technology aggressively and build many Stamp-charged Coke Oven batteries in India. Due to its cost-effectiveness and faster implementation of projects, the Chinese technology suppliers were naturally preferred by all Indian companies. Even, a top charged battery was built on LUET, China design at DSP during the period. Anyway, the Indian plant operators showed a clear-cut preference for Stamp Charged Coke Ovens over top charged batteries.

The period also witnessed the installation of many non-recovery stamp charge batteries by private companies. Almost all of them were based on Chinese design. Only the Sesa plant in Goa was based on technology supplied by Kembla, Australia. The liberal economic policy of the Government also favoured the import of complete technology along with full plant & equipment.

7. Bhushan and Paul Wurth

In the first phase of the plant in both Bhushan companies, Chinese stamp charged batteries were installed on economic consideration. While the Rengali Jharsuguda (BPSL) plant had non-recovery batteries of SSIT design, the Meramandli Angul (Bhushan Steel Ltd) plant had recovery type batteries of Shougang Design Institute design.

But the expansion program of both plants saw an altogether different approach by the management of both companies. The management of both companies decided to entrust the design of coke batteries to a European company Paul Wurth as a package along with high-capacity blast furnaces. While Paul Wurth was an already established technology supplier having more than 100 years of experience in blast furnace design, coke making was a comparatively new area at that time. Paul Wurth was assigned to design the very best high-capacity battery incorporating all modern and state of art features. The oven dimensions chosen were also very much unconventional. Paul Wurth responded to the challenges and came up successfully at both the plants with two entirely new designs. While the Angul plant had a top charge jumbo battery of 7.6 m height, the Rengali plant had a stamp charge battery of 5.5 m height. In fact, the jumbo battery of

Angul is the second-largest battery in the world today. Both these batteries established the credential of Paul Wurth as a leading designer of coke ovens.

8. Future Ahead

Table 3.1 is given for quick reference indicating the technology know-how suppliers of coke oven plants including non-recovery plants operating in India.

In view of the above-described historical development and adoption of various designs and technology in Indian Coke Oven plants, the current trends and expected trends in near future can be visualised. Because of total dependence on imported coking coal and the high price differential between premium coking coal and inferior coal suitable for stamp charging, the stamp charging technology will always be a winner. The Stamp charging technology will continue to attract the investors in near future also.

MECON will continue to hold its fort in top charge batteries mainly for rebuilding projects in PSUs. Bhilai Engineering Corporation (BEC), with support from Giprokoks, has also developed good capability in the technological area and will be a major challenger in this area. Larsen & Toubro (L&T) has also tried to develop technological capabilities in Coke making field by associating with various Chinese designers and Paul Wurth in Coke Oven projects.

Chinese companies can offer both non-recovery and recovery type stamp charge batteries at a comparatively cheaper price and hence will continue to attract Indian investors unless a significant change in Government policy or a change in political scenario takes place.

Among European companies, the two companies that can offer the very best of technology are Uhde and Paul Wurth. Uhde has not been able to penetrate the Indian market so far, mainly due to the company's internal policy. Paul Wurth can offer both top charge Jumbo ovens and large capacity stamp charge batteries. They will certainly be preferred by investors looking for premier technology.

9. Conclusion

The major challenges before the coke making industry today are threefold to reduce the cost of production of coke; to produce coke in a clean and safe environment and to meet the quality expectation of blast furnace operators.

Development in coke making technology had been gradual and steady with stories of success and failures. Much technological development and innovative ideas by various designers have been incorporated in modern coke ovens and these are meeting the challenges of the pollution-free and economic scale of operation. Though many coke ovens in Western World have closed down due to environmental and economic reasons, new coke ovens will continue to be built and existing coke ovens will continue to be re-built and technology will not die until mankind finds an alternative material or process.

10. Technology & Know-How Suppliers for the Indian Coke Oven Plants as per July 2021

TABLE 3.1 - TECHNOLOGY & KNOWHOW SUPPLIERS FOR THE INDIAN COKE OVEN PLANT

PLANT	CAPACITY (MTPY)	SIZE & TYPE of OVENS/BATTERY	TECHNOLOGY SUPPLIER	REMARKS
RSP, Rourkela	2.4	Old 4.5.New 7M	OTTO. GPK MECON	Top Charge.Recov.
DSP, Durgapur	2.2	All 4.45 M	Koppe OTTO, LuetMECON	Top Charge.Recov.
BSL, Bokaro	3.6	All 5 M	GPK &MECON	Top Charge.Recov.
BSP, Bhilai	4.8	Old 4.3M.New 7M	GPK, OTTO,MECON	Top Charge.Recov.
IISCO, Burnpur	1.6	Old 4.45M.New 7M	MECON, GPK	Top Charge.Recov.
VSP, Vizag	3.2	All 7M	GPK.MECON	Top Charge.Recov.
NINL, Duburi	0.8	7M	MECON.	Top Charge.Recov.
Tata Steel, Jamshedpur	3.8	4.57M	OTTO.ACRE	Stamp charge Recov.
Tata Steel, KPO, Orissa	3	5M	ACRE	Stamp charge recov.
BHUSHAN Steel, Angul	2	4.3 & 7.6 M	Shougang, PaulWurth.	SCR +TCR
BHUSHAN Power, Jsgd	1.4	5.5M	PaulWurth.SSITChina	Stamp charge recov.
JSW, Bellary	4.25	NR+4.3M	Sinosteel.Sesa	NR+ SCR
JSW, Siscol	0.5	NR	Sinosteel.Sesa	NR
JSW, Dolvi	4	5.5 &6.2M	Sinosteel, ACRE.	Stamp charge recov.
JSL, Jajpur	0.45	4.3M	SHOUGHANG	Stamp charge recov.
VISA	0.4	NR	SSIT	NR
TISCO, Haldia	1.6	NR	SSIT	NR
Uttam Galva, Wardha	0.5		CHINA	Recovery

PLANT	CAPACITY (MTPY)	SIZE & TYPE of OVENS/BATTERY	TECHNOLOGY SUPPLIER	REMARKS
Usha Martin, Jmsdpr	0.4	NR	SINOSTEEL	NR
DPL, Durgapur.	0.6	4.3M	OTTO	Top Charge.Recov.
NMDC, Nagarnal	1.7	7M	BEC+GPK	Top Charge.Recov.
ESSAR	1.35	5.5M	SHOUGHANG	Stamp charge recov.
NECO Jayswl, Raipur	0.4	NR	SSIT	NR
Electrosteel, Haldia	0.4	NR		NR
Electrosteel, Bokaro	1	NR	CHINA	NR
Sesa, Goa	0.3	NR	KEMBLA	NR
Bengal Energy, KGP	0.6	NR	SESA-KEMBLA	NR
JSPL, Raigarh	0.8	NR	SSIT	NR
JSPL, Angul	1.7	4.3M	SINOSTEEL	Stamp charge recov.
Marchent Cokeries.	2.85	NR	CFRI, Dasgupta etc	NR
LEGEND: NR: NON-RECOVERY.RECOV: RECOVERY SC: STAMPCHARGE.				
SCR:S TAMP CHARGE RECOVERY.TCR: TOP CHARGE RECOVERY.				

About the Author

Suresh Kumar Sinha was the topper of the 1977 batch of Chemical Engineering at Bihar Institute of Technology (BIT), Sindri and is the ex-GM of MECON. He completed his master's in chemical engineering from IIT Kanpur in 1981 and started his career in Coke oven as a Graduate Engineer in MECON, Ranchi, where he worked for 33 years. He also worked in Paul Wurth India for 5 years in a senior position. Presently, he is a freelance consultant.

III.2. India's Self Reliance as Coke Oven Equipment, Machinery & Refractory Supplier and The Technology Providers.

Suresh Kumar Sinha and Tarun Chatterjee

1. India's Self Reliance as Coke Oven Equipment, Machinery & Refractory Supplier.

Over the years Indian Coke making industry has become almost self-reliant in terms of vendors and supplies required for building the plants. Building the coke oven plant requires different types of engineering materials. The main plant units of coke oven plants are:

1. Coal Preparation Plant
2. Coke Oven Battery Proper
3. Coke Oven Machines
4. Coke Quenching (Wet/Dry)
5. By-Product Plant (Gas Treatment Plant)

1.1 Coal Preparation Plant

For the Coal Preparation Plant equipment, many Indian vendors have developed expertise and excellence to supply high quality and reliable products required for the process of receiving and storage of huge stock of different grades of coals, blending the coals in required proportions, crushing the coal blends to the required fineness (depending on the method of charging) and conveying system for carrying coal blends to coke oven battery for coal charging/coal cake preparation. While many suppliers for the conveying system are available over the years, there are few vendors available for specialized heavy material handling equipment. Elecon, HEC, McNally Sayaji, Metso, TRF and Thyssen Krupp, FL Smidth etc. have supplied very reliable wagon tippling equipment to Indian industries. Similarly, for yard machines viz. stacker, reclaimer etc. and crushing equipment also, the above-mentioned suppliers have been supplying quality systems and equipment.

1.2 Coke Oven Battery

The major material of construction for coke oven batteries is refractory and castings. Together, these two constitute about 60-70% of the construction tonnage of coke oven battery proper. The entire performance of the battery depends on the quality of these materials and the project schedule depends heavily on the delivery schedule of these materials.

About 25000-30000 tonnes of refractory material is required for the construction of one battery of about 0.8-0.9 Mt/yr capacity. Refractory material required is categorized in three broad categories viz. silica refractory, fireclay refractory and insulation refractory. Silica refractory for coke ovens needs to be very carefully chosen, because of its characteristic high thermal expansion and allotropic modifications at various temperatures. Additionally, silica refractory bears major thermal and mechanical stresses in operating coke ovens. Over the years, two indigenous manufacturers viz. OCL India Ltd. and TRL Krosaki Refractories Ltd. have genuinely emerged as quality suppliers of silica bricks for coke ovens meeting the global standards. Silica refractory for coke oven is also characterized by complexity of shapes and strict tolerance on dimensions. The capacity of both these manufacturers are limited and sometimes these are not able to meet the delivery schedule matching to the project schedule. In such cases and at times of commercial considerations (total project execution by foreign parties) imported silica bricks mainly from China and CIS countries are used. Another big consideration in the procurement of silica bricks is that the total tonnage of about 15000 tonnes for one battery is to source it from a single manufacturer having confirmed a single raw material base. Splitting of procurement is undesirable as it may lead to uneven expansion of bricks from different sources.

Unlike silica refractory, plenty of good manufacturers for both high and low alumina fireclay bricks indigenously. The manufacturers are geographically located across the country and can easily meet any requirement of fireclay refractory for coke ovens. Some notable suppliers of fireclay refractory suppliers in the country are Associated Ceramics, Brahma Refractories, Mahakoshal Refractories, Champion Ceramics, Manishri Refractories, Maithan Ceramics, Premier Refractories, Rajhans Refractories, Industrial Associates, SAIL Refractories, OCL and TRL.

Insulation bricks and castables have been supplied by Achint Chemicals, Associated Ceramics, Carborundum Universal, Premier Refractories, Maithan

Ceramics and other companies. Specialised sources for the supply of lightweight mica-based insulation bricks are located in the Bhilwara area of Rajasthan.

The second most important category of material required for the construction of coke oven battery is castings-majority being heat resistant type. Coke oven castings viz. doors, flash plates and door frames are characterized by a long length in the range of 5-8m and comparatively narrow width of weights around 3-5 tonnes apiece Over the years, Bhilai based companies viz. Simplex Castings, Beekay Engineering and Bhilai Engineering Corporation have specialized in the supply of major castings for Coke ovens. Simplex Castings have developed leak-proof doors with support from IKIO, a Japanese company. There is a need to develop more vendors for the supply of coke oven castings from large no of foundries available in the country.

Reputed Indian engineering companies are supplying all other materials required for coke oven construction viz. technological structures, pumps, pipes, valves and other items. There is some issue regarding the quality electro-of indigenously available helical springs required for battery anchorage. Because of the criticality of the item, imported German make springs have been used in many projects.

1.3 Coke Oven Machines

Coke Oven Machines is one area, where Indian companies need to improve their capabilities in designing and supplying modern hydraulic single spot oven machines. Traditionally, HEC has been supplying old generation electro-mechanical type pusher cars, charging cars, guide cars and quenching cars. HEC has also supplied modern pusher cars in collaboration with Schalke, Germany. Bhilai Engineering Corporation (BEC) and Beekay have established a good credential of supplying oven machines for top charged ovens in collaboration with CIS based engineering companies. For stamp charged ovens of Tata Steel, Tata Growth Shop (TGS) have supplied the stamping charging pushing machine (SCP)-the most complicated and heaviest machine in the coke oven complex. Except for Tata Steel stamp charged batteries, all other stamp charged batteries in the country are equipped with SCP supplied from Chinese companies as a part of the package from the technology supplier. For all upcoming stamp charging projects, Chinese machines have an edge on cost consideration. Chinese manufacturers viz. Dalian, Anshan etc. have also made a good reputation of supplying quality machines. Though Indian engineering companies have the required capability of manufacturing and assembling oven machines, no company In India has developed the desired skill, experience and capability to design the oven machines. In this respect, Indian companies are dependent on Chinese, CIS and German companies viz. Schalke, Hude, Koch, Thyssen Krupp, JM Henderson etc. These companies have mastered the art of integrating complicated mechanisms in a compact layout of moving machines.

1.4 Coke Quenching Process

Nowadays wet quenching is not a favoured option of coke quenching, though many improvements on account of pollution control viz. vapour spray and grit arrestors have been incorporated in the process. For dry quenching, all the equipment except the high-capacity mill fan is indigenously available. The high-pressure waste heat boilers can be supplied by Thermax and Thermal System. However, since a majority of CDQs are supplied by Chinese technologists, the equipment is also being supplied by Chinese companies.

1.5 By-Product Plant

The by-product plant (gas treatment plant) mainly consists of pumps, pipings, valves, tanks, heat exchangers and columns. They can be very well supplied by indigenous vendors. However, certain equipment and technologies still need to be imported, because of their criticality. So, exhausters are still imported, either of Chinese or German origin. The new plants are equipped with the de-sulfurization process to produce elemental sulphur. This entire technology and required equipment are imported including the reactor and the catalyst.

As can be seen above, the majority of equipment and material are available indigenously. However, Chinese suppliers have entered the Indian market in a big way, because of cost considerations. This trend may continue until there is a policy shift from the Government side to render the "MAKE IN INDIA" program more effective and successful.

2. The Technology Providers

Indian Coke Oven Plants for both Top Charge Coke Ovens and Stamp Charge (Recovery and Non-Recovery routes) continue to depend on the Design Knowhow from foreign sources like GIPROKOKS of Ukraine, ACRE, SINO STEEL, SHOUGANG, SSIT etc from PR of China and Paul Wurth of Germany. and GIPROKOKS, ACRE, NSSME of Japan for different capacities of CDQ. (Refer Table no 1 &2). The first CDQ plants came in RINL-Vizag Steel under a license agreement between Giprokoks and MECON and had many technical limitations as this was based on the first experimental CDQ in USSR. MECON got liberated from the license agreement and made several changes in Design & Engineering for the NINL project in 1997 and such changes were later incorporated in all CDQ Projects executed in SAIL, RINL & NMDC plants. The plants under private sectors chose higher capacity with state-of-the-art technology from NSSME. Paul Wurth, ACRE etc.

In the Merchant Coke Making Sector, some indigenous innovative work has been done by introducing the concept of Stamped Pellets charging which is working very well in many plants. This dependence on foreign know-how should not be a great matter of concern for example countries like Japan or Finland with a long history of Metallurgical Coke Making have also not entered in this field and wisely invested and patented the automation, computerized heating control and other allied activities.

TABLE 1: TECHNOLOGY & KNOW-HOW SUPPLIERS FOR THE INDIAN COKE OVEN PLANTS.

PLANT	CAPACITY MTPY	SIZE & TYPE OF OVENS/BATTERY	TECHNOLOGY SUPPLIER	REMARKS
RSP, Rourkela	2.4	Old 4.5.New 7M	OTTO. GPK MECON	Top Charge Recovery
DSP, Durgapur	2.2	All 4.45 M	Kopper OTTO, MECON	Top Charge Recovery
BSL, Bokaro	3.6	All 5 M	GPK &MECON	Top Charge Recovery
BSP, Bhilai	4.8	Old 4.3.New 7M	GPK	Top Charge Recovery
IISCO, Burnpur	1.6	Old 4.3.New 7M	OTTO. GPK	Top Charge Recovery
VSP, Vizag	4.2	All 7M	GPK	Top Charge Recovery
NINL, Duburi	0.8	7M	MECON	Top Charge Recovery
TISCO, Jamshedpur	3.8	4.57M	OTTO.ACRE-CHINA	Stamp Charge Recovery
TISCO, KPO, Orissa	3	5M	ACRE-CHINA	Stamp Charge Recovery
BHUSHAN Steel, Angul	2	4.3 & 7.6 M	Shougang, PaulWurth.	SCR +TCR
BHUSAN PowerJsgd	1.4	5.5M	PaulWurth.SSITChina	Stamp Charge Recovery
JSW, Bellary	4.25	NR+4.3M	Sinosteel.Sesa	NR+ SCR
JSW, Siscol	0.5	NR	Sinosteel.Sesa	NR
JSW, Dolvi	4	5.5 &6.2M	Sinosteel.ACRE.	Stamp Charge Recovery
JSL, Jajpur	0.45	4.3M	SHOUGHANG	Stamp Charge Recovery
VISA	0.4	NR	SSIT	NR
TISCO, Haldia	1.6	NR	SSIT	NR

PLANT	CAPACITY MTPY	SIZE & TYPE OF OVENS/BATTERY	TECHNOLOGY SUPPLIER	REMARKS
Uttam Galva, Wardha	0.5		CHINA	Recovery
Usha Martin, Jmsdpr	0.4	NR	SINOSTEEL	NR
DPL, Durgapur.	0.6	4.3M	OTTO	Top Charge Recovery
NMDC, Nagarnal	1.7	7M	BEC+GPK	Top Charge Recovery
ESSAR	1.35	5.5M	SHOUGHANG	Stamp Charge Recovery
NECO Jayswl, Raipur	0.4	NR	SSIT	NR
Electrosteel, Haldia	0.4	NR		NR
Electrosteel, Bokaro	1	NR	CHINA	NR
Sesa Goa	0.3	NR	KEMBLA	NR
Bengal Energy.KGP	0.6	NR	SESA-KEMBLA	NR
JSPL, Raigarh	0.8	NR	SSIT	NR
JSPL, Angul	1.7	4.3M	SINOSTEEL	Stamp Charge Recovery
Marchent Cokeries.	2.85	NR	CFRI, Dasgupta etc	NR
LEGEND: NR: NON-RECOVERY, SC: STAMP CHARGE				
SCR: STAMP CHARGE RECOVERY.TCR: TOP CHARGE RECOVERY.				

TABLE 2: TECHNOLOGY PROVIDERS FOR CDQ/ CDCP

PLANT	CDQ UNIT	NO OF CHAMBERS	CHAMBER CAPACITY T/HR	TECHNOLOGY PROVIDER	YEAR OF COMMISSIONING	REMARKS/ FEATURES.
NON-SAIL						
RINL	1	4	55	GPK	1989	First Generation CDQ of seventies design with Horizontal shaft boilers, electro-mech drives, 4 drive motors for lifters.
	2	4	55	GPK	1992	Vertical shaft boilers, hydraulic drives, single drive motor for lifter.
	3	4	55	GPK	1994	
	4	4	55	MECON	2010	
	5	4	55	GPK-BEC	2017	
SAIL						
IISCO	1	4	55	GPK-BEC	2013	Vertical shaft boilers, hydraulic drives, single drive motor for lifter
BSP	1	4	55	GPK-BEC	2015	
RSP	1	4	55	GPK-BEC	2013	
TATA STEEL						
Jamshedpur	1	1	135	NSSME.	2011	In Operation

PLANT	CDQ UNIT	NO OF CHAMBERS	CHAMBER CAPACITY T/HR	TECHNOLOGY PROVIDER	YEAR OF COMMISSIONING	REMARKS/ FEATURES.
	2	2	100	Paul Wurth	2018	In Operation
Kalinganagar	1	2	200	NSSME	2016	In Operation
Jharsuguda.	1	1	150			Under discussion.
JSPL, Angul	1	1	120	NSSME		
	1	1	120	ACCRE.		
NINL						
Jajpur	1	3	55	MECON	2008	MECON with improved features eg. Vertical shaft boiler, all hydraulic drives, a single drive motor for lifter and complete automation. Presently closed and can be restarted.
NMDC, Nagarnal	1	4	55	GPK-BEC	2017	ready for commissioning
	2	4	55	GPK-BEC	2018	

About the Author

Suresh Kumar Sinha was the topper of the 1977 batch of Chemical Engineering at Bihar Institute of Technology (BIT), Sindri and is the ex-GM of MECON. He completed his master's in chemical engineering from IIT Kanpur in 1981 and started his career in Coke oven as a Graduate Engineer in MECON, Ranchi, where he worked for 33 years. He also worked in Paul Wurth India for 5 years in a senior position. Presently, he is a freelance consultant.

IV. THE PHILOSOPHY OF COAL BLENDING AND QUALITY OF COKE

IV.1. Developments In Coal Blending for Prediction of Coke Quality

Surya Kishore Gurram

1. Introduction

Some of the life experiences in the field of Coal blending by the author during his journey in Coke oven in India and abroad are depicted here. These developments took place out of necessity for the survival of People/Plant/Businesses. These turned out to be small inventions and yielded great beak through in the operations of the organizations.

2. Development No.1

This is a story of a small merchant coke oven (GIKIL) in Europe in 2006. Before 2006 the coke oven used to work as captive coke ovens for the neighbouring blast furnace and thus not much stringent spec. for coke. However, with the recession in 2008, the blast furnace has been shut down and the merchant coke oven has to look for other global customers to sell its coke. The customer specs. were for very high-quality coke and the coke oven did not have the know-how of such a high-grade quality coke. The coke oven was also not equipped with Lab. Facilities to analyse and certify that what they produced is what the customer wanted. Further, the coke oven does not have any funds to spend to buy lavish laboratory equipment. At the same time job threat for 1000 employees of the coke oven is faced if the coke as per spec. is not produced.

In this background of the challenge of making coke as desired by Global customers, the management of the coke oven adopted SIX SIGMA methodology on the production data they have. They have identified the dependant variable of source coal quality for each product quality parameter. From the dependant variables of source coal quality, reliable prediction equations were developed for each coke quality parameter. With these model equations, blends were designed with the expected quality of coke. With the prediction equations, Coal selection has become very important and new guidelines were drawn.

As the coke oven was not having pilot oven testing facilities, Box testing of decided blends was done in the running coke ovens to be sure of the coke quality they produce. The coke obtained from Box testing was at variance to prediction equations. However, the blends were tried on an industrial scale and the results were encouraging.

At the time of this challenge, the coke oven did not have even a CSR testing apparatus and the sample used to be sent to DMT for process control. Such a result was just post-mortem of the process. Even that small help from DMT has been quite useful to perform by the coke oven to produce the required high-grade quality coke.

Thus, following this methodology GIKIL achieved the know-how of producing a very high-quality coke having a CSR of 68 in 2010 compared to 44 in 2006 when no prediction equations were available.

All this success story was presented at the AIST2011 conference held in Indianapolis, USA under the subject "Coal Blend Modelling and Coke Quality Prediction Studies – GIKIL's Success Story by Gurram Surya Kishore, Executive Director (Operation), Global Ispat Koksna Industrija"

These prediction equations were patented under Google patents - WO2012103600A2 - Coal blend models for determining coke quality - Google Patents: Inventor: Gurram Surya Kishore.

3. Development No.2

New coke ovens with stamp charging technology are under operation in India since 2012. They do not have linked up coal sources and all coking coal needs to be imported. This has thrown a challenge in the field of blending for optimum utilization of costly coals to produce coke with cheaper coals, but at the same time produce good quality coke. However, the prediction equations developed earlier were for top charged battery and they cannot be used for stamp charged batteries. Further, the coke quality from Box blends was also not consistent. The coke ovens installed were also not having pilot ovens for testing of blends. This has thrown further challenges to develop models for stamp charged. Batteries. A similar methodology of finding out the dependant variables of source coal and building reliable equations from data of stamp charging blends was done.

These were used to predict the coke quality for designing blends and put them into practice.

However, a necessity for a reliable tool for blend testing from the designed blends of equations was felt to improve the accuracy of blends and optimisation. This has led to the developing box which can give coke equal to the quality of industrial scale. Many prototype designs were fabricated and tried for producing the coke. However, all designs failed as the value of CSR obtained was at quite a good variance from the industrial scale. However, the author did not give up his efforts of development. He worked relentlessly to remove the shortcomings in design and ultimately by 2016 a Box that can produce coke exactly like an industrial scale was developed. The invention of Box has replaced the use of the expensive pilot oven for Blend testing as the results of Box testing are very near to Industrial scale compared to Pilot oven testing.

3.1 Methodology Of Box Testing

The coal blend is prepared for a quantity of 10kg and charged into the box. Bulk density is adjusted in Box as per requirement. The Box is charged into the oven after normal charging of the oven. The box is also pushed out along with the coke of a normal oven. The box is collected from the wharf. Coke in the box is tested for I20/I10/CSR/CRI.

A comparison of coke quality from this testing with Industrial-scale Blend is given below:

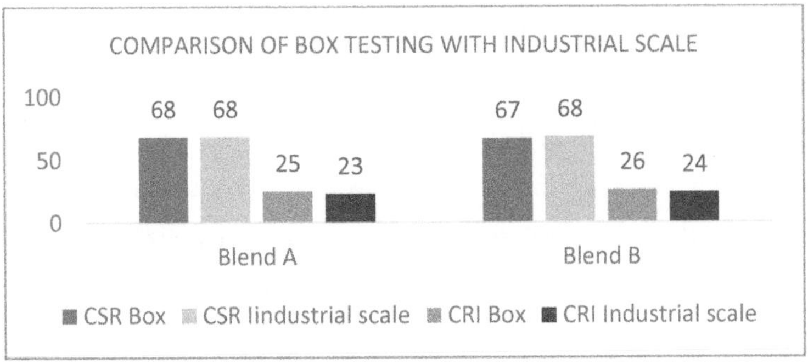

3.2 Advantages

i. Blend test can be conducted with a small number of source coals (Maximum 10kg) compared to 40kg for Pilot oven testing.

ii. Conduct blend test in the house without pilot oven and get results equivalent to industrial scale.

iii. Blend cost optimisation by implementing the results of such tests on an industrial scale.

iv. Investment for this testing is minimal and affordable for a Coke oven complex compared to investments in a Pilot oven.

v. Faster to conduct many variants of blends due to easier method of preparing blends.

vi. New source coals are selected for procurement after successful testing and thus give guarantee the performance.

vii. Gives vast scope for optimization of Coal blend cost with trials of different options of cheaper coals.

viii. Cheapest coal blend testing compared to Pilot oven testing.

ix. Results are near the value of industrial scale and are more reliable than Pilot oven.

Thus, this blend testing has resulted in optimizing the blend to utilize more cheap coals and achieve a reduction in coke cost and thus survive in the business. This technique of blend testing is being successfully done for designing blends, coal blend optimization and source coal quality assessment etc since 2016 in the coke ovens in India where the author has worked.

4. Development No.3

Ash chemistry of coal plays a vital role in achieving CSR of coke. However, many a time ash chemistry varies due to the ingress of contaminants in the process route and may result in sudden dips of CSR and affect Blast furnace production. Such dips have been observed with contamination of coal in its travel from port to Coke oven end in one Coke oven where the author has worked. To ensure that the coal charged into the oven does not have any contaminants which affect CSR it is necessary to know the ash chemistry of charge coal before it is charged into the oven. Calculation of ash chemistry is normally done by volumetric method or spectrophotometer. Both these methods are time-consuming and probably the coal might have been consumed by the time result is available.

To overcome this problem of the long lead time of ash chemistry measurement, a prediction method for ash chemistry was developed to know the ash chemistry of source coal within 4 to 5 hours of sample preparation. This prediction technique is based on the variance of the colour of ash with its ash chemistry. Coal with good ash chemistry will be white/creamy in colour whereas coal with adverse ash chemistry will be brown. Data of ash chemistry with photographs are collected and a prediction chart is developed based on the ash colour. The chart is used to predict the ash chemistry after making ash from a coal sample. Such prediction values are accurate enough for normal process control. With this tool in hand, the Coke ovens act immediately in case of ingress of contaminants in the system and can ensure that the quality as required is achieved.

5. References

1. **"Coal Blend Modelling and Coke Quality Prediction Studies – GIKIL's Success Story by Gurram Surya Kishore, Executive Director (Operation), Global Ispat Koksna Industrija"- Proceedings of AIST conference -2011.**
2. Google patents; WO2012103600A2 - Coal blend models for determining coke quality - Inventor: Gurram Surya Kishore

About the Author

Surya Kishore Gurram is the author of "Developments in Coal blending for prediction of Coke quality". He was a gold medallist of 1974 batch chemical engineers from Andhra University. He was a topper of the 1977 batch of MTech chemical engineers from IIT Kanpur. He started his career in Coke oven as a Graduate Engineer from the Durgapur Steel plant (1977 to 1983). Since then, he has worked in Vizag steel (1983 to 1996), Ispat Industries (1996 to 2003), GIKIL-Bosnia (2003 to 2012). Since returning home from his foreign assignment he worked as a consultant for coke ovens of BSL, JSPL(Angul), JSPL(Raigarh), TSBSL, and JSW (Dolvi).

IV.2. Coal Blending in An Age of Transition

Dr Rob Boyd

1. Background

There is a worldwide movement, increasingly strident in many places, demanding the cessation of anthropic CO_2 emissions in order to minimize the impact of "climate change". The complexities of the issue in terms of science, politics, solutions (real versus perceived) and the range of long-term impacts (recognised and unrecognised) are not the subject of this paper. But suffice to say, the general public in many parts of the world wants to stop coal mining as part of the decarbonisation of human activities.

As a result of this socio-political pressure, "BHP and Anglo appear to have embarked on a strategy to naturally deplete their metallurgical coal resource base as they shy away from capital spending on new projects and seek divestment of non-core coal assets amid a global push to decarbonization" (IHS Markit Inside Coal and Steel Raw Materials, 20th May 2020), and other miners may follow suit. Likewise, West Cumbria Mining in the UK has received severe opposition from green lobby groups (outside the local area) which has forced a government review of its permit; and despite the UK wanting to maintain domestic steel production and the indicative coal specification indicating good fuel rate savings.

According to a briefing to the European Parliament (EPRS_BRI (2020) 641552_EN), currently "the worldwide use of fossil fuels for electricity generation, transport, heating and industry leads to the release of ~37 gigatonnes (Gt) of CO_2 annually". About 3.3 Gt of CO_2 or about 9% of all CO_2 emissions can be attributed to the production of iron and steel.

However, the ongoing need for steel is well recognized, and some people are advocating an immediate change to hydrogen reduction. In reality, hydrogen steelmaking is still only at pilot scale testing and is a long way from technology commercialization. Figures from a recent conference (ESTAD 2019) indicated changing to hydrogen steel making would require: CAPEX ~€117bn; would require an entirely green source of large amounts of electricity (an additional 20% in the case of Germany; EPRS_BRI [2020]641552_EN); would mean steel costs ~200% of current costs; protectionist government policies would be required to avoid steel dumping. It seems to be generally acknowledged commercialization of hydrogen steelmaking is still "decades into the future" (BHP quoted in IHS Markit Inside Coal & Steel Raw Materials, 19th October 2020).

Meantime, around the world companies have very substantial installed capital assets for coke & ironmaking via the current blast furnace iron making route. These assets will not be easily replaced, even once new technologies such as hydrogen reduction mature, and particularly in developing economies such as India where virgin iron units are urgently required for infrastructure. (Compared with the utilization of electric arc furnaces in developed economies where recycling of iron is significant due to the high steel intensity in the economy).

The environmental imperatives imply a need to minimize the coke rate through good blend design and operational performance to minimize the CO_2 emissions per tonne of hot metal. But failure to replace metallurgical coal assets in certain geopolitical situations indicates the potential for a shortage of coking coals, driving a need to utilise a more diverse portfolio of potentially non-ideal specifications, analogous to the situation in China where government policies mandate the use of a wide variety of domestic coals supplemented by the import of premium material.

And therein lies a tension in designing a blend to minimize fuel rate (i.e., minimize CO_2 emissions) with a more restricted portfolio of coals. The core principles of designing a blend to minimize fuel rate will be the key topic of this paper, in the hopes the principles will assist in using a wider range of coal yet still make a high-quality coke.

2. Technical Details

The major drivers of fuel rate in a blast furnace are coke strength and coke ash.

1. Coke strength is a fundamental requirement to be achieved by the coke maker. In situations where the coke strength is too low, there is an increased risk of collapse of the burden and a chilled hearth. As coke strength improves, the fuel rate decreases (Figure 1). However, it is apparent the rate of fuel rate reduction appears to flatten when using cokes with a CSR greater than ~62. And there is some anecdotal evidence that suggests if the CSR is too high, the coke doesn't burn in the raceway, builds up in the deadman, and reduces furnace permeability resulting in an increased fuel rate.

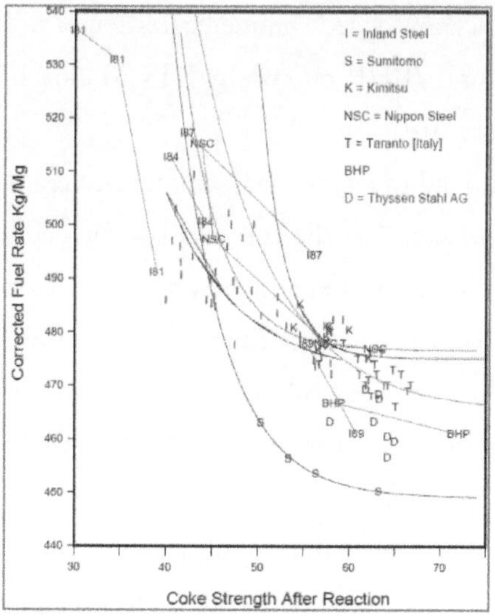

FIGURE 1 - CHANGE IN COKE RATE WITH CSR

2. Coke ash increases lead to increased coke rate due to:
    a. 1. use of fuel to melt the ash in the coke rather than for reduction
    b. increased flux rate to lower the melting point of the ash
    c. increased slag volume leading to reduced permeability in the blast furnace.

Coke makers around the world are well aware of the pre-eminent requirement for coke strength and have methods of blend design to achieve this. Many of the blend design schemes from around the world essentially have a rank parameter and a binding parameter (for example Figure 2).

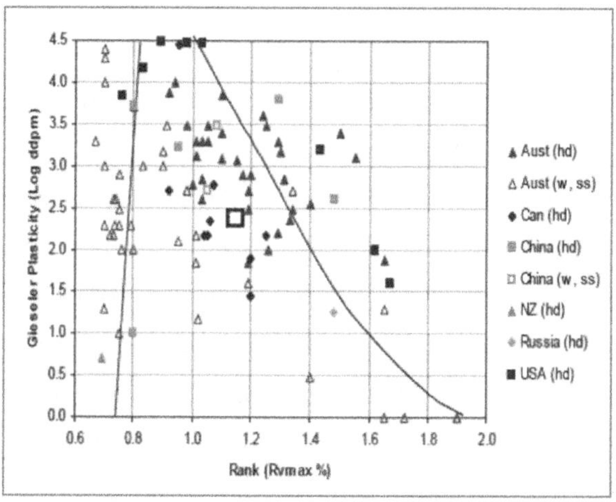

FIGURE 2 - FIGURE FROM ALS COAL QUALITY COURSE

As demonstrated in Figure 3, the primary driver of coke strength is coal rank (as inferred by vitrinite reflectance). Many design models use vitrinite reflectance as the rank parameter, but some use volatile matter on a dry ash-free basis. Coal rank drives the ordering of the coal structure, with aromaticity increasing as the rank increases. Larger polyaromatic units are inherently less reactive and hence the coke reactivity index is lower and CSR higher.

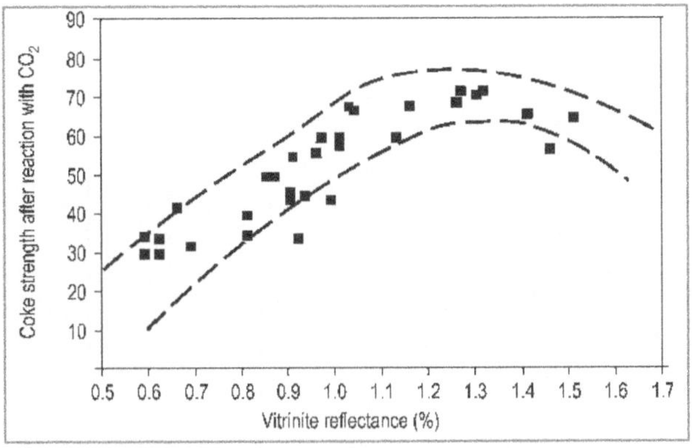

FIGURE 3 - INFLUENCE OF COAL RANK ON COKE STRENGTH
FROM DIEZ, ALVAREZ & BARRIOCANAL (2002)

In conjunction with the increased aromaticity, oxygen decreases in coal, and hydrogen and carbon increase as a proportion. Given the right proportion of hydrogen and carbon, coal exhibits thermoplastic properties, i.e. the ability to soften and swell on heating. Many parameters are used as indicators of thermoplasticity, including the proportion of reactive macerals, gieseler fluidity, Audibert-Arnu or Rhur dilatometer, Chinese G Index & Gray-King coking index. The importance of thermoplastic properties is in indicating a coal's ability to bind with other coals in a blend. Insufficient binding properties in a blend means the coke won't be well integrated and hence weak.

Premium grade coking coals often have a good combination of rank and thermoplastic properties (and good ash chemistry if high CSR). However, in China, coals ranging from semi-anthracite to high volatile coking coal may be used in a coke blend as dictated by centrally governed availability. Such experience suggests that, in the absence of "ideal" coals, a wider range of coals can be used as long as they are targeted into a "window" of rank and thermoplasticity for the total blend.

Usefully for India (NB also China), where the use of stamp charging of coke ovens is widespread, stamp charging offers a particular benefit in facilitating the use of a wide range of coals. As illustrated in Figure 4, the range of both volatile matter (rank parameter) and G-Factor (binding parameter, derived from dilatometer analysis) that can be used in stamp charging is much wider than for a gravity charge oven. Of most note is the wider tolerance for lower binding/ caking properties in a blend because coal grains are being mechanically pushed together, lowering the need for high swelling/ expansion to facilitate intergrain cohesion. Notably, many lower swelling coals are lower volatile matter, hence also improving coke yield.

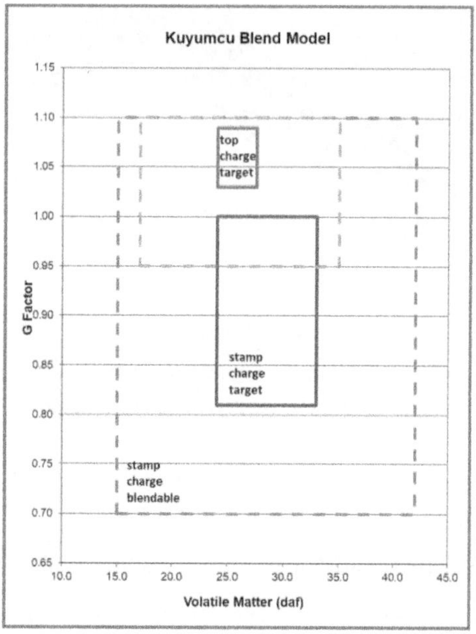

FIGURE 4 REPRODUCTION OF THE KUYUMCU COKE BLEND CHART
FIGURE FROM KUYUMCU, MPI PRESENTATION ON STAMP CHARGING

Theoretically, the right blend of rank and caking properties complying with a blend target will provide a good drum index and a coherent coke for CSR testing. For the CSR test, the other factor of importance is the chemistry of the ash. Ash chemistry high in alkalis will tend to catalyse the boudouard reactions that attack the coke structure (i.e. increase CRI), thereby weakening the coke and reducing the CSR. Numerous ash chemistry indices are available to assess this issue, however, the better indices take account of the ash yield (i.e. higher alkalis in low ash coal may still deliver fewer alkalis to coke than moderate alkalis in high ash coal) & coke yield. One useful index in this regard is the Modified Basicity Index (MBI) developed by Canmet Resources: 100 x

Ash(db) x (Fe_2O_3 + MgO + CaO + K_2O + Na_2O)/((SiO_2 + Al_2O_3) x (100 – Volatile Matter (db))

The second method of lower fuel rate is lowering the ash in the coke. Modelling ash is less difficult, with the blend ash being additive depending on the proportion of the various coals in the blend, and then recalculated to coke ash by taking account of the yield loss. As noted above, in general terms lowering ash will assist with lowering alkalis delivered into the coke, thereby lowering coke CRI and increasing the CSR.

2.1 Techno-Economic Modelling of Coke Blends

In future, if a diverse range of coking coals of various properties are available rather than large volumes of premium coals, a coke maker may be tasked with designing a blend using a more restricted portfolio of coals. In India, given some Indian government announcements, this may include a higher proportion of domestic Indian coals. To the author's knowledge, Indian coals tend to be higher ash, lower rank and high in inert macerals, implying a need for complementary import coals to optimize the blend and achieve the two key blend objectives.

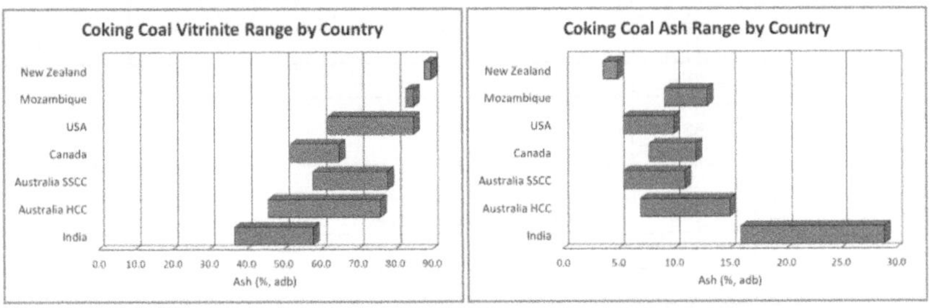

FIGURE 5A & 5B, NEW ZEALAND COAL ATTRIBUTES COMPARED TO WORLD COALS

The following discussion is New Zealand coal focused due to the author's background, but the concepts can equally be used with other coals when evaluating coal for use in a blend.

New Zealand coal has been sold into India for over thirty years because of two key quality attributes: high vitrinite (~90%) providing reactives for coke binding, and very low ash. New Zealand coal is also higher in sulphur, more comparable to some US type coals.

The current New Zealand product sold into India is Alpine coking coal, a semi-hard type coking coal. Semi-hard coking coals (defined here as approx. Ro(max) 0.85 to 1.00%) are also available from Australia, the USA, Russia and Indonesia. Buyers are faced with a choice of where to procure coal. But the purchase price is only one part of the decision tree; cheap coal that increases fuel rate in the blast furnace may add cost, and certainly will increase CO_2 emissions per tonne of hot metal.

A Generic Average SHCC specification has been calculated for coals of 0.85 to 1.0% Ro(max) from Australia, the USA, Russia and Indonesia using a database of coal qualities for these coals. A comparison of key specifications for comparison is shown in Table 1. Note MCI is another ash chemistry index, similar to the MBI Index; the higher the number the more alkaline the ash.

TABLE 1 - COMPARISON OF GENERIC AVERAGES FROM BY COUNTRY

	NZ ALPINE	AUST GENERIC	INDONESIAN GENERIC	RUSSIAN GENERIC	USA GENERIC
ASH (%, ADB)	3.6	8.6	7.8	9.7	7.4
VOLATILE MATTER (%, ADB)	34.0	31.0	36.2	29.2	35.8
SULPHUR (%, ADB)	1.60	0.57	0.72	0.39	1.05
TOTAL REACTIVE	95.6	85.5	96	84.4	68.8
RO (MAX) %	0.93	0.91	0.85	0.90	0.93
FSI	7.5	7.4	8.5	6.0.	7.5
FLUIDITY (DDPM)	300	2650	4420	10880	22700
TOTAL DILATATION (%)	53	100	175	141	190
MCI	0.015	0.022	0.075	0.062	0.032
INDICATIVE CSR	57	49	45	44	49

As shown in Table 2, in terms of specifications that drive coke strength, Alpine is superior overall, and this is reflected in the higher CSR relative to coals from other regions (although note, in particular, the Indonesian semi-hard coals are lower Ro(max), inherently detrimental to coke strength).

In terms of binding properties, while other coals are lower vitrinite, they may have higher fluidity and total dilatation. However, the Gray-King numbers of G7-G10 confirm the excellent binding capability of the Alpine coal.

Clearly, the ash of the Alpine coal is markedly lower than coal from other regions. A blend model has been constructed by the author over a number of years to model the change in coke quality and infer the consequential changes in coke/ fuel rate in the blast furnace. The model is comparative; it compares a base blend with a new blend. For parameters such as coke yield and sulphur partitioning in coke making, numerous equations have been gathered and an average of the calculated fuel rate changes is used.

TABLE 2 - COMPARISON OF COKE STRENGTH & COKE BONDING DRIVERS BY COUNTRY

	NZ ALPINE	AUST GENERIC	INDONESIAN GENERIC	RUSSIAN GENERIC	USA GENERIC
COKE STRENGTH DRIVERS					
RO (MAX) %	0.93	0.91	0.85	0.90	0.93
TOTAL REACTIVES	95.6	85.5	96	84.4	68.8
ASH (%, ADB)	3.6	8.6	7.8	9.7	7.4
MCI	0.015	0.022	0.075	0.062	0.032
INDICATIVE CSR	57	49	45	44	49
BLENDING BONDING DRIVERS					
TOTAL REACTIVES	95.6	85.5	96.0	84.4	68.8
FSI	7.5	7.4	8.5	6.0	7.5
FLUIDITY (DDPM)	300	2650	4420	10880	22700
TOTAL DILATION (%)	53	100	175	141	190

NEGATIVE VS ALPINE
STRONGLY NEGATIVE VS ALPINE
POSITIVE VS ALPINE

For drum indices and CSR, there are numerous equations available but none of them works universally. Accordingly, drum index and CSR predictions that appear outside the range of other results are discarded so that all results end up (for CSR) within +/-5 points of an average. It has been found over time that the predicted results are generally close to the results of coke testing of a blend by the customer. The model is considered to work well for predicting coke qualities, particularly noting the CSR test has a repeatability of +/-3.3%, reproducibility up to 8.5% (depending on the standard used).

Thereafter, the coke rate impact of changes in drum index, CSR, coke ash, slag bulk etc is evaluated using "rules of thumb" developed by integrated steel plants around the world. Where graphical data for parameters such as CSR as shown in Figure 1 are available, the data has been modelled with a best-fit curve and an equation used. Again, there are numerous rules of thumb for each parameter & sometimes equations, therefore an average figure for fuel rate impact is used.

Table 3 shows the result of blend modelling and the comparative value of replacing the Alpine product with the other "generic" semi-hard coals from various countries. As can be seen, only the Generic Indonesian specification is showing a significant lowering of predicted CSR, undoubtedly due to a combination of the lower Ro(max) and the relatively poor ash chemistry. The key differences in value are driven by the lowering of the coke ash, which as above drops fuel rate to melt the coke ash and also reduces the slag bulk.

TABLE 3 – RESULTS OF BLEND MODELLING

NOMINAL BLEND	BASE INC NZ ALPINE COAL	REPLACE NZ WITH AUST	REPLACE NZ WITH INDO	REPLACE NZ WITH RUSSIAN	REPLACE NZ WITH USA
GOONYELLA	30%	30%	30%	30%	30%
CARBOROUGH DOWNS	20%	20%	20%	20%	20%
HAIL CREEK	10%	10%	10%	10%	10%
TATA ALPINE	15%	0%	0%	0%	0%
BLACKWATER WEAK	25%	25%	25%	25%	25%
AUSTRALIAN GENERIC		15%			
INDONESIAN GENERIC			15%		
RUSSIAN GENERIC				15%	
USA GENERIC					15%
COKE YIELD CHANGE (%)		+0.70	+0.11	+0.85	+0.49
COKE ASH CHANGE (%)		+0.81	+0.74	+0.96	+0.66
COKE SULPHUR CHANGE (%)		-0.14	-0.12	-0.15	-0.08
HOT METAL SULPHUR (%)	0.041	0.035	0.036	0.034	0.037
M10	6.2	6.3	6.3	6.3	6.2
M40	84.7	83.3	83.2	83.2	83.4
CSR	65.8	66.1	61.3	63.8	65.3
COKE USE CHANGE KG/THM		+7.97	+12.85	+11.99	+6.71
COST CHANGE/THM		$0.66	$1.70	$1.39	$0.73
COST/TONNE OF NEW COAL		$7.63	$19.21	$15.94	$8.50
NB ASSUMES COST OF COKING COAL ~$118/T					

Of note, the hot metal sulphur (Table 3) is predicted to decrease slightly when replacing Alpine with the other generic coals, but not in proportion to the change in coke sulphur. There are two key reasons for this:

1. When the fuel rate per tonne of hot metal decreases, so does the total sulphur loading in the blast furnace.
2. When coke ash is decreased, if the flux rate is not reduced the slag becomes more basic, capturing a higher proportion of the sulphur in the furnace.

Note from Point 2 above, if hot metal desulphurization is installed, the flux rate could be reduced leading to further fuel rate savings through lower slag bulk, but with increased hot metal desulphurization reactants and potentially treatment time. The cost-benefit could be evaluated, depending also on the plants specific hot metal sulphur target.

3. Suggested Evaluation Process

The author is not aware of any coke plant that does not have a model to allow it to formulate blends, and likely many plants will have an integrated model capable of assessing the impact of changes in coke quality. It is suggested here such integrated modelling gives the excellent capability to assess the potential to formulate blends using a diverse portfolio of coals as some coal brands disappear. Not only can the ability of a blend to achieve key blast furnace targets be assessed, but optimization of coke strength and reduction of coke ash as means to reduce fuel rate (economic benefit) and thereby CO_2 emissions per tonne of hot metal can be evaluated as part of the decision process.

Obviously, modelling cannot fully replace coke testing of a proposed blend via a pilot oven that mimics the coke plant conditions or using a box charge method (as outlined in the accompanying paper by Shri Kishore). But the modelling stage has good potential to reduce the number of coke blend trials to a few key options.

Further, modelling may provide integrated steel plants with a means of communicating how the plant is working to minimize emissions in the years before new lower emitting steel making technologies become fully commercialized.

4. Conclusions

Given the current socio-political pressure against the mining and use of coal, it seems likely production from many countries will decline as mines are exhausted. However, the timeframe to fully commercialise & implement hydrogen steel making is decades from today. As a result, there appears to be potential for a period of difficult transition, when coke and integrated steel plants are forced to utilize a diverse portfolio of coals while also under pressure to minimising fuel rate for both economic and CO_2 emission considerations.

One approach to formulate an effective blend using a diverse coal portfolio is through coke blend modelling integrated with fuel rate change assessment. Using such a model allows plant staff to assess a range of potential blends that fit within a blend target to see how key blast furnace coke quality requirements can be met while also minimising fuel rate for both economic and emission reasons. Thereafter, a targeted series of coke tests can be performed to validate the modelling results ahead of full-scale oven trials.

5. References

- ACIRL Mine Skill Australia (7th Edition) (2011) Understanding Coal Quality. ALS Laboratories Coal Quality Training Manual.
- Diez, M, Alvarez, R. and Barriocanal, C. (2002) Coal for Metallurgical Coke Production: Predictions of Coke Quality and Future Requirements for Cokemaking. Int Jnl of Coal Geology V50, p389-412
- Kuyumcu, H (2002). MPI presentation on Stamp Charging

About the Author

Dr Rob Boyd is currently Technical Marketing Manager for BT Mining, New Zealand. He holds a Post Grad Diploma in Environmental Science and an MSc (Hons) & PhD in Geology. He started his career as a geologist in the Greymouth Coalfield in New Zealand, ending as the Senior Resource Geologist for the Greymouth & Reefton coalfields with Solid Energy. In 2006 he moved into the marketing team in Solid Energy & latterly BT Mining and has remained in that role for the last 15 years. During this time, he has visited 29 coke and integrated iron plants around the world, always learning something new.

IV.3. Evaluation Of Coal for Coke Making - A Perspective as Input Raw Material for Seven-Meter-Tall Top Charged Battery

Sanjit Kumar Seet

1. Abstract

A coking coal valuation mechanism based on several vital coal properties is proposed for categorizing single coals or coal blends. It is suggested that the proposed system fulfils the requirements of the coke-maker in that it is sensitive to small changes in coal composition and the parameters used can be correlated with coke quality, impact on battery life and cost- economics.

2. Introduction

At any integrated steel plant having a BF-BOF route, coal selection has been one of the thrust areas to ensure coke making process rendering coke suitable to blast furnace and avoid unsafe battery operation. In this paper, several perspectives of coal evaluations are discussed which incorporates BF coke quality, safe battery operation and maximization of value at coke oven and BF end.

3. Coal Composition and Structure

Coal science and technology of coke making has made great advancement for the last several decades; however, the views of experts in this field still carry divergent opinions regarding the structure-property correlation of coal and coal to coke transformation mechanism. Below is a pictorial diagram illustrating the basic composition of coal.

<table>
<tr><td colspan="6">COAL</td></tr>
<tr><td colspan="2">MINERAL MATTER</td><td colspan="2">PURE COAL
(C,H,N,O,S,N, TRACE ELEMENTS)</td><td colspan="2">MOISTURE</td></tr>
<tr><td>ASH</td><td>LOI</td><td>VOLATILE MATTER</td><td>FIXED CARBON</td><td>INHERENT MOISTURE</td><td>ADHERENT MOISTURE</td></tr>
</table>

FIGURE 1 SHOWING COMPOSITION OF COAL

Pure coal consisting of carbon, hydrogen, nitrogen, sulphur & other trace element bearing material is the prime part, which needs to be understood w.r.t., its structure-property correlation. Even after extensive efforts to understand the coal structure, the prediction of product properties after high-temperature pyrolysis such as carbonization has remained unresolved. This unpredictable nature of the coking mechanism and resultant coke quality from the originally understood coal structure forced commercial coke makers to resort to a long list of empirical test parameters for characterization.

4. Concept of Coal Characterization

Characterization of coal for its relative merits & demerits towards the metallurgical usage revolves around three major aspects i.e., Property of coal, influence on coke making and finally value obtained at the blast furnace end. This idea is pictorially depicted below:

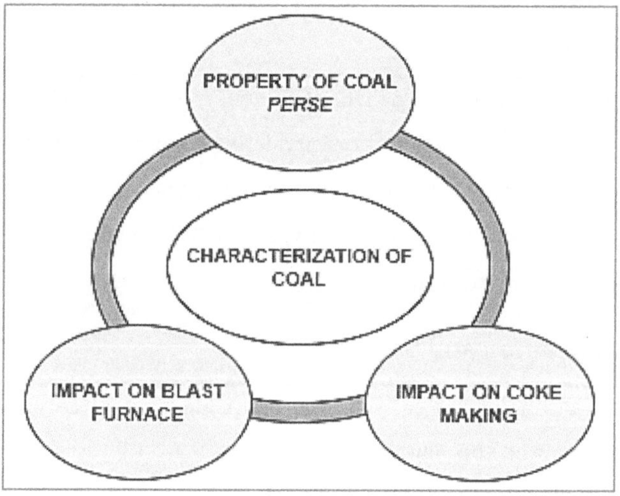

FIGURE 2 - THE THREE FACETS OF COAL CHARACTERIZATION

4.1 The Coal Property

Characterization of coking coal is done based on three aspects i.e., **Grade, Rank & Type.** The ***Grade*** of coal indicates the level of various impurities that can adversely affect the coke quality, the ***Rank*** of coal is the maturity of coal indicated by parameters such as MMR, Volatile Matter content & the ***Type*** of coal is related to the plastic or binding property during coking.

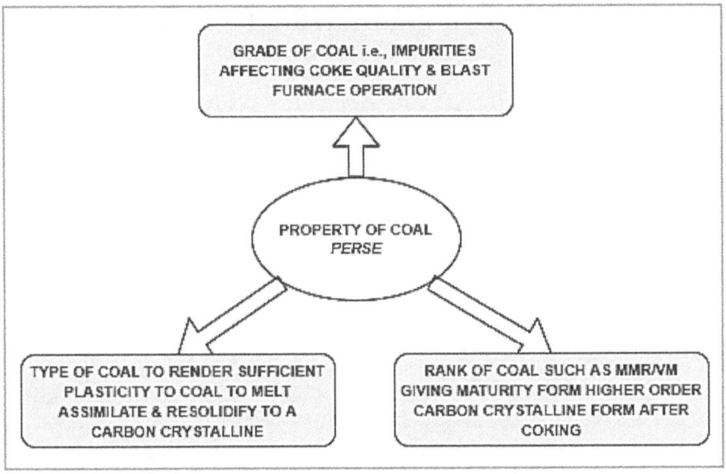

FIGURE 3 ELEMENTS OF COAL PROPERTY

4.2 Coal Assessment in Line with International Classification of Hard Coal

Based on the International Classification of coking coal by Type, the code of the coal and thereby interpreting its quality such as coking, strongly coking or non-coking is done as shown below.

A comparison of coke quality from this testing with Industrial-scale Blend is given below:

ESTIMATION OF CODE OF THE CANDIDATE COAL AS PER ECE CLASSIFICATION

Feed coal property		
IM	0.5	COAL A Tech Coal
VM db	21.0	
Ash db	9.5	
C	82.0	
H	4.3	
S	0.40	
LTGK	G 4	
CSN	7.00	
Max Dilation	13.21	

Output Coal Data for coal Classification as per Economic Commission of Europe (ECE)

			Candidate coal Code No.
Vmdaf	23.6		
GCV	8093		
CSN	7.00		
LTGK	G 4	LTGK	
Max Dilation	13		
1st figure	2nd figure	3rd figure	431
Class	Group no	Sub Gr. No.	
4	3	1	

FIGURE 4 ASSESSMENT OF COKING PROPERTY AS PER ECU

4.3 Prediction Of Mean Maximum Reflectance (MMR) From The H/C Ratio by The Regression Analysis

Immediately when coal is received at the laboratory MMR analysis cannot be done to know its rank value. Thus, based on the proximate and ultimate analysis data MMR prediction formula was developed using extensive data analysis using the H/C ratio.

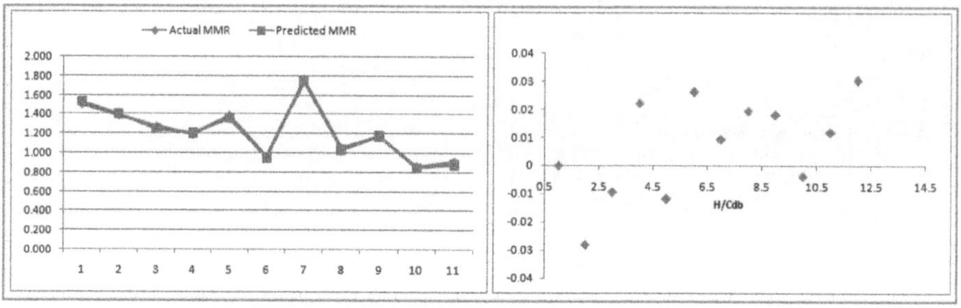

FIGURE 5 - COAL RANK PREDICTION MODULE FROM PROXIMATE AND ULTIMATE ANALYSIS ALONG WITH REGRESSION RESIDUAL

4.4 Prediction Of Coke Strength from Ultimate Analysis

Data analysis is performed to develop a regression model to predict the CSR of coke from the %carbon data of the coal. After extensive data analysis, it was found that aromatic carbon (derived from total carbon C_{daf}) correlates well with CSR values.

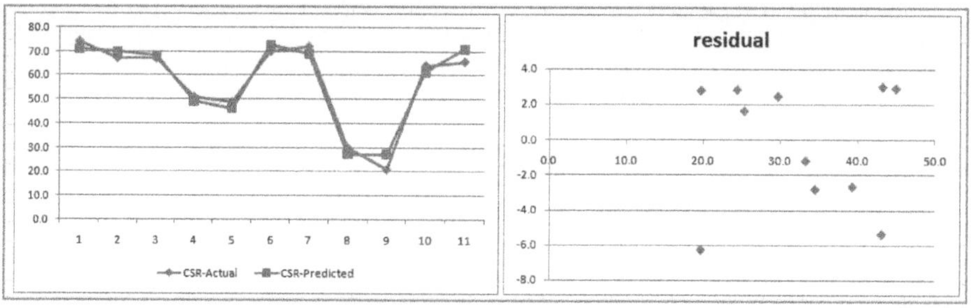

FIGURE 6 - COAL RANK PREDICTION MODULE FROM PROXIMATE AND ULTIMATE ANALYSIS ALONG WITH REGRESSION RESIDUAL

TABLE 1. ACTUAL AND CALCULATED CSR FROM CARBON OF DIFFERENT COALS

Coal Brand	C_{daf}	$CSR_{-Actual}$	$CSR_{-predicted}$
Australian High Rank-1	92.6	74	73.12
Australia Hard-1	90.8	67	67.16
Australia Hrad-2	89.7	67	65.58
Australian Soft	88.1	51	51.2
Canadian Soft	88.6	49	49
Canadian Hard	91.8	70	71.52
Australian High Rank-1	92.7	72	73.24
Indian Medium coking coal(MCC)	83.6	30	36.58
Newzealand semi-soft	85.6	21	36.6
Newzealand Hard	89.7	64	60.02
Indonesian Hard	92.7	65.7	73.1

4.5 Implication At the Coke Making Stage

The candidate coal should in the first place produce coke most suitable to the quality required by the blast furnace in terms of chemical & physical property. Coal also should provide a maximum yield of coke & by-product. Finally, the coal during carbonization should not generate coking pressure unsafe to the structure of the battery. Selection of proper MMR value, petrography constituents, volatile matter content, charge granulometry, moisture & heating rate will control the wall pressure within the safe range.

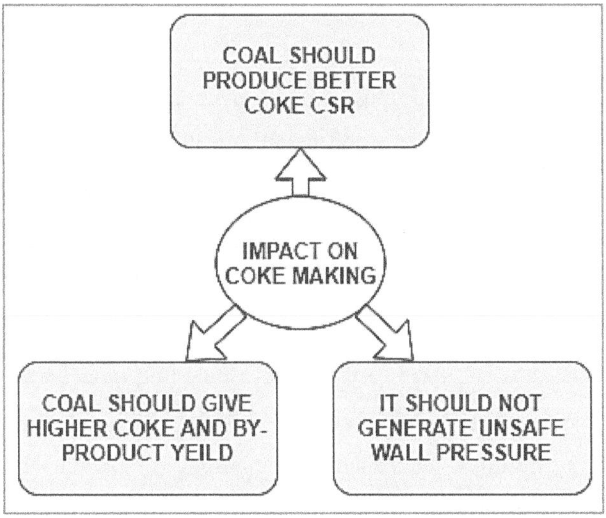

FIGURE 7 - REQUIREMENT FROM THE CANDIDATE COAL AT THE COKE-MAKING STAGE

4.6 Blast Furnace Coke Quality Requirement

Production of high-quality BF grade coke requires a good understanding of coal quality, coal to coke transformation mechanism and stipulation of coke quality criteria in the coke making stage.

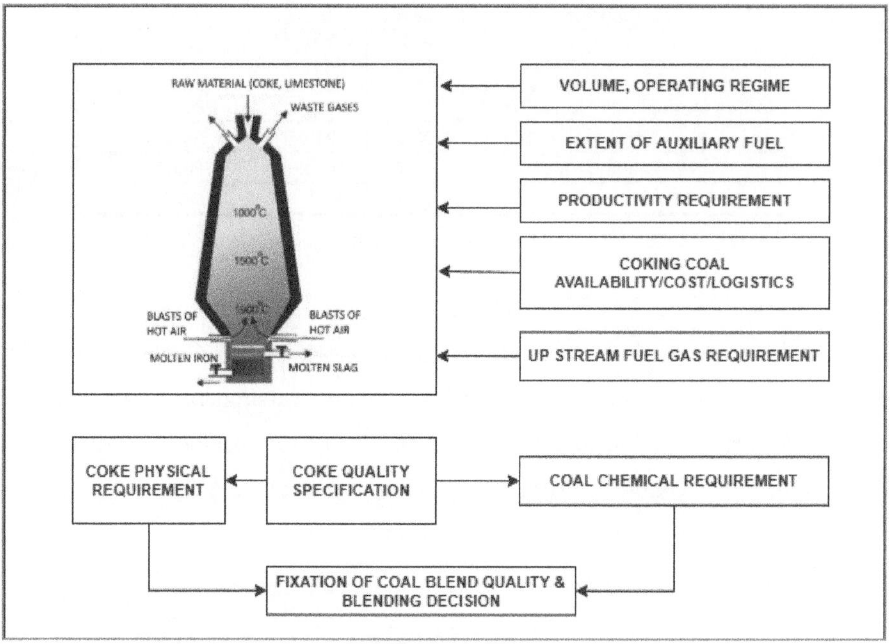

FIGURE 8 - THE EVOLUTION OF COKE QUALITY SPECIFICATION

4.7 Blending Process

Development of blending philosophy at coke oven has been an outcome of long operating experience of seven-meter-tall batteries with varied coal sources, extensive literature survey, analysis of plant data and trial run data of several carbonization studies world over.

4.8 Use Of Linear Programming Technique in Coal Blending

Different coals are blended to formulate the coke oven charge. For example, a quaternary blend of the above idea is utilized to formulate a blending model which will give the blend composition having the minimum cost fulfilling the laid down blend design criterion. Table-2 shows an equivalent compilation of coal quality variables and associated blending criteria.

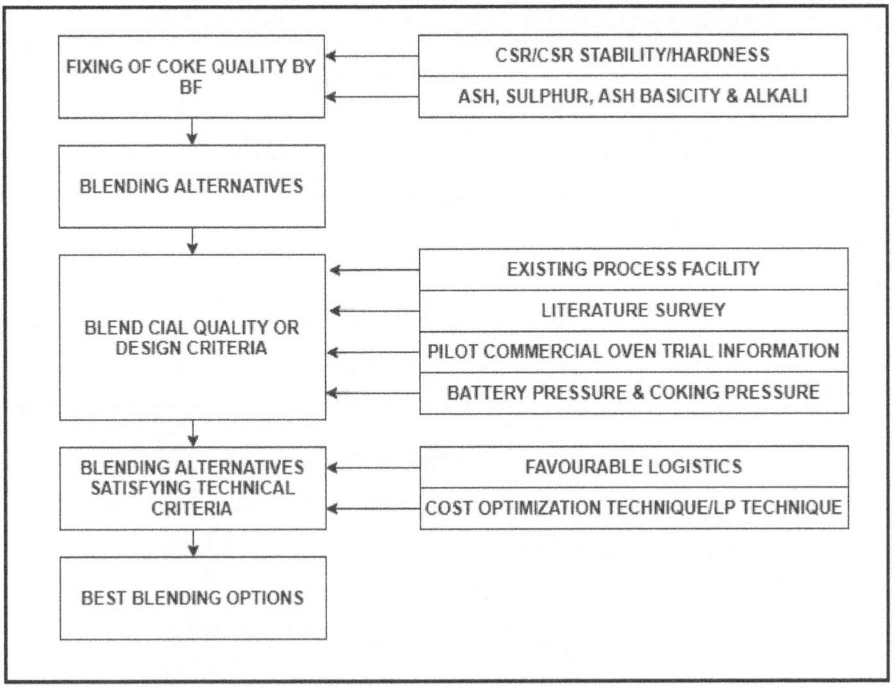

FIGURE 9 PROCESS OF EVALUATION OF COAL BLENDING OPTIONS

TABLE 2 - THE INPUT COAL PARAMETERS FOR THE BLENDING MODEL

COAL PARAMETER	COAL	COAL	COAL	COAL	COAL	BLEND DESIGN CRITERION	
Under consideration	1	2	3	4	5	Lower limit	Upper Limit
Blend fraction (Output of model)	x1	x2	x3	x4	X5		
Volatile Matter	Vm1	Vm2	Vm3	Vm4	Vm5	VM cbl	Vmcbu
MMR	MMR1	MMR2	MMR3	MMR4	MMR5	MMRcbl	MMRcbu
ASH	ASH1	ASH2	ASH3	ASH4	ASH5		ASHcbu
% Sulfur	S1	S2	S3	S4	S5		Scbu
Cost	C1	C2	C3	C4	C5	Optimum Cost of blend (Output of the model)	

The Suffix "cbl" & "cbu" are the coal blend lower and upper limits of the coal parameters of blend in the criterion. coke of a normal oven. The box is collected from the wharf. Coke in the box is tested for I20/I10/CSR/CRI.

The objective function is; Minimize; $Z(\text{Cost})=c_1x_1+c_2x_2+c_3x_3\ldots\ldots c_nx_n$(4)

A typical process of development of the constraint equations, to solve a linear programming problem is illustrated below in Table 3.

TABLE 3 - A TYPICAL PROCESS OF DEVELOPMENT OF THE CONSTRAINT EQUATIONS

SL NO	PARAMETER	CONSTRAINTS	CONSTRAINTS EQUATION	SLACK/ SURPLUS VARIABLE
1	Blend VM	VM cbl	vm1x1+vm2x2+vm3x3+vm4x4-s1+A1= VM cbl	s1/A1
		Vmcbu	vm1x1+vm2x2+vm3x3+vm4x4+s2= VM cbu	s2
2	Blend	MMRcbl	mmr1x1+mmr2x2+mmr3x3+mmr4x4-s3+A2= MMRcbl	s3/A2
	MMR	MMRcbu	mmr1x1+mmr2x2+mmr3x3+mmr4x4+s4= MMRcbu	s4
3	Blend Ash	ASHcbl	ash1x1+ash2x2+ash3x3+ash4x4+s5= ASHcbl	s5
4	Blen Sulphur	Scbu	S1x1+S2x2+S3x3+S4x4-s6+A3= Scbu	s6/A3

The objective function is; Minimize; $Z(Cost)=c_1x_1+c_2x_2+c_3x_3......c_nx_n$(4)

A typical process of development of the constraint equations, to solve a linear programming problem is illustrated below in table 3.

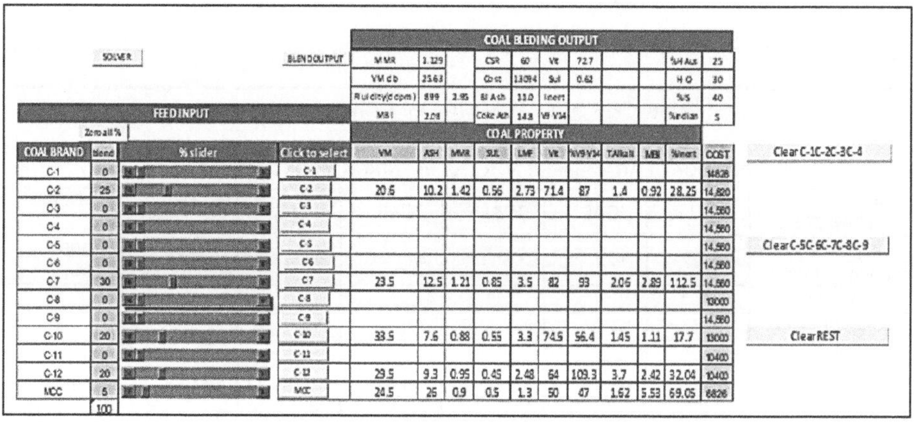

FIGURE 10 - THE INPUT-OUTPUT SCREEN OF THE BLENDING PROGRAM WITH LP OPTIMIZATION

4.9 Predicting The Value Rendered at Blast Furnace by the Candidate Coal (Kg Per Ton of Hot Metal Produced)

The value rendered at the blast furnace end is calculated by a parameter called Net Effective Carbon (NEC). Below in figure no-11, aspects connected to value at blast furnace end by the coke are illustrated.

A comparison of coke quality from this testing with Industrial-scale Blend is given below:

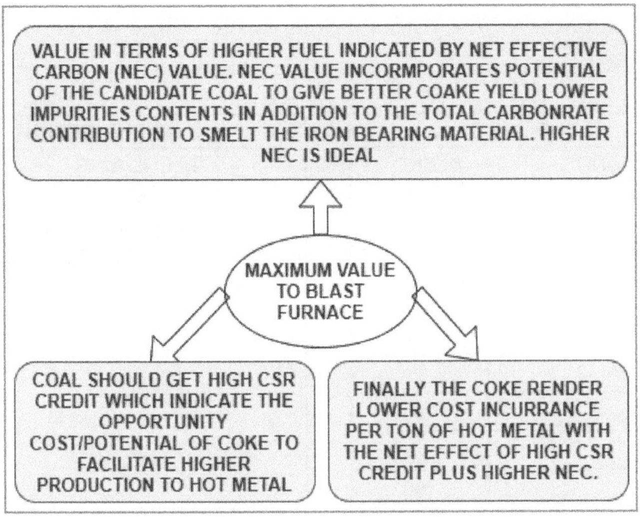

FIGURE 11 - FINAL VALUE AT BLAST FURNACE END.

The rationale of estimating NEC is to examine the chemical potential of candidate coal against a standard one. Below in figure no-12 the typical process calculation of NEC and value of coal used is illustrated as an example.

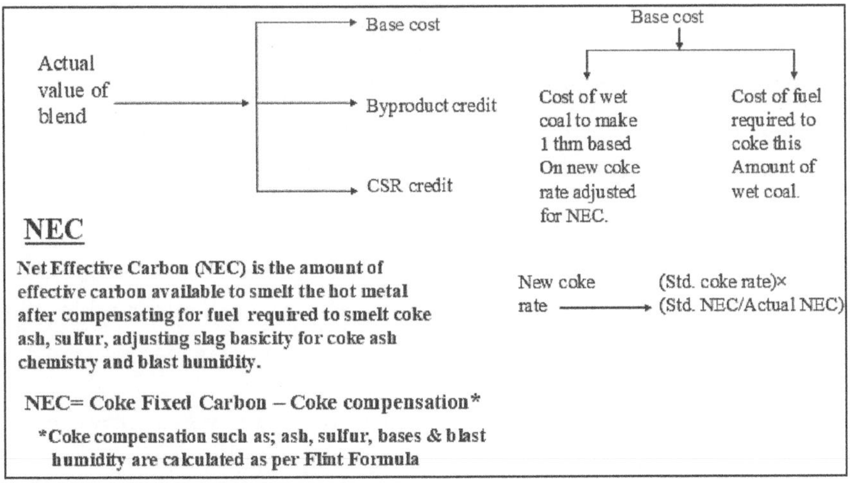

FIGURE 12 - EXPLANATION ON NET EFFECTIVE CARBON (NEC)

4.10 Estimation And Analysis of Coking Pressure During Blending

The pressure developed during the carbonization of coal in the slot oven plays a very crucial role in the selection of coal for blending and formulation of the coke oven charge. An important study was made in this regard by John J Grimley& Colin E Radley, Coke and Iron Technical of British Steel, by assessing 142 carbonization blends consisting of 33 different coal sources in tall coke ovens.

BLEND INPUT	
COAL BRAND	BLEND%
HIGH RANK COAL	30
HARD COAL-1	15
HARD COAL-2	15
	0
SOFT COAL-1	15
SOFT COAL-2	15
	0
MCC	10
SUM	100

BLEND OUTPUT	
BLEND MMR	1.154
VM%	24.73
BLEND FLUIDITY (DDPM)	426
PREDICTED WALL PR	
WALLPR (PSI)	0.02

FIGURE 13 - THE TYPICAL INPUT-OUTPUT MODULE FOR PREDICTION OF WALL PRESSURE OF BLEND FOR A 7-METER-TALL BATTERY

They finally developed an empirical model to predict coking pressure from the following property of the coal blend.

1. % Of high-rank coal (MMR>1.35)
2. % Of Medium rank coal (MMR 1.1-1.35)
3. MMR of high rank, medium rank & low rank
4. Log Maximum Fluidity of high rank, medium rank & low rank
5. The % moisture content of the blend
6. % Additives to blend such as breeze and oil

Using the above study result, the prediction of carbonization pressure for blends formulated at a 7-meter-tall battery is estimated using the empirical formula as shown below.

4.11 Characterization Of Candidate Coal Through Scoring –Evaluation System

To characterize a candidate coal regarding its relative suitability towards the production of metallurgical coke has been done to fine-tune the process of evaluation so that the coal under consideration is examined with respect to all related angles of coking & coke quality. The input-output screen is shown below in Table no-4.

TABLE 4 - PREDICTION OF COKING PROPENSITY AS PER SEVERAL INTERNATIONAL COAL CLASSIFICATIONS WITH SCORING SYSTEMS

SELECT NAME OF CADIDATE COAL FROM THE DROPDOWN MENU	
Candidate Coal	COAL-5

OUTPUT- EVALUATION OF CANDIDATE COAL- COAL-5 FROM BMA

Candidate Coal	COAL-5
Supplier Name	BMA

Rank Parameter	Rank Value	Rank score	Maximum score
MMR	1.180	80	140
VM db	22.30		
VMdaf	25.45		

Type Parameter	Type Value	Total Type score	Maximum score
LMF	2.75	88	203
Coal V9-V14	49.0		
CSN	9		
CBI	1.39		

Grade Parameter	Grade Value	Total Grade score	Maximum score
Blend Alkali	0.14	-9.3	12
MBI	0.58		
Sulfur	0.54		

Compatibility Parameter	Value	Total compa score	Maximum score
Reflectogram Code	0	76.5	79.5
Reflectogram stdev	0.05		
Plastic Range	80.00		

Effect on Coke Oven Battery life of coal			
Negative coal factor	Effect	Battery life score	Maximum score
Clinkering?	No	-0.8	0
Wall presure-MMR	No		
Wall presure-VM	Yes		

Total score	Total Max Score
234	435

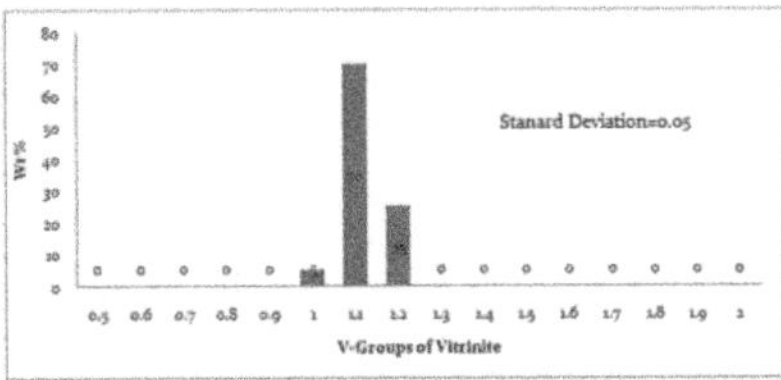

Comment on Rank & classification of the coal

Classification of coal by rank as per ASTM D 388-99(2004)	Medium Volatile
International class No.in classification of coal by Type(ECU)	4
International code No (ECU)	434
International coding (ECU) & assessment	Very Strong Coking
U.K.Based classification	Very Strong Coking
USA Based	High Vol Biyuminus A
Classification of coal by rank as per Coke Makers	100% battery chargeable

Comment on coking potential

Not thermoplastic dominant

sufficient with coking V9-V14 vitroids

Inert-reactive not in optimum range

Comment on blending compatibility of the coal

straight coal	straight coal&Good plastic range & straight coal
Good plastic range & straight coal	
Good plastic Range	

Critical Parameter	Norm	Value for the Coal	
MMR	1.12-1.155	1.18	Strong coking
VMdmmf		22.3	
LMF	>2.8	2.75	
Coal V9-V14	>40%	49.0	
CBI	0.7-1.3	1.39	
Blend Alkali	<0.25%	0.14	
MBI	<1.6%	0.58	
Plastic Overlap(deg C)	>35	80	
Wall Pr/Clinkering	0	-0.8	
CSR	>65	73	

Comment on impurities level

Medium SulfurLow iron oxideLow alkai

Comment on battery effect of the coal

Noclinkering & Normal wall pressure

The assessment summary of the coal

As per RINL inhouse assessment the coal is -Strong coking and classification as per rank by coke maker is-100% battery chargeable The ASTM Classification wise it is -Medium Volatile, its International classification by Type no-4 Code Number is-434,which is interpreted as -Very Strong Coking

The coal is-Noclinkering & Normal wall pressure typeMedium SulfurLow iron oxideLow alkai.

The coal is- with reflectogram stdev of-0.0509901951359278 hence it can be termed as straight coal The coal having Good plastic Range is individualy - Good plastic range & straight coal

The validation of the scoring is done by relating the estimated scores of the coal blend used in the battery with that of the actual CSR value, as depicted below in figure no-15.

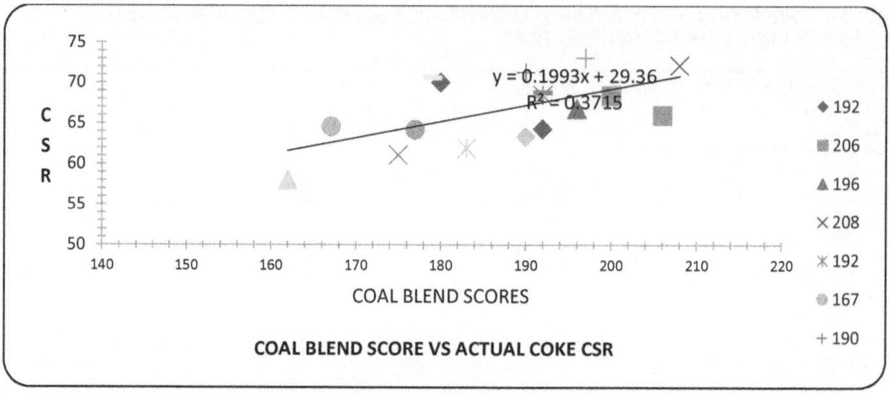

FIGURE 15 - CORRELATION BETWEEN ACTUAL CSR OBTAINED AND COAL SCORE

4.12 Prediction Of Coke Strength Parameter

Based on coal quality the prediction of coke strength parameters such as M-10 & CSR provides an effective tool in the hand of coke makers to control the process of coke making.

The regression model of M-10 is related to Volatile matter in dry mineral matter basis (VM_{ddmm}) and crushing fineness of the feed coal to the battery (% -3.20 mm)

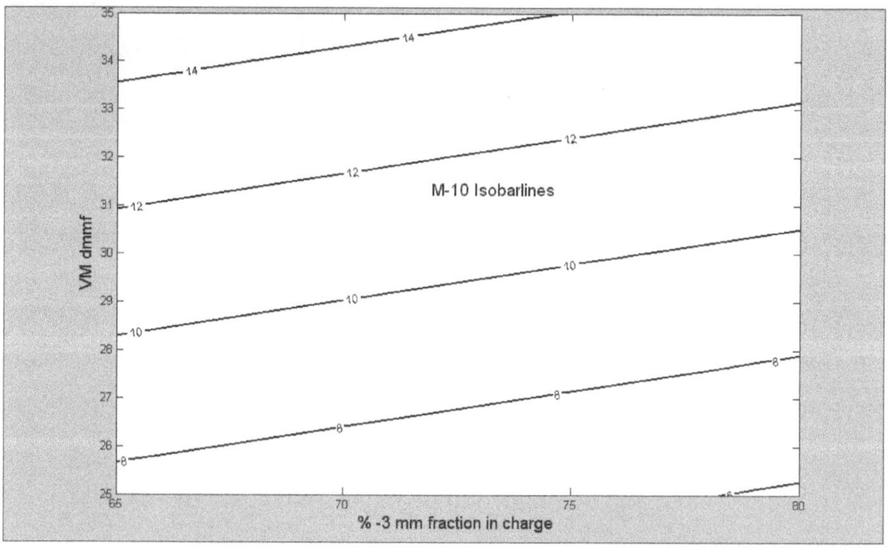

FIGURE 16 - PREDICTION OF M-10 FROM RANK PARAMETER (VM DMMF) & PROCESS PARAMETER (% -3 2 MM FRACTION CRUSHING INDEX)

4.13 Prediction Of CSR

Based on the available data on several coal parameters such as MMR, maximum fluidity, proximate analysis & ash chemistry most influencing factors are correlated to develop a regression model.

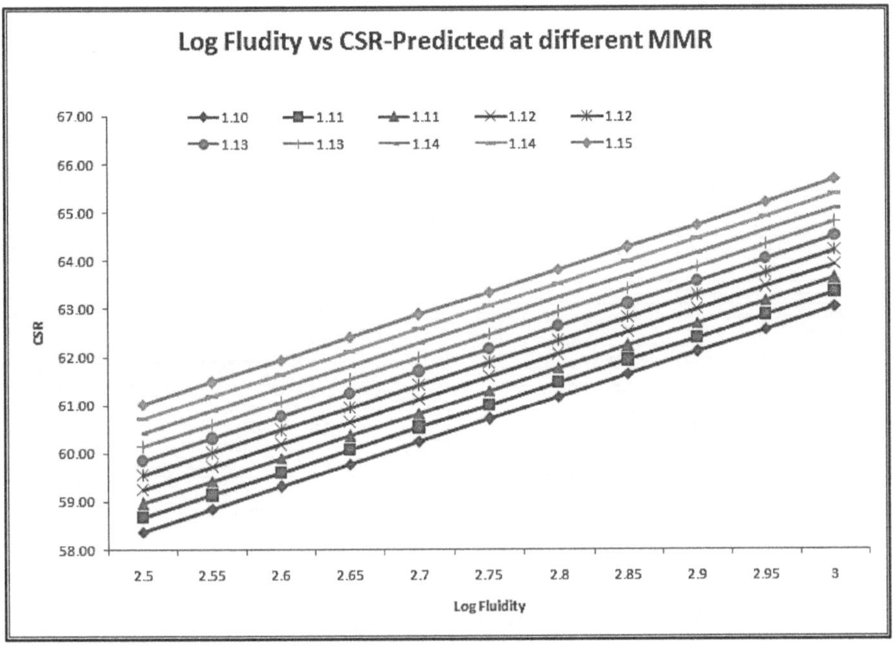

FIGURE 17 - PREDICTION OF CSR FROM RANK PARAMETER (MMR) & TYPE PARAMETER (MAX.FLUIDITY)

5. Final Evaluation Report for Coal & Coal Blend

A final evaluation report is generated if the input coal brand name is typed in the space provided as "FEED COAL BRAND BELOW". The program will retrieve the basic coal data from the database, process it for its scores and have several other important implications during coking. Once the brand name of the candidate coal is selected from the drop-down menu, all other above mention parameters and graphs are automatically populated and renewed for visualization. The screenshot of the program with a typical candidate coal brand "Coal-1" is shown in figure no-18.

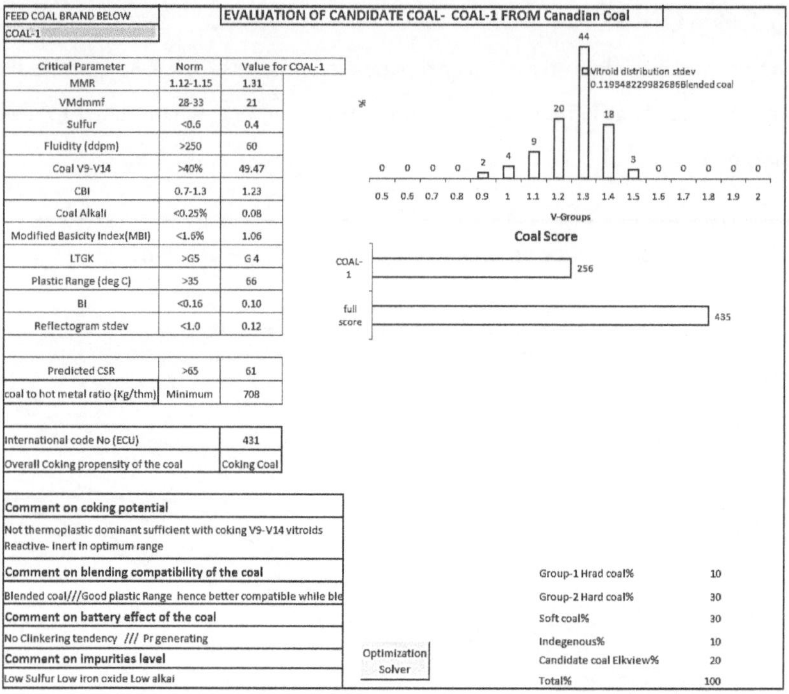

FEED COAL BRAND BELOW

COAL-1

EVALUATION OF CANDIDATE COAL- COAL-1 FROM Canadian Coal

Critical Parameter	Norm	Value for COAL-1
MMR	1.12-1.15	1.31
VMdmmf	28-33	21
Sulfur	<0.6	0.4
Fluidity (ddpm)	>250	60
Coal V9-V14	>40%	49.47
CBI	0.7-1.3	1.23
Coal Alkali	<0.25%	0.08
Modified Basicity Index(MBI)	<1.6%	1.06
LTGK	>G5	G 4
Plastic Range (deg C)	>35	66
BI	<0.16	0.10
Reflectogram stdev	<1.0	0.12
Predicted CSR	>65	61
coal to hot metal ratio (Kg/thm)	Minimum	708

International code No (ECU)	431
Overall Coking propensity of the coal	Coking Coal

Comment on coking potential
Not thermoplastic dominant sufficient with coking V9-V14 vitroids Reactive- inert in optimum range
Comment on blending compatibility of the coal
Blended coal///Good plastic Range hence better compatible while ble
Comment on battery effect of the coal
No Clinkering tendency /// Pr generating
Comment on impurities level
Low Sulfur Low iron oxide Low alkai

Group-1 Hrad coal%	10
Group-2 Hard coal%	30
Soft coal%	30
Indegenous%	10
Candidate coal Elkview%	20
Total%	100

FIGURE 18 - FINAL COAL EVALUATION REPORT

Using any of the above characterization and prediction methodology, the % usage of premium hard coking coal can be reduced significantly with cost savings on account of costlier coal consumption at coke oven battery without hampering the coke quality. The prior assessment will be useful while introducing new coal into the battery.

6. Conclusion

Coal science and technology of coke making has made great advancement for the last several decades; however, the views of experts in this field still carry divergent opinions regarding the structure-property correlation of coal and coal to coke transformation mechanism. As a result, the standards such as ASTM etc. is still using the conventional parameters such as volatile matter, calorific value & carbon content to classify coal for its relative merits & de-merits against a particular use. The proposed mechanism however to a fair degree is able to predict coke quality, battery life and evaluation at blast furnace end along with assessment on international coal classification by rank and type. A comprehensive coal evaluation sheet (figure no-18) is generated from the program giving details about the coal under examination. It was also found that the coal scores obtained against any coal or blend is relating well with the actual CSR of coke. The evaluation scheme enumerated above has helped coke oven managers in reducing the cost of blend and increase its supplier base to tackle any emergency.

7. Acknowledgements

The Author extends his regards to Sri Ravindra Ranjan, Ex. Executive Director (Works), Head of the Works division of RINL-VSP & Sri R Nagrajan, Ex Executive Director Services & Ex Executive Director Projects & Expansion of RINL-VSP, who have rendered critical guidance for the above study.

8. References

1. Robert J Marshalla a Process Metallurgy Department, British Steel Corporation, Scunthorpe Division, P.O. Box 1, Scunthorpe, South Humberside, UK
2. Coke quality & production- Roger Loison, Pier Foch & Andre Boyer
3. UNECE International Classification of Hard Coals "International classification of coal by type".
4. HS Valia, DD Kaegi, VI Addes, DA Zeeke, "Production and use of high CSR coke at Inland Steel" ISS 1990, pp. 59-72.
5. Y Ishikawa, M Kase, Y Abe, K One, M Sugota & T Nasi, "Influence of post-reaction strength of coke on blast furnace" ISS 1990, pp. 37-48
6. HS Valia "Prediction of CSR with CO2 from coal analysis at Inland Steel Company" ISS 1990, pp.109.119
7. Pearson, David E, "Influence of geology on CSR" Coal petrography
8. Richard Sakurovs, "Interaction between coking coal in blend" CSIRO Division of Energy Technology, PO Box 136 North Ryde NSWI670 Australia.
9. G.M. Latshaw,H.R.Mccollum,R.W.Stanley "Wall pressure determination by measurement and interpretation of internal gas pressure in a six-meter battery" United States Steel Corporation. ISS 1992 pp. 51-58.
10. H.S. Valia" Effect of coal oxidation on coke making" Iron Making Conference Proceedings -1990, pp. 199-207.
11. John, price T, "Coal petrography and characterization" Mac Master -2003 Volume-1.
12. H.S. Valia" Coal Blend Design" Mac Master -2003 Volume-1.
13. Van Drew, S.L (1987) "Coal preparation and blending techniques based on Linear Programming" Mineral Research Management By personal Computer, Society
14. The UNECE International Codification System for Medium and High-Rank Coals.

About the Author

Sanjit Kumar Seet at present is General Manager, Quality Assurance & Technology Development at Visakhapatnam Steel Plant, RINL. He has thirty-two years of rich experience in quality control activity in an integrated steel plant.

V. SAFETY

V.1. Process Safety Management

Ashish Chakraborty

Unexpected releases of toxic materials/gases, reactive substances, flammable or melted liquids & other harmful gases in processes involving highly hazardous industries have been reported for many years.

PROCESS SAFETY MANAGEMENT (PSM) is an analytical tool focused on preventing releases of any substance defined as a "highly hazardous chemicals" by the EPA or OSHA. PSM refers to a set of inter-related approaches to managing hazards & risks associated with the process industries and is intended to reduce the frequency and severity of incidents resulting from releases of chemicals and other energy sources.

OCCUPATIONAL SAFETY

PROCESS SAFETY

1. Occupational Safety

1. Effects: Minor damage: max. 1-2 victims
2. Approach: People-orientated man-machine interactions
3. Investigation: relatively easy to find root causes
4. To all workplace scenarios
5. Focused on personal Safety (... to keep people safe)
6. Unsafe act
7. Unsafe condition
8. PPE's

9. On-site hazard ids-
    a. Noise/Dust
    b. Slip/Trip/Falls/open pit
    c. Radiation,
    d. Metal sparks,
    e. Hot water/surface
    f. moving machines/conveyor

2. Process Safety

1. Effects: Major Material damage / Potential multiple victims
2. Approach: Process orientated. Potential threats of consequences for safety, environmental, reputation, product and cost.
3. Investigations: More complex to find the direct and root causes
4. Mechanical Integrity & Quality Assurance
5. Process Design conditions
6. Critical Equipment Maintenance
7. Identification of hidden process hazards & analysis
8. Consequence Analysis
9. Emergency Preparedness
10. Human Factor Analysis
11. Training & competency

3. Safety Triangle

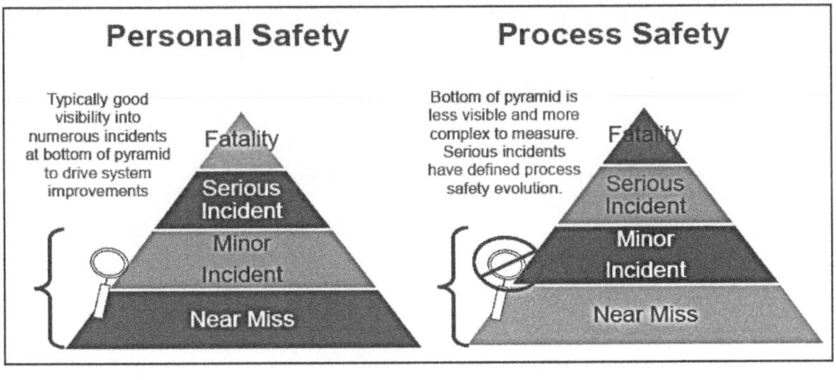

4. Need For Process Safety

Experts say up to ten million gallons of 'missing oil' from the 2010 BP disaster (pictured) has been discovered at the bottom of the Gulf of Mexico

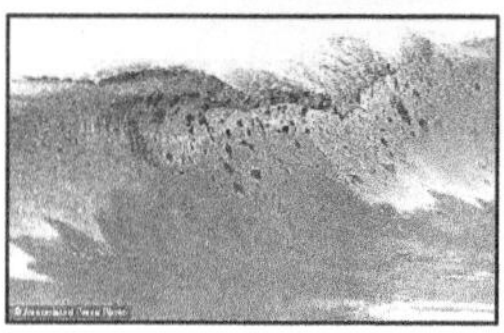

The explosion at BP's Deepwater Horizon rig killed 11 workers, injured 17 more and caused nearly 50,000 barrels of oil per day to spew into the gulf.

The well went unplugged for nearly three months after the explosion on April 20, 2010, damaging fishing and tourism as well as marine and wildlife habitats

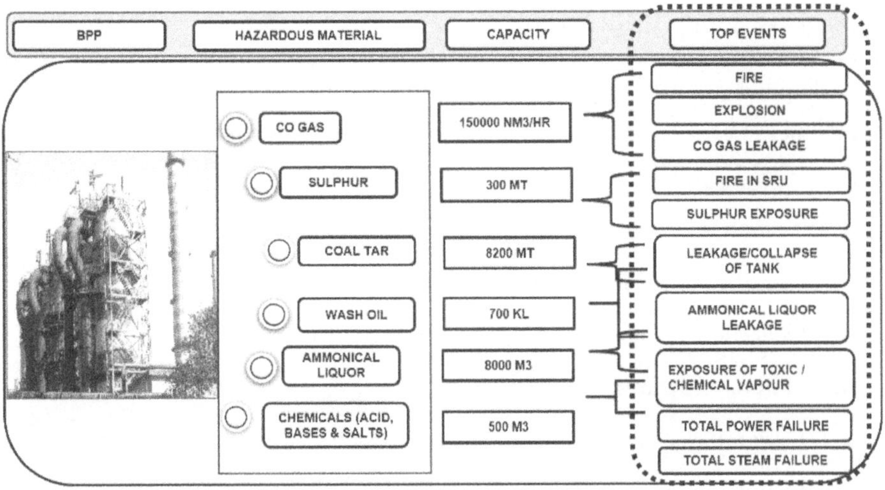

FIGURE 6: COKE OVEN BY-PRODUCT PLANT PROCESS HAZARDS - TOP EVENTS

5. Elements Of Process Safety Management

1. Process Safety Wheel is divided into three broad sections "**Technology, Facilities & Personnel**"
2. Centre of the Wheel (axel) is **Management Leadership & Commitment**
3. Altogether, there are **14 elements** of Process Safety Management
4. The first step of PSM is collecting **Process Safety Information**

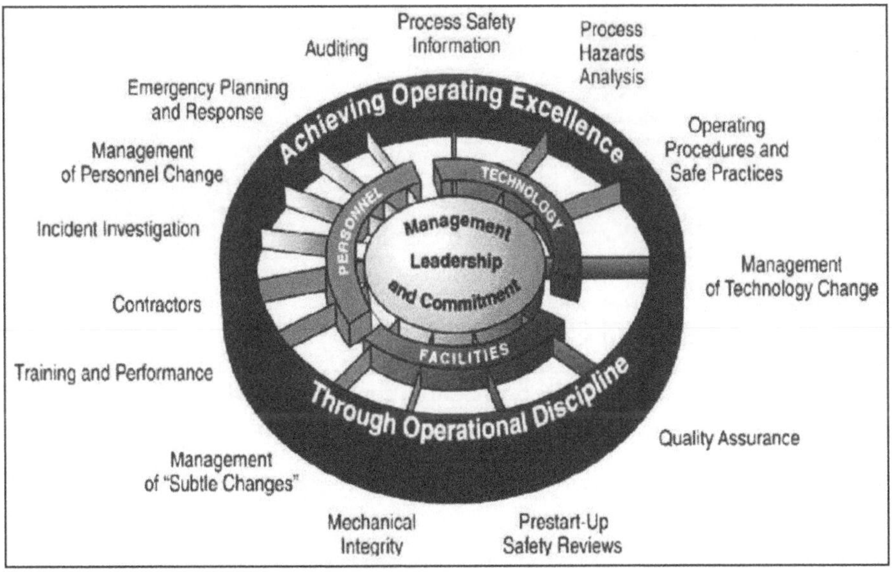

6. Process Safety Information

A compilation of written process safety information that helps to identify and understand the hazards posed by those processes involving highly hazardous materials.

6.1. PSI Elements

1. Team Formation
2. Identification & listing Areas/ Sub-Processes (Process Selection Criteria)
3. Collection/Preparation of P &IDs, BDs & PFDs
4. Hazard of the Material
5. Design Basis for Safe Operating of the Process
6. Identification of Equipment Critical to Process Safety (Categorization)
7. Equipment Design Basis
8. Maintaining Library

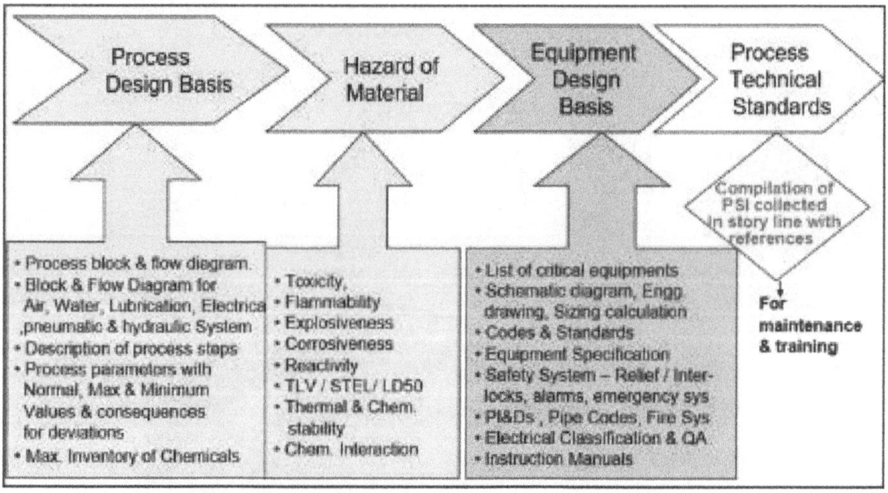

FIGURE 7: STEPS FOR PROCESS SAFETY INFORMATION

7. Process Selection Criteria

Identify processes-sub processes & categorization in terms of (e.g., Uncontrolled release, Explosiveness, Flammability, high corrosivity, Thermal instability/latent heat. Table presented on the next page.

TABLE 1 - CRITERIA FOR SELECTION OF PROCES FOR PSRM (Form No.: PSRM/PSI/PSC/001 Rev. No.:01 Eff. Dt.: 01.11.2015)

Department: By-Product Plant	Process Name:		
Document No:	Rev No:	Effective Date:	

Department	Section	Process	Involving Hazardous Substance having/causing: - Acute Toxicity	Explosivity	Flammability	Corrosivity	Thermal Instability	Latent Heat or Compression	Consequences: Significant Property damage (> 50 lakhs)	Potential for Fatality	Significant Environmental Impact	HHP	LHO	Ranking	Category
By product Plant	Old by-product plant	A. Gas Cooling	Y	Y	Y	Y	N	N	Y	Y	N	Ö		9	B
		B. Detarring	Y	Y	Y	Y	N	N	Y	Y	N	Ö		8	B
		C.Exhauster	Y	Y	Y	Y	N	N	Y	Y	Y	Ö		4	A
		D.Scrubbing (Ammonia & Naphthalen)	Y	Y	Y	Y	N	N	Y	Y	Y	Ö		3	A
		E. Stripping (Ammonia liquor) & Incineration	Y	Y	Y	Y	N	N	Y	Y	Y	Ö		6	A
		Naphthalene Distillation	Y	Y	Y	Y	N	N	Y	Y	Y	Ö		2	A
		Tar & Liquor separation	Y	N	Y	Y	N	N	Y	Y	Y	Ö		7	A
		Pre-treatment process	Y	N	N	Y	N	N	Y	Y	Y	Ö		1	A
		BOT process	Y	N	N	Y	N	N	N	N	Y	Ö		10	C
		CO gas boosting	Y	Y	Y	Y	N	N	Y	Y	Y	Ö		5	A

*If any one of the consequences is applicable, then the process is considered as High Hazard Process (HHP).
*If any of the criteria mentioned above is applicable, then the process is considered as Low Hazard Operation (LHO).

Ö	Applicable
X	Not Applicable
Y	Yes
N	No

A- when consequence is having 3 nos of yes
B- When Consequence is having 2 nos of Yes
C- When Consequence is having 1 no of yes
Priority to be given in the order of Human, Environment & Property

8. Hazard Of Materials

A hazardous material is a material that, because of its quantity, concentration and physical or chemical characteristics, may pose a physical, environmental or health risk of acute or chronic nature.

The example includes materials that are toxic, corrosive, flammable, reactive, explosive and/or emitting ionizing radiation. Hazardous material can be a solid, liquid, gas and may exhibit one or more potentially dangerous physical or chemical properties.

Form No.: PSRM/PSI/HOM/001 Rev. No.:02 Eff. Dt.:01.01.2016														
Hazard of Material														
Department : By Product Plant								**Process Name : Napthalene Distillation Process**						
Document No : CP/PSRM/BPP/PSI/002								**Rev No: 1.01**			**Effective Date: 01.04.2017**			
Sl. No.	**Hazardous Substance**	**Reactivity**	**Toxicity**			**Explosivity**		**Flammability**				**Other Process Hazards (Like Stored Energy, Asphyxiants, radioactive etc)**	**Inventory (If Available)**	
			TLV	**STEL**	**LD50 / LC50**	**LEL**	**UEL**	**Flash point**		**LFL**	**UFL**		**Total Inventory**	**Statutory Norm**
1	Naphthalene		10 ppm	15 ppm	1250 ppm			79 c	Auto ignition: 567 c Boiling pont: 218 c Melting point : 80.2 c	**0.90%**	5.90%			
2	Wash Oil				**2000 mg/kg**	no data available		151 c	Auto Ignition: 500 c Boiling range: 230-320 C	no data available		Stored energy	191 KL	NA
3	I.P. Steam		**NA**	NA	**NA**	NA	NA	NA	NA	**NA**	NA	Stored energy		
4	Water		NA	NA	NA	NA	NA	NA	NA	NA	NA			
5	AIR		NA	NA	NA	NA	NA	NA	NA	NA	NA			
6	Nitrogen		**NA**	NA	**NA**	NA	NA	NA	NA	**NA**	NA			

9. Process Design Basis (Safe Operating Limits & Safe Operating Conditions)

SOL (Safe Operating Limit) & SOC (Safe Operating Condition)-

Describe the process steps with process parameters for each step, their limits (max., min. & desired values) and process capabilities to manage deviations. The table has been presented in the next page.

<table>
<tr><td colspan="9"></td><td>Process Design Basis</td></tr>
<tr><td colspan="9">Department : By Product Plant</td><td></td></tr>
<tr><td colspan="9">Document No : CP/PSRM/BPP/PSI/002</td><td></td></tr>
<tr><th rowspan="2">Sl No.</th><th rowspan="2">Parameter</th><th rowspan="2">UoM</th><th rowspan="2">PSM Critical</th><th colspan="2">Safe Operating Limit (SOL)</th><th colspan="2">Standard Operating Condition (SOC)</th><th colspan="2">Control Measures to avoid deviation (SOC)</th></tr>
<tr><th>Min</th><th>Max</th><th>Min</th><th>Max</th><th>Identification</th><th>Action</th></tr>
<tr><td rowspan="2">1</td><td rowspan="2">Rich W.Oil Flow FCV2101</td><td rowspan="2">kg/hr</td><td rowspan="2">yes</td><td rowspan="2">500</td><td rowspan="2">6000</td><td rowspan="2">3500</td><td rowspan="2">4700</td><td rowspan="2">Max & Min: Alarm in DCS (FCV2101) at low & high limits</td><td>Min- Increase the flow through FCV2101 manually from DCS. In case of failure of contoller, adjust by the by-pass valve of controller manually.</td></tr>
<tr><td>Max- Decrease the flow through FCV2101 manually from DCS. In case of failure of contoller, adjust by the by-pass valve of controller manually.</td></tr>
<tr><td rowspan="2">2</td><td rowspan="2">Rich wash oil Temp after pre heater (TC2632)</td><td rowspan="2">Deg C</td><td rowspan="2">yes</td><td rowspan="2">130</td><td rowspan="2">160</td><td rowspan="2">135</td><td rowspan="2">155</td><td>Max : Oil temp after Preheater > 155°C , an alarm generated for corrective action.</td><td>Max :- To decrease steam flow by TCV2632 manually from DCS. In case of failure of controller operate by-pass valve of controller from field.</td></tr>
<tr><td>Min: Oil temp after Preheater < 135 C , an alarm generated for corrective action.</td><td>Min:- To increase steam flow by TCV2632 manually from DCS. In case of failure of controller operate by-pass valve of controller from field.</td></tr>
<tr><td rowspan="2">3</td><td rowspan="2">Level of still LCV2317</td><td rowspan="2">%</td><td rowspan="2">yes</td><td rowspan="2">30</td><td rowspan="2">80</td><td rowspan="2">40</td><td rowspan="2">60</td><td rowspan="2">Max & Min: Alarm in DCS (LCV 2317) at low & high limits</td><td>Max- Decrease the level through FCV2101 manually from DCS. In case of failure of contoller, adjust the flow by the by-pass valve of controller manually.</td></tr>
<tr><td>Min- Increase the level through FCV2101 manually from DCS. In case of failure of contoller, adjust the flow by the by-pass valve of controller manually.</td></tr>
</table>

Process Name : Napthalene Distillation				
Rev No: 1.01		**Effective Date:**		
Consequence of deviation (SOL)		**Control/Measures to avoid deviation (Existing Safegaurd) From SOL**		**References**
min	**max**	**min**	**max**	
1. Low level in still , rise in oil temp & pressure inside Preheater causing gasket and tube damage due to which high pressure steam ingress will take place in tube resulting in hunting and damage of pipelines & loss of containment. 2. Cavitation in outlet pump & pump damge may occure.	1.High level, can lead to flooding of naphthalene distillation column	Steam flow to be reduced as per SOP-CP/SOP/BPP/OPN/122	WO & Steam flow to be reduced as per SOP-CP/SOP/BPP/OPN/122	Operating Manual
Improper distillation , quality of CO gas will affect. No significant consequence	Chances of Over pressureization of naphthalene column	Temperature controller TCV 2632 with alarm followed by operator action as per SOP CP/SOP/BPP/OPN/151	Temperature controller TCV 2632 with alarm followed by operator action as per SOP CP/SOP/BPP/ OPN/151	Operating Manual
1. Low level , rise in top temperature & pressure of the still may lead to overpressurisation. 2. Cavitation in outlet pump may damage the pump.	1.High level, can leads to flooding of naphthalene distillation column. 2. Decrease in top temperature	Level control valve LCV2317, interlocking of outlet pump with Low low level of 30%	Level control valve LCV2317,	Operating Manual.

9.1 Barriers Or Layers of Protections

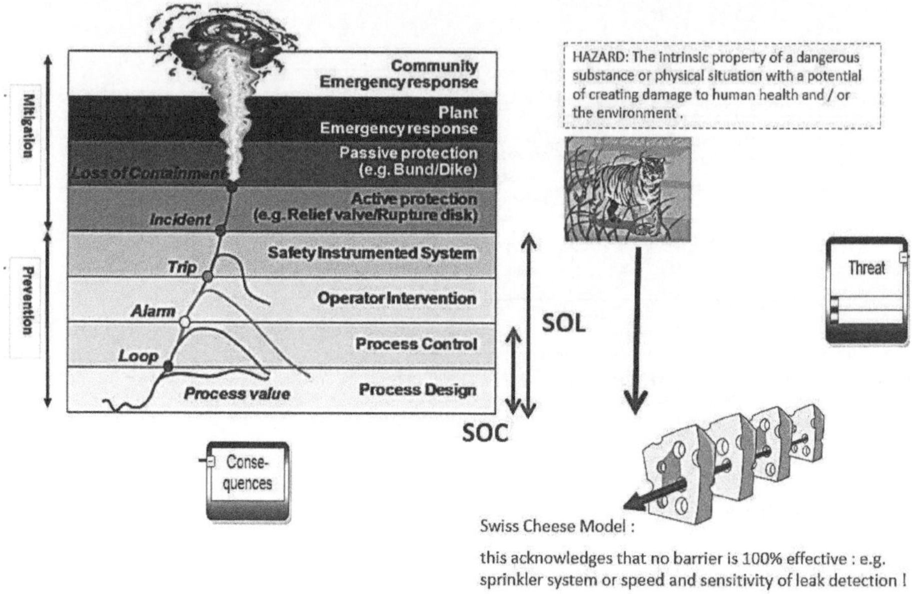

9.2 Material Energy Balance

M&E balances are an integral part of the overall PDB to clearly define the process and the relationships between process variables and materials, and to identify where energy can accumulate in hazardous quantities

Naphthalene Stripper

<u>Streams:</u>

Input:

RWO-HOT – Rich wash oil (hot)

Steam – Steam

TOTAL	UNITS	IN	OUT	REL. DIFF
Mole-flow	kmol/hr	236.52	236.52	0.00
Mass-flow	kg/hr	8500.00	8500.00	0.00
Enthalpy	Gcal/hr	-14.33	-16.75	0.14

Output:

VDIST – Vapour distillate

LDIST – Liquid distillate

LWO – Lean wash oil

BLOCK: STRIPPER MODEL: RADFRAC

INLETS -	RWO-HOT	STAGE 2
	STEAM	STAGE 16
OUTLETS -	VDIST	STAGE 1
	LDIST	STAGE 1
	LWO	STAGE 16

PROPERTY OPTION SET: RK-SOAVE STANDARD RKS EQUATION OF STATE

	RWO-HOT	STEAM	VDIST	LWO	LDIST
Temperature, °C	140	190	96.1	151.6	96.1
Pressure bar	1.275	1.471	1.079	1.079	1.079
Vapor Frac	0	1	1	0	0
Mole Flow kmol/hr	14.484	222.034	2.052	14.698	219.768
Mass Flow kg/hr	4500	4000	99.87	4440	3960.13
Volume Flow l/min	92.747	96230	953.16	92.549	71.553
Enthalpy Gcal/hr	**-1.818**	**-12.529**	**-0.065**	**-1.853**	**-14.85**

10. Equipment Design Basis – PSM Critical Equipment List

Form No.: PSRM/PSI/EDB/002 Rev. No.:01 Eff. Dt.:01.11.2015

Department: By Product Plant

Process Name: Naphthalene Distillation

Document No: CP/PSRM/BPP/PSI/002 Rev No: 1.01

EQUIPMENT DESIGN BASIS						
SL NO	EQUIPMENT	RESPONSIBILITY	BASIS OF SELECTION			PSM CRITICAL
			CONSEQUENCE BASED PSRM CRITICAL	PREVENTION & MITIGATION EQUIPMENT	PRESCRIPTIVE PSM CRITICAL EQUIPMENT	
1	Pre-Heater 2E102A	MECH	Consequence Based PSRM Critical			Yes
2	Pre-Heater 2E102B	MECH	Consequence Based PSRM Critical			Yes
3	Pre-Heater 2E102C	MECH	Consequence Based PSRM Critical			Yes
4	Naphthalene Still 2V101	MECH	Consequence Based PSRM Critical			Yes
5	Pressure relief valve: PRV-NOV-100-1776 (set point:1.5 kg/cm2)	MECH		Controlled release equipment/ systems		Yes

SL NO	EQUIPMENT	RESPONSIBILITY	BASIS OF SELECTION			PSM CRITICAL
			CONSEQUENCE BASED PSRM CRITICAL	PREVENTION & MITIGATION EQUIPMENT	PRESCRIPTIVE PSM CRITICAL EQUIPMENT	
6	Stripped wash oil outlet pump: P-101A	MECH	Consequence Based PSRM Critical			NO
7	Stripped wash oil outlet pump: P-101B	MECH	Consequence Based PSRM Critical			NO
8	Condenser 2E 101 A	MECH	Consequence Based PSRM Critical			Yes
9	Condenser 2E101 B	MECH	Consequence Based PSRM Critical			Yes
10	Vapour inlet valve to the condenser (2E101 A)	MECH	Consequence Based PSRM Critical			No
11	Vapor inlet valve to condenser (2E101 B)	MECH	Consequence Based PSRM Critical			No
12	Pressure relief valve: (Condenser 2E101 A) Set point: 1.5kg.cm2 (PRV-NOV-50-1814)	MECH		Controlled release equipment/ systems		Yes

11. Equipment Design Basis – List of Critical Equipment

Document No : CP/PSRM/BPP/PSI/002						Rev No: 1.01						
S No	Hazardous Substance (Equipment is related with)	Sub Process / Sub System	Equipment		SAP ID	Basis of Selection Note#1	Manufacturer / Supplier	Model No / Type	References Documents (vendors Manual, FD, Operating/maint. manual details)	Schedule for Maint./ Calibration /inspection	Location of the Documents / Drawings etc	Remark
1	Wash Oil	Napthalene Distillation Process	Pre -Heater	2E102A	10333725	Consequnce Based PSRM Critical	Bcock Woodall Duckham Pvt. Ltd.		Tata Drawing No. 331948	1 Year	Drwaing Mgmt Systrem MEG	
2	Wash Oil	Napthalene Distillation Process	Pre -Heater	2E102B	10333726	Consequnce Based PSRM Critical	Bcock Woodall Duckham Pvt. Ltd.		Tata Drawing No. 331948	1 Year	Drwaing Mgmt Systrem MEG	
3	Wash Oil	Napthalene Distillation Process	Pre -Heater	2E102C	10333727	Consequnce Based PSRM Critical	Indigenous		Tata Drawing No. P/5652	1 Year	Mgmt Systrem MEG	
4	Wash Oil	Napthalene Distillation Process	Naphthalene Still	2V101	10333728	Consequnce Based PSRM Critical	Bcock Woodall Duckham Pvt. Ltd.		Tata Drawing No. 331943,327653, 327655,	1 Year	Drwaing Mgmt Systrem MEG	
5	Wash Oil , Naptahalene Vapour	Napthalene Distillation Process	PRV- Napthalene still	PRV-NOV-100-1776	10333729	Controlled release equipment/ systems	Bcock Woodall Duckham Pvt. Ltd.		Tata Drawing No. 331943,327653, 327655,	1 Year	Drwaing Mgmt Systrem MEG	
6	Wash Oil , Naptahalene Vapour	Napthalene Distillation Process	PRV - Condenser A	PRV-NOV-50-1814	10333730	Controlled release equipment/ systems	Bcock Woodall Duckham Pvt. Ltd.		Tata Drawing No. 331948	1 Year	Drwaing Mgmt Systrem MEG	
7	Wash Oil , Naptahalene Vapour	Napthalene Distillation Process	PRV - Condenser B	PRV-NOV-50-1819	10333731	Controlled release equipment/ systems	Bcock Woodall Duckham Pvt. Ltd.		Tata Drawing No. 331948	1 Year	Drwaing Mgmt Systrem MEG	
8	Steam	Napthalene Distillation Process	PRV - cooling water - Condenser A	PRV-CWR-25-1840	10333732	Controlled release equipment/ systems	Bcock Woodall Duckham Pvt. Ltd.		Tata Drawing No. 331948	1 Year	Drwaing Mgmt Systrem MEG	
9	Steam	Napthalene Distillation Process	PRV - cooling water - Condenser B	PRV-CWR-25-1847	10333733	Controlled release equipment/ systems	Bcock Woodall Duckham Pvt. Ltd.		Tata Drawing No. 331948	1 Year	Drwaing Mgmt Systrem MEG	
10	Wash Oil	Napthalene Distillation Process	Regenerator	2V102	10333734	Consequnce Based PSRM Critical	Bcock Woodall Duckham Pvt. Ltd.		Tata Drawing No. 331963,327651		Drwaing Mgmt Systrem MEG	

12. Process Hazard Analysis-HAZOP

Hazard and operability study (HAZOP) is a technique for the identification of hazards that are involved in the process, based on guide words and recommend control of risk. Identification of hazards in the coal chemical department of the steel industry is done by using the HAZOP study technique. The studies were carried out by a multidisciplinary team.

The hazard and operability study is the identification of Hazards in process of any industry and finds out the lack of safety. It suggests preventive controls to eliminate all potential hazards in the process industry like Coal chemical plant i.e., the by-product for coke oven gas.

Scenario	Protected Risk			Recommendations	RESIDUAL RISK		
	Severity	Frequency	Risk		SEVERITY	FREQUENCY	RISK
Rupture/Leakage due to over pressurization of Ammonia still.	C-3	F-2	R-2	1. Revive of Pressure Transmitter in inlet of thermo compressor. 2. Reliability of PRV AS VS 220. & PRV AS VS 221. 3. Maintenance of safeguards should be accordingly to standard maintenance practices.	C-3	F-4	R-3

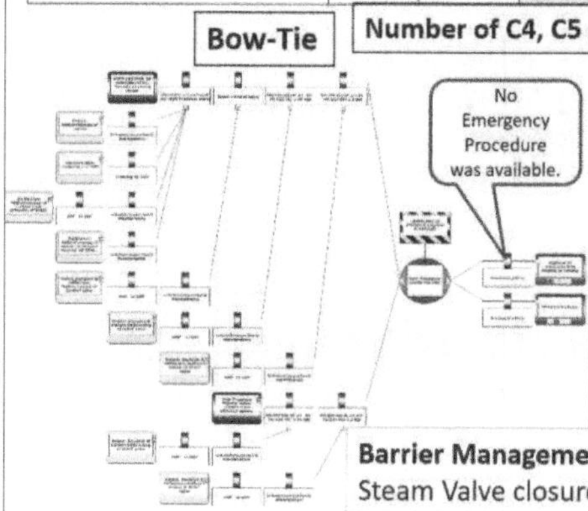

Number of C4, C5 scenarios identified:

Risk Matrix

F-1	3	2	2	1	1
F-2	3	3	2	1	1
F-3	4	3	3	2	1
F-4	4	3	3	3	2
F-5	4	4	4	3	3
	C-1	C-2	C-3	C-4	C-5

Recommendations on Mitigation Safeguard:

1.Remote Stop to be provided.

2. Emergency Procedure to be made.

Barrier Management helped us to visualize the importance of Steam Valve closure during emergency.

13. Management Of PSM Critical Equipment

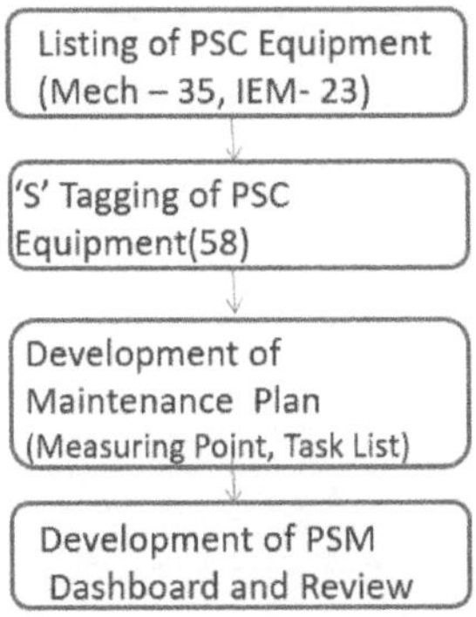

'S' Tagging of PSC Critical Equipment

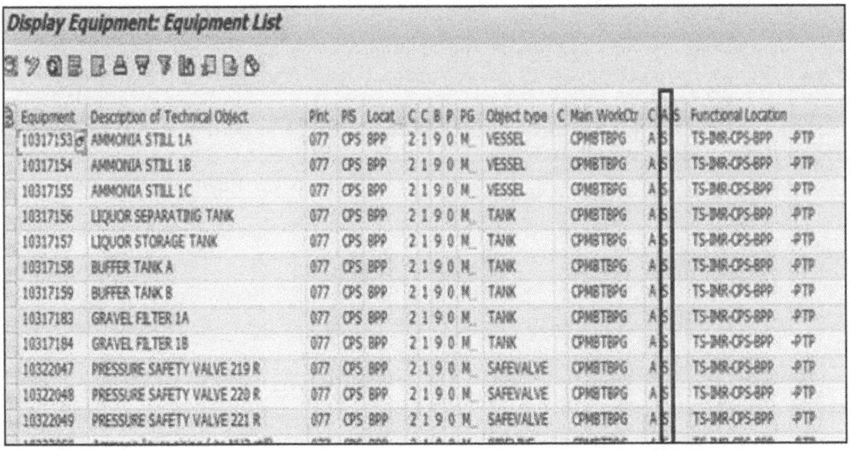

Display Equipment: Equipment List

Equipment	Description of Technical Object	Plnt	PS	Locat	C	C	B	P	PG	Object type	C	Main WorkCtr	C	A	S	Functional Location
10317153	AMMONIA STILL 1A	077	CPS	BPP	2	1	9	0	M_	VESSEL		CPMBTBPG	A	S		TS-JMR-CPS-BPP -PTP
10317154	AMMONIA STILL 1B	077	CPS	BPP	2	1	9	0	M_	VESSEL		CPMBTBPG	A	S		TS-JMR-CPS-BPP -PTP
10317155	AMMONIA STILL 1C	077	CPS	BPP	2	1	9	0	M_	VESSEL		CPMBTBPG	A	S		TS-JMR-CPS-BPP -PTP
10317156	LIQUOR SEPARATING TANK	077	CPS	BPP	2	1	9	0	M_	TANK		CPMBTBPG	A	S		TS-JMR-CPS-BPP -PTP
10317157	LIQUOR STORAGE TANK	077	CPS	BPP	2	1	9	0	M_	TANK		CPMBTBPG	A	S		TS-JMR-CPS-BPP -PTP
10317158	BUFFER TANK A	077	CPS	BPP	2	1	9	0	M_	TANK		CPMBTBPG	A	S		TS-JMR-CPS-BPP -PTP
10317159	BUFFER TANK B	077	CPS	BPP	2	1	9	0	M_	TANK		CPMBTBPG	A	S		TS-JMR-CPS-BPP -PTP
10317183	GRAVEL FILTER 1A	077	CPS	BPP	2	1	9	0	M_	TANK		CPMBTBPG	A	S		TS-JMR-CPS-BPP -PTP
10317184	GRAVEL FILTER 1B	077	CPS	BPP	2	1	9	0	M_	TANK		CPMBTBPG	A	S		TS-JMR-CPS-BPP -PTP
10322047	PRESSURE SAFETY VALVE 219 R	077	CPS	BPP	2	1	9	0	M_	SAFEVALVE		CPMBTBPG	A	S		TS-JMR-CPS-BPP -PTP
10322048	PRESSURE SAFETY VALVE 220 R	077	CPS	BPP	2	1	9	0	M_	SAFEVALVE		CPMBTBPG	A	S		TS-JMR-CPS-BPP -PTP
10322049	PRESSURE SAFETY VALVE 221 R	077	CPS	BPP	2	1	9	0	M_	SAFEVALVE		CPMBTBPG	A	S		TS-JMR-CPS-BPP -PTP

Tagging at Shop floor

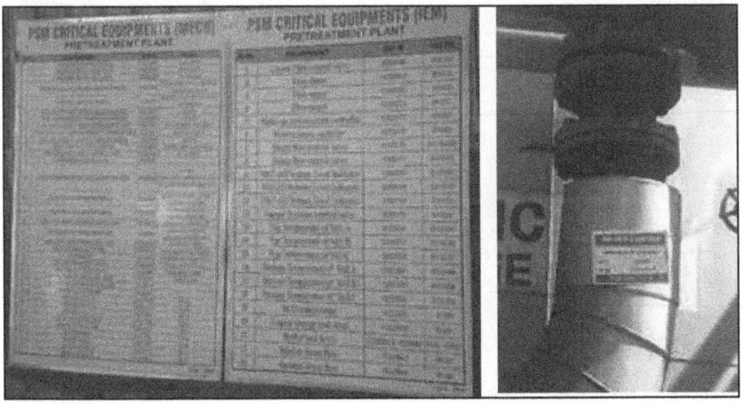

Maintenance Plan

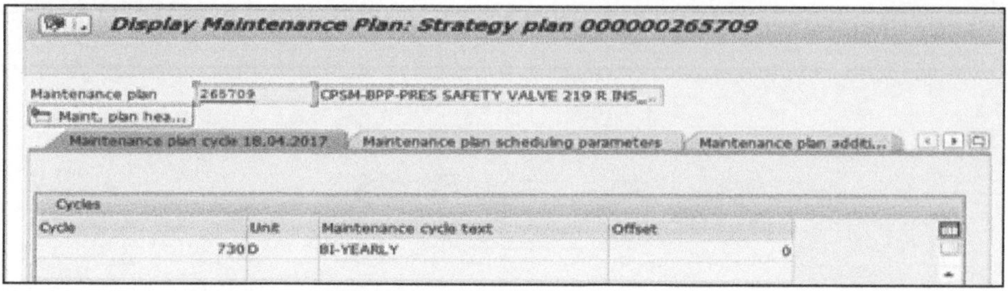

Maintenance Order

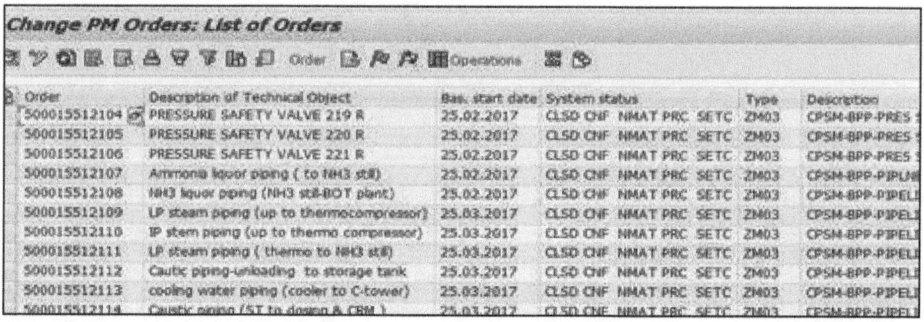

Order	Description of Technical Object	Bas. start date	System status	Type	Description
500015512104	PRESSURE SAFETY VALVE 219 R	25.02.2017	CLSD CNF NMAT PRC SETC	ZM03	CPSM-BPP-PRES
500015512105	PRESSURE SAFETY VALVE 220 R	25.02.2017	CLSD CNF NMAT PRC SETC	ZM03	CPSM-BPP-PRES
500015512106	PRESSURE SAFETY VALVE 221 R	25.02.2017	CLSD CNF NMAT PRC SETC	ZM03	CPSM-BPP-PRES
500015512107	Ammonia liquor piping (to NH3 still)	25.02.2017	CLSD CNF NMAT PRC SETC	ZM03	CPSM-BPP-PIPLI
500015512108	NH3 liquor piping (NH3 still-BOT plant)	25.02.2017	CLSD CNF NMAT PRC SETC	ZM03	CPSM-BPP-PIPELI
500015512109	LP steam piping (up to thermocompressor)	25.03.2017	CLSD CNF NMAT PRC SETC	ZM03	CPSM-BPP-PIPELI
500015512110	IP stem piping (up to thermo compressor)	25.03.2017	CLSD CNF NMAT PRC SETC	ZM03	CPSM-BPP-PIPELI
500015512111	LP steam piping (thermo to NH3 still)	25.03.2017	CLSD CNF NMAT PRC SETC	ZM03	CPSM-BPP-PIPELI
500015512112	Cautic piping-unloading to storage tank	25.03.2017	CLSD CNF NMAT PRC SETC	ZM03	CPSM-BPP-PIPELI
500015512113	cooling water piping (cooler to C-tower)	25.03.2017	CLSD CNF NMAT PRC SETC	ZM03	CPSM-BPP-PIPELI
500015512114	Caustic piping (ST to dosing & CBM)	25.03.2017	CLSD CNF NMAT PRC SETC	ZM03	CPSM-BPP-PIPELI

PSM Dashboard — Form No.: PSRM/COE/Dashboard/1

	Process Safety Incidents (Number)			SOC Deviation	SOL Deviation	No of PSM Critical equipment failed	Barrier Audit conducted		Barrier Health (C4/C5 Scenario)			Compliance of PSM Critical Equipment Maintenance Plan (Number of MOs generated / completed)				Delayed Compliance of MOs (Number)		Z8 Notification for PSM Critical Equipment (Number)			
												IEM		Mechanical				IEM		Mech	
	Red	Yellow	Green	Nos	Nos	Nos	Plan	Actual	Total No of barriers	Assessed	Red/Amber Nos	Generated	Completed	Generated	Completed	IEM	Mechanical	Open	Closed	Open	Closed
For the Month	0	0	0	11	0	0	1	1	9	9	0/1	51	51	10	10	0	0	0	0	0	0
Year till date	0	0	0	27	0	0	3	3	23	23	0/2	159	159	36	36	0	0	0	0	0	0

Take Away:

1. Enhanced Focus of PSV
2. System developed to validate Process PSV through SSTG

14. Sustenance Of COE (Centre of Excellence)

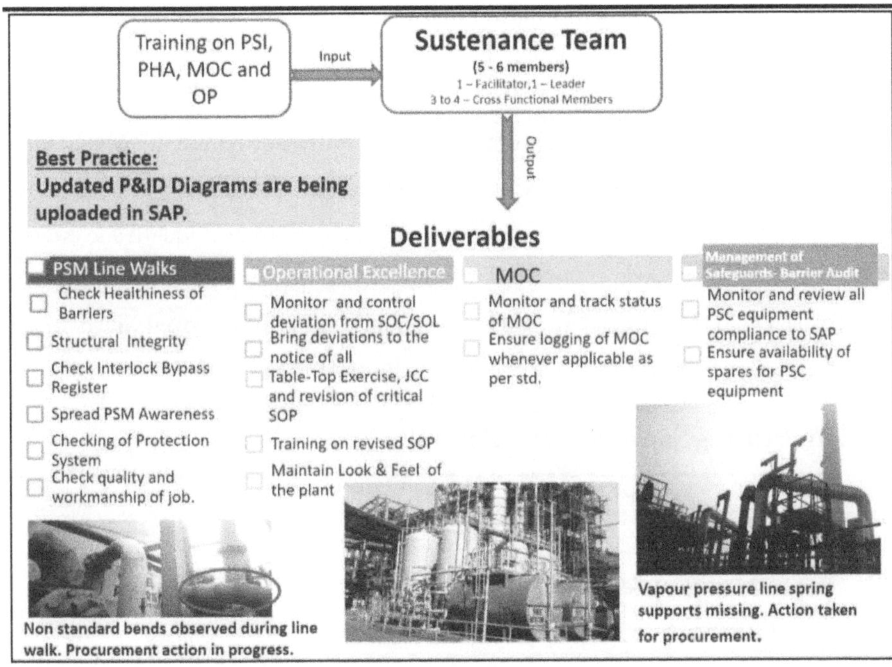

About the Author

Ashish Chakraborty possesses a master's in chemical engineering from NIT, Durgapur. He brings with him 38+ years of Experience in Premier Organisations. Ashish started his career in Hindustan Lever Ltd. After a short stint, he moved to Durgapur Projects Ltd., a State Govt undertaking, where he worked in different capacities (Coke Oven, Coal Chemicals). Later, he joined TATA Steel and further extended his exposure in Operations of Coke Oven By-Product, Process Safety and various Audits and is still associated with it. Apart from actively participating and leading in various Projects, Ashish played a strategic role in a particular project related to "Quality of Coke Oven Effluent", which got qualified in TATA Innovesta, earning lots of accolades and recognitions.

V.2. Hazard Identification and Risk Mitigation in Chemical Recovery Coke Oven

Susanta Kumar Halder

Since the early era of coke making on an industrial scale, the coke plant in totality is considered as one of the most polluting industries having a lot of dangerous processes related to safety hazards carrying a high level of risk for the plant installations, working people and also the surrounding living habitats.

Coke making process gives rise to "air pollution and water pollution" which apart from creating work zone safety hazards also affects the surrounding area with air, water and land pollution causing a health safety risk to the people around, aquatic life around and vegetation around.

Safety hazard in the form of "fire and explosion" also exists in coke oven which carries a lot of risk for the working crew and the installations within the coke plants.

1. Main Process Sources of Air Pollution

1. Process of oven pushing and charging
2. Coal crushing, storage, blending and transportation
3. Coke cutting, screening and transportation
4. Crude tar production, storage, distillation etc
5. BTX production, storage, distillation etc
6. Ammonium sulphate/sulphur pastil production
7. Naphthalene production and storage
8. Coke dry quenching
9. Chimney emissions
10. Flare stack
11. Wet quenching of hot coke

All the above processes give rise to a generation of fugitive emissions containing dust, gaseous fumes, suspended liquid particles etc in the plant proper as well as far surrounding areas which possess considerable personal health risk for the exposed people.

The respirable SPM (< 10 microns) present in the emissions beyond the permissible exposer limits and prolonged continuous exposer may lead to "lung cancer, kidney cancer, skin cancer, asthma "to the working crew directly and may cause health problems to the surrounding habitats.

The most dangerous constituent in the coke oven emission is the "benzene soluble fraction "which is carcinogenic and may cause a very high level of a safety hazard to the working crew.

2. Risk Mitigation Plan & Practices to Ensure SPM In Ambient Air < 50 Mg/Nm^3

1. Pushing & charging an oven should have a proper emission control system to ensure visible emissions < 16 sec. Zero leak oven doors, water seal ap cap, peakless charging adopting individual oven pressure control system, land-based de-dusting system for coke pushing on coke side, emission control unit in pusher car to take care of pushing and oven levelling/coal cake charging emissions.
2. Effective de-dusting / dust suppression units in all types of crushing, cutting & screening units in coal and coke plants.
3. all the storage tanks in the coal chemical areas should have nitrogen flushing at the free space above the liquid level and bleeder valves of the tanks should be connected to the inlet of primary coolers to arrest the obnoxious gaseous fume bleeding to the atmosphere.
4. Clause process to be adopted for removal of sulphur and ammonia from coke oven gas. Ammonium sulphate plant for removal of ammonia should be avoided on account of the very hazardous and risk-prone process.
5. The Flare stack should have an automatic online firing system to avoid clean coke oven gas bleeding to the atmosphere.
6. Fugitive emission control system in wet quenching of hot coke.

Even with all the above risk mitigation technology in place, there exists health-related personal safety risk to the working people in the certain regulated area of working where prolonged exposer cannot be avoided on account of technological requirements. To mitigate the personal safety risk, the following work practices are suggested to be adopted.

A. **Permissible exposure limit:** 150 micrograms per cubic meter of air determined as an average over 8 hrs period.

B. **Regulated areas:** coke oven battery including oven top, pusher side, coke side, battery ends, coke wharf, coal coke plant, tar plant, BTX plant, clause plant, naphthalene plant etc.

2.1 Health Hazard Data

Prolonged exposure to coke oven emissions beyond permissible exposure limits is a cause of lung cancer and kidney cancer in humans. Although there has not been an excessive number of skin cancer cases in humans.

2.2 Use Of Protective Clothing and Equipment

1. Respirators: suitable respirators to be used for working in areas where the level of coke oven emissions are over the PEL, and the work practice controls are not feasible.
2. Protective clothing: use of suitable protective clothing like "flameproof jackets & pants, flame-resistant hand gloves, heat-insulated helmets with face shield & safety shoes etc" to save the body from coke oven emissions and the heat generated during the coking process.

2.3 Hygiene Facilities and Practices

Eating, drinking, smoking, chewing gum or tobacco are strictly forbidden in regulated areas except for drinking water. All these activities are permitted in a designated place only. Use eye washing and emergency showers in case of emergency need.

2.4 Signs, Slogans and Labels

All required safety-related slogans, warning signs & labels for protection, hazards involved etc to be displayed in regulated areas. Strict entry restrictions are to be enforced in areas where coke oven emissions are above the PEL.

2.5 Medical Examinations

Workmen working in regulated areas for more than 30 days per year are to undergo periodic medical examinations. The examination must include medical history, chest x-ray, pulmonary function test, weight comparison, skin examination, urine analysis and urine cytology for early detection of urinary cancer.

2.6 Training And Education

All workmen in the coke oven are to be educated and informed on these items and to be trained on hazards of coke oven emissions and engineering and work practice controls associated with every kind of job done in regulated areas including proper use of PPE.

All works in these regulated work zone should be done with certified work instruction (sop) by authorised persons only.

3. Water Pollution and Risk Mitigation

1. The source of water pollution is in a chemical recovery coke oven is the generation and treatment of excess ammonia liquor during coal carbonization which contains highly toxic elements like "cyanide, ammonia, phenol etc. "and that requires elaborate chemical and biological treatment to make it safe for reuse/discharge outside. As per the CPCB guidelines, all coke oven plants are now zero discharge plants to ensure safety to the aquatic life, agriculture, animals and people surrounding the plant.
2. To achieve the CPCB norms in the processed wastewater outlet, the following measures are a part of the standard plant practices: -
3. Separate drainage system for the process wastewater system and the stormwater drainage
4. Bod plant to treat the excess ammonia liquor.
5. Large groundwater pit to collect the process drain water for pre-treatment and feed to bod plant for biological treatment.
6. All stormwater drainage is to be collected in large pits for settlement and treatment before being fed to the plant reservoir for reuse.
7. Part of the water to be used as make up for wet quenching of hot coke.
8. Treated water can be used for gardening, floor and road washing, SMS slag yard etc.

Very careful and safe practices are required to be adopted to handle this liquor in the process and during subsequent use after treatment. This can cause harm to people who come in direct contact.

4. Fire And Explosion Hazard Associated with Coke Making Process

The entire coke oven complex is a fire and explosion prone area due to the presence of "gases, fumes, chemicals and fine coal coke dust. Fire and explosion hazards exist in all the areas having a different degree of occurrence and risk. Any incident of fire or explosion may cause permanent damage to the installations/equipment and can cause loss of life due to a burning injury.

Zone classification of the fire and explosion hazards.

Gases and flammable vapours	Zone 0	An environment where the explosive/fire-prone atmosphere is present continuously or for long periods.
Gases and flammable vapours	Zone 1	An environment where the probability of existing a fire cum explosive atmosphere is likely to occur in normal operation
Gases and flammable vapours	Zone 2	An environment where a fire cum explosive atmosphere is not likely to occur in normal operation and if any that will be for a short period.

Dust	Zone 20	An environment where flammable dust is present continuously or frequently under normal operating conditions in enough amount to generate an explosive concentration of mixed dust with air and /or areas where may occur excessive amount of dust with no further control.
Dust	Zone 21	An environment where the probability of existing a fire cum explosive atmosphere mixed with the dust in the air is likely to occur in normal operation.
Dust	Zone 22	An environment where the probability of existing a fire cum explosive atmosphere mixed with the dust in the air is not likely to occur in normal operation or if any, this will be for a short period.

Areas of coke oven as per the explosive zone classification:

1. Zone 0: battery cellar, chemical storage pits with storage tanks and areas which are confined in nature and handling gas / flammable liquid vapours. Slight leakages can lead to the concentration of explosive fumes leading to fire/explosion.
2. Zone 1: entire coal chemical plant and gas pip-lines.
3. Zone 2: clean coke oven gas and bf gas pipelines. Gas preheaters etc.
4. Zone 20: coal storage silos, coal tower, coke bunkers
5. Zone 21: coal yard, crushers etc.

4.1 Zone Wise Risk Mitigation

Zone 0, zone 1, zone 2

- All electrical installations are of explosion/flameproof type.
- Leakage free atmosphere
- Use of gas monitors “CO, BTX vapour, ammonia vapour etc” (permanent and portable)

- Use of non-sparking tools for maintenance work
- All maintenance work with proper protocol.
- All operations following a certified **sop.**
- All work with certified-trained personnel only.
- Strick observance of gas safety rules.
- Surface colling of chemical storage tanks with nitrogen sealing of the tank inside free space.
- All tank safety valves, bleeders, vents to be connected to raw cog line before primary cooler.
- Blanking and de-blanking of gas lines with proper protocol and with oxygen < 0.8 % inside the pipeline after flushing with nitrogen.
- Use of all necessary personal safety gear at work.

Zone 20,21,22

- Nitrogen flushing in coal silos and coal tower
- Water mist spray in the coal yard
- Proper quenching of hot coke
- Efficient de-dusting system in crushing and screening houses
- Dust-free MCC installations

An environment impact assessment register has to be developed and maintained at the shop floor for information and training which shall be upgraded every month based on the study and observations to ensure timely corrective actions on the highly adverse activities scoring high weightage.

Format for Impact Aspect										
Sl No	Activity	Aspect	Impact	Type	Condition of Significance				Qty A	Prob. Of Occur B
					LC	RSP	IPC	Emer		
1	Battery Heating	Waste Gas	Adverse	Air Pollution	√		√		4	5
2	Coal Charging	Fugitive Emission	Adverse	Air Pollution	√		√		1	4
3	Coal Carbonisation	Fugitive Emission	Adverse	Air Pollution	√		√		1	5
4	Oven Pushing	Fugitive Emission	Adverse	Air Pollution	√		√		1	4
5	COG Cleaning	Excess Coal Water	Adverse	Water Pollution	√	√	√		2	2
6	Coal Preparation & Conveying	Fugitive Emission	Adverse	Air Pollution	√	√	√		2	4
7	Coke Sizing & Transportation	Fugitive Emission	Adverse	Air Pollution	√	√	√		2	4
8	Crude Tar Processing	Solid waste (Sludge)	Adverse	Land Pollution	√	√	√		3	3
9	Crude Tar Processing	Tar Vapour	Adverse	Air Pollution	√	√	√		2	5
10	Excess Coal Water Treatment	Solid waste (Sludge)	Adverse	Land Pollution	√	√	√		1	1
11	Benzol Storage	Benzol Vapor	Adverse	Air Pollution	√	√	√		1	5
12	Solid Sulpher Production	Waste Water	Adverse	Water Pollution	√	√	√			

Evalatio

Note:
LC: Legal Concern
RSP: Resource saving potential
IPC: Interested party concern
EMER: Emergency

1. Quantity	
Scale:(1 -5)	
Excessive	5
Very High	4
High	3
Moderate	2
Low	1

2.Occurance
Scale:(1 -5)
Continious
Several Times in a Day
Once in a Day
Once in a Week
Once in a Month /Less frequency

4.Legal Requirment	
Meeting legal	1
Not Meeting legal requr.	5

5. Detection	
Preventive Mechanisim	
Scale:(1 -5)	
More than 24 Hours	5
More than 12 Hours	4
More than 06 Hours	3
More than 01 Hour	2
Immeditely	1

Note:if total score

ts Register

Impact C	Legal Requ. D	Ctrl E	D'Tecn F	Score G(A*B*C*D*E*F)	SIG /. Non Aspect	OCP/ EMP
2	1	1	1	40		
2	5	1	1	40	16 sec Visible emmission	
2	1	1	5	50		
2	1	1	5	40		
3	1	1	4	48		
1	5	1	1	40		
1	5	1	1	40		
3	1	2	1	54		
2	1	2	1	40		
3	1	1	1	3		
4	1	1	1	20		
				0		

on Criteria:

5
4
3
2
1

3.Impact	
Scale:(1 -5)	
Fatal to Human Life	5
Human Health EFFECT	4
Kills Marine Life Flora/Fauna/Global issues/ Resource Cojumption	3
Cause Discomfrt/Acid Rain/Nuisance	2
Slight Impact/negligible Vissual Impact	1

e is 50 then it should be consider as significant.

5. Hira (Hazard Identification and Risk Abetment)

Plans are to be prepared for identifying the hazardous processes associated with the coke ovens for continuous monitoring and improvement in the risk factors.

Below is the model format for some identified process activities/works is attached as an Appendix for illustration purposes.

About the Author

Susanta Kumar Halder is Ex Sr. VP, BSL TSL, B. Tech (Metallurgy), IIT KANPUR. He has forty years of total working experience in the field of Coke Making Comprising of "Project Conception, Project Engineering, Project Execution, Heating & Commissioning, Repair & Rebuilding, Operation and Maintenance. For 26 years he has worked in Chemical Recovery Top Charge Coke Oven of SAIL Plants (Bhilai and Durgapur) having Batteries of 4.3 m, 4.5 m and 7.0 m tall ovens of "Russian, British and German Design. He has 4 years of experience in Heat Recovery Horizontal and Vertical Coke Ovens of Electrosteel with Stamp Charging (Indian & Chinese Design) and Waste Heat Base Power Generation as head of coke ovens. For 10 years he worked as head of Coke Oven in 4.5m Stamp Charge Chemical Recovery Battery of Chinese Design and 7.6m top charge battery of Paul Wurth Design in TATA Steel BSL Limited. 180 tph Single Chamber Coke Dry Quenching Plant of Nippon Steel Design. State of the Art BOD Plant of 130 Cu.m/hr by Mutagg, Germany. He has special Expertise in Coking Coal Evaluation, Procurement and Cost-Effective Blend Making for Carbonization Since September 2019 providing Technical and Engineering Consultancy to different industries related to Coke Making comprising of "Greenfield project, O&M, Repair, DD, Health Assessment and usage of Coking Coal etc.

6. Appendix

INDENTIFICATION OF OH&S HAZARD, RISK ASSESMENT AND RISK MANAGEMENT MEASURE

Activity / Services	OHS Hazard / Associated Risk Identification								Risk Assesment						LC/DC If any	Risk Management Measure Required
	Hazards	Condition			of Measure		Risk			Severity : S2 = L X C						
		D/I	R/NR	N/AN/E	Existing	Gaps if any	Likelyhood	Consequence	Scale (S1)	Likely hood (L)	Level of Consequence (C)	Delectability (D)	Total = S1+S2+D	Risk Classification		
Oven Pushing	Wrong Pushing(Hot Coke Spillage)	D/I	NR	E	√		HUL	Severe Property Damage	1	1	6	1	8	Substantial		Strict Compliance to SOP. No bypass of Machine to Oven Interlocking system.Use of PPE.Operation by trained manpow er.
	Exposer To Heat & Rediation	D/I	NR	E	√		HUL	Superficial Burn	1	1	2	1	4	Moderate		Strict compliance to use of PPE. Emergency Show er
	Pusher Beam Stuck inside the Oven	D/I	NR	AN	√		UL	Property Damage	1	1	4	1	6	Moderate		To maintain the emergency w inch alw ays in w orking condition.Operation by trained manpow er as per SOP.
Oven Charging	Exhauster Failure	D/I	NR	E	√		UL	Severe Property Damage/Burn Injury	1	1	4	1	6	Moderate		Stop Pushing/Charging. Strict compliance to SOP for pow er failure. Use of PPE. Trained manpow er. No further operation.
	Exposer To Heat,Raw Gas & Rediation	D/I	NR	E	√		UL	Superficial Burn / Suffocation	1	1	2	1	4	Moderate		Strict Compliance to SOP . Use of PPE.Trained manpow er.
	SOPRECO/HPLA Failure(Smokless Charging)	D/I	NR	E	√		UL	Severe Property Damage	1	1	4	1	6	Moderate		Strict Compliance to SOP.Use of PPE. Trained manpow er. No charging w ithout SOPRECO.
Heating Gas Changeover in cellar	Gas Leakage / Explosion/Fire	D/I	NR	AN	√		UL	Severe Property Damage/burn Injury/Fatal	1	1	6	1	8	Substantial		By Protocol only. Strict Compliance to Gas Safety.Use of PPE.Use of portable CO monitor.Use of Cellar Ventillation system. Cellar equipped w ith portable Fire Extinguisers and fire hydrant line.
Blanking/De-blanking of Gas Lines	Gas Leakage / Explosion/Fire	D/I	NR	AN	√		UL	Severe Property Damage/burn Injury/Fatal	1	1	6	1	8	Substantial		By Protocol only. Strict Compliance to Gas Safety.Use of PPE.Use of portable CO monitor.Use of non sparking tools. Standby fire Tender.
Exhauster Changeover	Gas Leakage / Explosion/Fire	D/I	NR	AN	√		UL	Severe Property Damage/burn Injury/Fatal	1	1	6	1	8	Substantial		Strict Compliance to SOP. Use of PPE . To be done by PLC control only. By high skilled trained manpoer.

LC : Legal Concern , if any concerning activity / associated hazard is covered under Indian OHS legislation.

DC : Domino Concern , if concerning activity / associated hazard can trigger series of accident.

Any Other : Any Other relevent factor if applicabe related to / of affecting the significance of OHS risks.

D: Direct , **I** : Indirect , **R**: Routine , **NR**: Non Routine , **N**: Normal , **AN**: Abnormal , **E** : Emergency

FR - Frequent , **L**-Likely , **UL** : Unlikely , **HUL** : Highly Unlikely

SOP : Standard Operating Practice

INDENTIFICATION OF OH&S HAZARD, RISK ASSESMENT AND RISK MANAGEMENT MEASURE

Activity / Services	OHS Hazard / Associated Risk Identification								Risk Assesment						LC/DC If any	Risk Management Measure Required
	Hazards	Condition			of Measure		Risk		Scale (S1)	Severity : S2 = L X C						
		D/I	R/NR	N/AN/E	Existing	Gaps if any	Likelyhood	Consequence		Likely hood (L)	Level of Consequence (C)	Delectability (D)	Total = S1+S2+D	Risk Classification		
Crude Tar Storage and dispatch	Spillage / Leakage / Fire	D/I	NR	E	√		UL	Severe Property Damage/Fatal	1	1	6	1	8	Substantial		Strict Compliance to SOP for storage and Tanker Loading. Use of PPE. Earthing of Tankers. Proper overflow control. Emergency Show er . Eye Spary.Dyke around the tank. Operation by Trained Manpow er.
BTX Production	Spillage / Leakage / Fire	D/I	NR	E	√		UL	Severe Property Damage/Explosion/burn Injury/Fatal	1	1	6	1	8	Substantial		Strict Compliance to SOP for Operation of the Stripper and Transfer of BTX to Tank. Installation of Foam Spray Required. Proper Earthing of all electrical installation w ith flame proof sw itches and motors. Overflow control. Emergency Show er . Eye spary. Use of proper PPE.Fencing around the stripper area. Use of LEL Sensors. Operation by Trained Manpow er.
BTX Storage & Dispatch	Spillage / Leakage / Fire	D/I	NR	E	√		UL	Severe Property Damage/Explosion/burn Injury/Fatal	1	1	6	1	8	Substantial		Strict Compliance to SOP for storage and Tanker Loading. Installation of Foam Spray Required. Earthing of Tankers. Overflow control. Emergency Show er . Eye spary. Dyke Wall around the tank area.Use of LEL Meter. Operation by Trained Manpow er.
Solar Oil Unloading & Storage	Spillage / Leakage / Fire	D/I	NR	AN	√		UL	Severe Property Damage/Explosion/burn Injury/Fatal	1	1	6	1	8	Substantial		Strict Compliance to SOP for storage and Tanker Loading. Installation of Foam Spray Required. Earthing of Tankers. Overflow control. Emergency Show er . Eye spary. Dyke around the storage tank. Operation by Trained Manpow er.
NaOH Unloading & Storage	Leakage / Spillage	D/I	NR	AN	√		UL	Superficial burn Injury	1	1	4	1	6	Moderate		Strict Compliance to SOP for storage and Tanker unloading. Installation of Foam Spray Required. Earthing of Tankers. Overflow control. Emergency Show er . Eye spary. Use of PPE. Dyke around the storage tank.
Storage of Liquid Sulpher	Spillage/ Leakage / Explosion/Fire	D/I	NR	AN	√		UL	Severe Property Damage/burn Injury/Fatal	1	1	6	1	8	Substantial		Strict Compliance to SOP for storage . Fire Hydrant. Dyke around the tank.Overflow control. Emergency Show er . Eye spary.Operation by Trained Manpow er.

LC : Legal Concern , if any concerning activity / associated hazard is covered under Indian OHS legislation.

DC : Domino Concern , if concerning activity / associated hazard can trigger series of accident.

Any Other : Any Other relevent factor if applicabe related to / of affecting the significance of OHS risks.

D: Direct , **I :** Indirect , **R:** Routine , **NR:** Non Routine , **N:** Normal , **AN:** Abnormal , **E :** Emergency

FR - Frequent , **L**-Likely , **UL :** Unlikely , **HUL :** Highly Unlikely **SOP** : Standard Operating Practices

INDENTIFICATION OF OH&S HAZARD, RISK ASSESMENT AND RISK MANAGEMENT MEASURE

Activity / Services	OHS Hazard / Associated Risk Identification								Risk Assesment						LC/DC If any	Risk Management Measure Required
	Hazards	Condition			of Measure		Risk		Scale (S1)	Severity : S2 = L X C						
		D/I	R/NR	N/AN/E	Existing	Gaps if any	Likelyhood	Consequence		Likely hood (L)	Level of Consequence (C)	Delectability (D)	Total = S1+S2+D	Risk Classification		
Sulpher Pastill bagging and storage	Spillage / Fire	D/I	NR	E	√		UL	Property Damage/Burn Injury/Suffocation	1	1	4	1	6	Moderate		Strict Compliance to SOP for storage and Baggigng. Fire Hydrant. Eye Spray. Emergency Show er. Operation by trained manpow er.
Methonol Unloading & Storage	Spillage / Leakage / Fire/Explosion	D/I	NR	E	√		UL	Severe Property Damage/burn Injury/Fatal	1	1	6	1	8	Substantial		Strict Compliance to SOP for storage and Tanker Unloading. Installation of Foam Spray Required. Earthing of tanker during unloading. Explosive License.Operation by trained manpow er
Storage & Handling of H3PO4.	Spillage / Leakage	D	NR	AN	√		UL	Superficial Burn Injury.	1	1	4	1	6	Moderate		Strict Compliance to SOP for storage and IBC Unloading. Installation of Eye Washer Required.Operation by trained manpow er.
Storage & Handling of FeCl3	Leakage / Spillage	D	NR	AN	√		UL	Superficial Burn Injury.	1	1	4	1	6	Moderate		Strict Compliance to SOP for storage and IBC Unloading. Installation of Eye Washer Required.Operation by trained manpow er.
Storage & Handling of Phosphoric Acid	Spillage/ Leakage	D	NR	AN	√		UL	Superficial Burn Injury.	1	1	4	1	6	Moderate		Strict Compliance to SOP for storage and IBC Unloading. Installation of Eye Washer Required.Operation by trained manpow er.

LC : Legal Concern , if any concerning activity / associated hazard is covered under Indian OHS legislation.

DC : Domino Concern , if concerning activity / associated hazard can trigger series of accident.

Any Other : Any Other relevent factor if applicabe related to / of affecting the significance of **OHS** risks.

D: Direct , **I :** Indirect , **R:** Routine , **NR:** Non Routine , **N:** Normal , **AN:** Abnormal , **E :** Emergency

FR - Frequent , **L**-Likely , **UL :** Unlikely , **HUL :** Highly Unlikely

SOP : Standard Operating Practice

V.3. Safety Protocol in Coke Oven Plants

Ramakrishna Batabyal

As a young boy growing up in a post-liberalization India, where the internet boom had opened big opportunities everywhere, I wished to follow the footsteps of my grandfather who worked in Bokaro Steel Plant. For many people, these dreams do not often come true, so it came as a shock when Tarun Chatterjee allowed me to present an engineering student's perspective on Safety Protocols within Coke Oven plants and I remembered my visits to Bokaro Steel as a high school boy to start making studies on the safety aspects of this most hazardous unit of the Steel plant, the Coke Oven and Chemical Plant; it was indeed a great learning experience for me.

I realized that if one has to learn the technical aspects of engineering details of an industry one must first learn the most important of all aspects, i.e., SAFETY. The Coke Oven is very unique in that it has moving machines, a coal storage yard, and conveyor system, Coke Oven Batteries handling red hot coke, Coke dry cooling plant and a Chemical plant handling toxic flammable chemicals in liquid and gaseous forms. I will introduce this plant to the reader through its Safety protocols.

In our everyday lives, we witness accidents galore, and property, jobs, relations, often even lives are lost. Human error is one of the largest triggers for accidents, and as an ever-advancing society, we must find ways to not err as such. People find it convenient to bypass safety rules, take shortcuts and carry out tasks without regard for the second person, so to enforce rules strictly in accident-prone areas like a Coke Oven plant is essential. Writing this paper through research on the internet and combing through government guidelines I found that the Ministry of Steel, Government of India had made very valuable recommendations identifying the Hazards and Risk Control Measures specific to the various units under a Coke Oven Plant. I have also written about Safety Protocols for HT & LT installations and the Heat Recovery Steam Generation units of Coke Oven Plants.

1. Coke Oven

Coke Ovens have 3 major sub-units: Coal Handling Plant, Coke Oven Batteries and Coke Sorting Plant.

TABLE-1: SAFETY PROTOCOL FOR COKE OVEN SUB-UNITS[1]

S.No.	AREA/SECTION	HAZARDS	RISK CONTROL MEASURES
1	COAL HANDLING PLANT	Hand getting pressed during coupling decoupling of wagons	1. Use scotch block below wagon wheel. 2. Proper Communication. PPE Usage
		People struck by moving Side arm charger	1. Proper personnel security. 2. Interlock between sidearm charger operation and ground working
		Moving Conveyor/shuttle conveyor	1. Shutdown/permit to work with electrical isolation. always operate the local emergency switch. Auto hooter blowing whenever shuttle shifting command is given. 2. No work should be done on the conveyor in running condition. Sufficient illumination. regular cleaning. Proper guard for both Head and Tail end
		Rotating machinery	1. Coupling guards to be in place. 2. Loose cloth to be prohibited.
		Fire hazard due to self-ignition of coal and due to rubbing of conveyors/ Idlers etc.	1. Proper Maintenance. 2. Zero Speed Switches. 3. No overload
2	PARTIAL BRIQUETTING	Moving conveyors	SOP for Conveyor to be developed & followed.
3	COKE OVEN BATTERIES	Wrong Pushing	Machine interlock to be provided & maintained
		Coke falling on a person from an open door	1. Restricted entry 2. FR Jackets to be worn. 3. Frame Cleaning 4. Use of coal cake Scrapper during breakage. 5. No frontal contact during breakage. 6. Charging level to be maintained and levelling shall be done.
		Person hit by moving machines	1. Siren, gong bell during movement of oven machines. 2. Auto announcement during any coke oven machine operation. 3. Before starting a standing car long hooter should be given to an operator to check.

[1] *Source:* *Document SG/28 – Safety Guidelines for Iron & Steel Sector, Ministry of Steel, Government of India*

S.No.	AREA/SECTION	HAZARDS	RISK CONTROL MEASURES
		Gas Leakage and backfire in cellar and gas pipelines	1. Online CO monitors with an alarm system to be installed for detecting gas leakage if any. 2. In addition to this, Portable CO monitors are to be made available at the site to detect gas leakage. 3. A person should not move without a detector inside the cellar. 4. Adherence to SOP for Gas change over operation 5. Check that all COG DP plugs are in position before opening CO gas cocks. 6. Close BF gas cock by observing the indication mark. 7. Use the gas monitor to check the concentration of CO. 8. Don't be on an empty stomach. 9. Ensure smooth operation of CO /BF gas butterfly valve. 10. Ensure positive pressure (>50mmwc) is always in the gas pipeline. 11. Interlock to be provided to bring reversing winch machine in the pause in case of very low pressure to prevent air infiltration inside the pipeline to protect explosion. 12. Fire extinguishers to be provided.
4	COKE SORTING PLANT	Moving Conveyor	Conveyor safety procedure to be followed. (Regulations for Fire control given above)

2. By-Product Plant

The Coke Oven gas has a Calorific value of 4400 kcals/NM3 and is used as fuel in the steel plant after recovering the chemicals. The gases that emerge from the oven are extracted, cooled off and then separated for different chemicals, such as Ammonia, or Benzol. The gas is later used as Fuel. The BPP comprises the following components and sub-units: Gas Condensation Plant, Ammonium Sulphate Plant, Benzol Recovery Plant, Benzol Rectification Plant, Tar Distillation Plant, Sulphuric Acid Plant.

TABLE-2: SAFETY PROTOCOL FOR BY-PRODUCT PLANT SUB-UNITS[2]

S.No.	AREA/SECTION	HAZARD	RISK CONTROL MEASURES
1	**GAS CONDENSATION PLANT**	Spillage of hot liquor from leaky flanges/pump gland	□ Regular inspection to observe flange/gland leakage before a changeover and rectify the same with safety protocol.
		Burn due to leaky steam line flange / hot condensate / touching hot steam line	□ PPEs to be used for leaks in glands/flanges
		Exposure to CO/ Mixed gas	□ Use of gas mask if CO % is above 50ppm. Maintain water seal.
2	**AMMONIUM SULPHATE PLANT**	Exposure to ammonium Sulphate powder/Sulphuric Acid Spray/Acidic Liquor	Regular Maintenance and housekeeping. Regular cleaning of pipes and tubes. Provision of Safety Shower
3	**BENZOL RECOVERY AREA**	Exposure to Benzol vapours	Check/ use of continuity jumpers over pipeline flange.
4	**BENZOL RECTIFICATION PLANT**	Exposure to BTX vapours	Usage of mechanical seals in pumps.
5	**TAR DISTILLATION PLANT**	Formation of Explosive mixture leading to an explosion	The explosive mixture is to be checked before starting work.
		Spraying/ Spillage of tar/ tar product	Isolation of line & draining of line before starting job. Regular cleaning of the area
		Exposure to cold wash oil	Exposure to cold wash oil
6	**SULPHURIC ACID PLANT**	Exposure to acid from the leaky flange, valve glands etc	Regular checking of pipelines, flanges and glands for possible leakages and attend leakages on priority. Keep a safe distance from acid lines while the pump is in operation,
		Overflow of acid from wagon	Develop & follow SOP for Acid loading / lifting. Keep a safe distance from the acid wagon.
		Exposure to any acid/chemical compound deemed hazardous	Ensure proper isolation and shut down before hand over of line for maintenance. Keep a safe distance from open flanges.

[2] *Source:* *Document SG/28 – Safety Guidelines for Iron & Steel Sector, Ministry of Steel, Government of India*

3. Coke Dry Cooling Plant

Red hot coke pushed out of the coke oven battery at 1050° C is cooled to 180° C in this unit and heat recovered in the form of high-pressure steam is used to generate power in the steam turbine.

TABLE-3: SAFETY PROTOCOL FOR COKE DRY COOLING PLANT[3]

1. ALL CDCP FACILITIES	
HAZARD	THE EXPLOSIVE MIXTURE ON THE CIRCULATING GAS ROUTE
RISK CONTROL MEASURES	Working of interlocks and automation shall be ensured. The percentage of oxygen content should not exceed 1 %. The percentage of hydrogen content should not exceed 4 %. The percentage of Carbon mono oxide should not exceed 8%. Availability of Nitrogen shall be ensured. Proper functioning of nitrogen valves shall be ensured. Proper functioning of the gas bleeder valve after mill fan shall be ensured. Proper functioning of pre-chamber bleeder valve shall be ensured Explosion flaps located in the circulating gas path shall be in properly cleaned condition and should be inspected from time to time. Proper functioning of air breather valve at pre-chamber shall be ensured. Proper functioning of the gas analyser shall be ensured.
HAZARD	PERSON HIT BY ARMS & HOOKS OF CRANE AT COKE BUCKET REPAIR GARAGE.
RISK CONTROL MEASURES	Siren gong bell and auto announcement during movement of Crane. Before starting, Crane long hooter should be given to an operator to check for any person doing any job. Development & adherence to SOP.
HAZARD	MOVING CONVEYOR AFTER DISCHARGING DEVICE
RISK CONTROL MEASURES	Shutdown/ permit to work with electrical isolation. No work should be done on the conveyor in running condition. Always operate Local emergency switch for approaching conveyor. Availability & working of Pull chord. Siren the system before restarting the conveyor. Loose clothes are prohibited. Area barricading if the material is removed from height.

[3] *Source:* ***Document SG/28 – Safety Guidelines for Iron & Steel Sector, Ministry of Steel, Government of India***

	Headend and tail end guard. Crossing the conveyor only through the designated place. SOP for Conveyor to be developed & followed
HAZARD	**BURN & INJURY DUE TO HP/ MP/ LP STEAM**
RISK CONTROL MEASURES	Regular inspection of the steam pipeline. IBR authorized/certified personnel should only be allowed for operation and maintenance of the Boiler. The periodic inspection of the Boiler inspector for certification & validity of license shall be ensured. Calibration of visible inspection gauge on regular basis. Proper inspection, testing and functioning of safety valves of Boiler drum & SH Steam pipeline. Operator/ technician should wear Safety jackets while working around the steam pipeline. Development & adherence to SOP.
HAZARD	**HOT COKE FALLING ON A PERSON FROM COKE BUCKET WHILE LIFTING/MOVING OF COKE BUCKET LIFTER. THE FREEFALL OF HOT COKE BUCKET.**
RISK CONTROL MEASURES	Visual inspection of coke bucket lifter rope on regular basis. Scheduled replacement of rope as per SOP shall be ensured. Visual inspection of brake and brake shoes of coke bucket lifter on regular basis. Nobody should pass in front/ below of Coke Bucket Lifter. Proper gripping of hooks at both side-arms of coke bucket shall be ensured while lifting hot coke bucket. Interlock to stop lifting of the bucket in case the hook is engaged only on the side. Also, it is preferred to install load cells on lifters to give feedback of load being lifted as well as individual hook load to ascertain proper load transfer on the lifter hooks. All watchmen/fieldmen of the lifter should wear FR jackets. Siren gong bell during lifting/moving of Coke Bucket Lifter. Working of interlocks and automation of coke bucket lifter shall be ensured.
HAZARD	**PERSON HIT BY COKE BUCKET CAR & ELECTRIC LOCO**
RISK CONTROL MEASURES	Siren gong bell during movement of Coke Bucket Car & Electric Loco. Auto announcement during Coke Bucket Car & Electric Loco operation.

	Before starting a standing Coke Bucket Car & Electric Loco long hooter should be given and operator to check for any person doing any job. Permit to work before undertaking any maintenance job around Coke Car Track & Bus bar. Climbing is prohibited on the running machine. Interlock to be provided to stop the machine if someone tries to climb on the running machine. Development & adherence to SOP.
HAZARD	CO GAS ACCUMULATION & FIRE IN CONVEYOR TUNNEL
RISK CONTROL MEASURES	Online CO monitors with an alarm system to be installed for detecting CO gas accumulation. Portable CO gas monitors are to be made available at the site to detect CO gas accumulation. A person should not move without a detector inside Conveyor Tunnel. Adherence to SOP for maintenance job in Conveyor Tunnel. Don't go to work on an empty stomach Fire Extinguisher to be provided. Ensure continuous running of ventilation fan on Conveyor Tunnel. Spot quenching system to be provided to avoid hot coke onto the moving conveyor. Working of interlocks and automation shall be ensured. Coke temperature detectors installed on conveyor belt / discharging device will give feedback to control the discharging rate to avoid hot coke discharge on Coke conveyor than the permitted value. Provide automatic fire detection cum alarm system (by LHS cable) along with automatic MVWS system.
HAZARD	WORK IN CONFINED SPACE/PITS
RISK CONTROL MEASURES	Permit to Work to be taken from the owner department. Safety training and do's/ don't instruction to workers. Check the concentration of oxygen, toxic, explosive gases, etc. before work. Proper emergency rescue arrangements.
HAZARD	ROTATING MACHINERY (ID FAN, MILL FAN, PUMPS, MOTORS, ETC.)
RISK CONTROL MEASURES	Coupling guards to be in place. Loose cloth to be prohibited. Motor fan cover to be in place.

2. WASTE HEAT BOILER	
HAZARD	EXPLOSION; LEAKAGE; STANDARD SAFETY REGULATIONS BASED ON THE INDIAN BOILER REGULATIONS ACT 1950
RISK CONTROL MEASURES	All Boilers must have IBR certification/approval. All pipes and valves must be of IBR defined standards The boilers must never be used beyond defined power and pressure limits. Regular (on the clock) maintenance to check for loose oil on the floor before flash-up, loose connections/valves, cracks allowing leakage of steam, clean pipes and lines, and more safety standards. Apart from the IBR defined regulations, only a trained professional should operate the boiler to ensure maximum precaution.

The possibilities for accidents are huge in number. To out plan such incidents, we need to ensure that basic safety guidelines, like those listed by the Ministry of Steel, Government of India, such as in Document SG/28, are followed without fail, without any lapses. This means PPE kits, Fire Extinguishers and Fire Hazard kits, caution around heavy machinery, safety shower sprays, minimal technological distractions etc., have to be strictly implemented across Steel Plants, in Coke Ovens.

4. Machines & Motors

There are, of course, several machines used inside a Coke Oven Plant. In investigating the safety hazards for such machines (which are quite large given their operations), we must briefly go through what those machines are; these are divided between HT and LT Motors. HT motors are High Tension or High Voltage devices. These drive much greater torques than LT motors or Low Voltage motors. The voltage ranges for HT motors are between 3.3kV, 6.6kV, and 11kV, while LT runs on ratings of 230V, 415V, and 690V. HT motors are very large due to the large insulation required and their large voltage ratings, while it is not the same for LT motors.

The machinery listed hereunder LT and HT motor classifications have to be handled with extreme care, have to be given power only when required, all connections must be secure, thoroughly insulated, and far from the risk of sparking (& fires), and should be run only within safe limits. The transportation of these machines needs to be safe, without much damage to the system, there should be no personal inflammable used around the equipment, and regular maintenance has to be carried to maintain sanitary and industry standards.

The following is a list of primary machines in use within a CO Plant:

TABLE-4

S.N.	MACHINE	USE
1	COKE GUIDE CAR	Door extractor, engage extraction and installation of the door, approach-withdrawal of coke guide cage to the oven, directed coke descent in coke quenching car, cleaning of doors and frames, collecting of dust-gas mixture, removal of coke spillages
2	COKE PUSHER CAR	The Pusher Car to a large extent operates the pushing and extraction of the door, levelling coal blend and coke pushing as is obvious.
3	COAL CHARGING CAR	Charging Cars mainly Charge Coal blend to Coke Ovens.
4	COKE BUCKET CARS	Coke bucket cars are designed for receipt of coke having the temperature 1000-1050°C from the chamber of coke oven battery, its transportation by means of Electric Loco, lifting and discharging of coke into quenching chambers of coke dry cooling plants (hereinafter CDCP).
5	COKE QUENCHING CARS	Coke Quenching Car is designed for receipt of hot coke from coke oven chamber, transportation of hot coke to wet quenching tower, transportation and discharging of cooled coke on coke wharf.
6	ELECTRICAL LOCO	The Electric Loco moves the Coke Quenching Car or Coke Bucket Car along the factory line of processes.
7	HOISTER (CDCP)	The Hoister transfers/lifts/hoists the red-hot coke we observe post-coking to the top of the CDC Plant.

5. References

- SG-25: Safety Guideline for Loco Operation & SG-18: Personal Protective Equipment
- (PPE) Management, Ministry of Steel, G.O.I.
- SG-03: Safety Guideline for Working in a Confined Space, Ministry of Steel, G.O.I.
- SG-04: Safety Guideline for Permit to Work (Operation & Maintenance), Ministry of Steel, G.O.I.
- SG-05: Safety Guideline for Illumination at the workplace, Ministry of Steel, G.O.I.
- SG-09: Safety Guideline in Equipment & machine Guarding, Ministry of Steel, G.O.I.
- SG-28: Safety Guidelines for Iron & Steel Sector, Ministry of Steel, G.O.I.
- SG-16: Safety Guideline for Fire Safety, Ministry of Steel, G.O.I

About the Author

Ramakrishna Batabyal is a student of Shiv Nadar Institute, Greater Noida, Uttar Pradesh. Currently, he is in the third year of electrical engineering

VI. ENVIRONMENT

VI.1. The Environmental Challenges for Indian Coke Making and Future Technologies

K D Choudhury

1. Introduction

We know that Coke is used in blast furnaces in the conversion of iron ore to iron, which can be further refined to produce steel. Coke plants produce coke from coal, using coke oven batteries. A battery consists of a group of ovens connected by common walls.

The air toxics reduced by the good practices will include polycyclic organic matter, polynuclear aromatic hydrocarbons, benzene, toluene and xylene. The health effects associated with exposure to these air toxins can include cancer, leukaemia, central nervous system effects, and acute health disorders such as respiratory and skin irritation. The Ministry of Environment, Forest and Climate Change, Government of India, finally came out with some rules. Today's final rule in India focuses on limiting air toxics from recovery type coke oven processes from charging, pushing, quenching and from battery stacks and effluents. Though there are no rules for non-recovery coke oven batteries, yet it is mentioned as one of the low emission processes. Non-recovery batteries are gaining acceptance in India by small coke manufacturers. Attempts have been made to address both in this article.

2. Environmental Impacts of By-Product Recovery Coke Oven

The process of Coke Making is driving out the volatile matter by heating the coal at high temperatures for a few hours. The red-hot coke is taken out, quenched and used for various purposes, mostly in blast furnaces and foundries. During coke making, lots of leakages take place from the battery. During coke quenching with wastewater, carcinogenic hydrocarbons are emitted when water is poured on red hot coke. The coke oven gas containing heavier hydrocarbons such as tar, pitch, naphthalene and lighter hydrocarbons are separated and cleaned in by-product plant. The cleaned coke oven gas (approx. 4200 kCals/Nm^3) is used as fuel. The ammonia in the raw coke oven gas is either recovered by treating with sulphuric acid thereby making ammonium sulphate or incinerated. A Typical Scheme of By-Product Recovery Type Coke Oven is shown in **Figure X.1**. The effluents coming out from the by-product plant contains ammonia, phenol and cyanide and are treated in Wastewater Treatment Plant, commonly known as BOD plant for reuse.

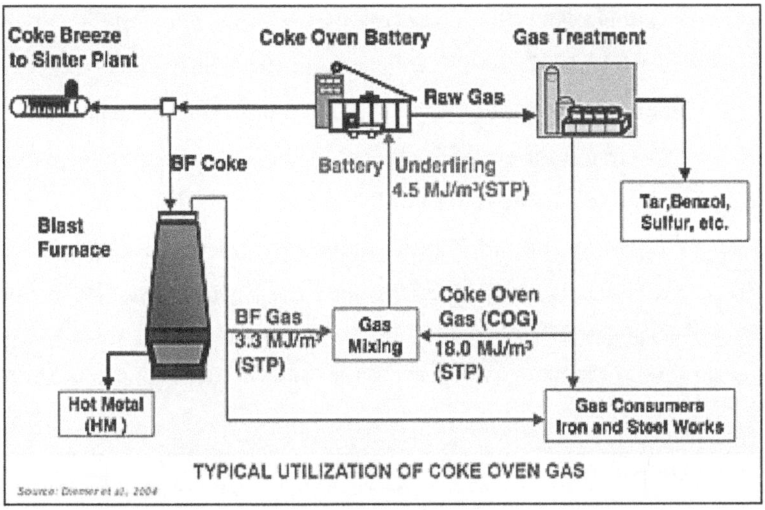

A TYPICAL SCHEME OF BY-PRODUCT RECOVERY TYPE COKE OVEN

Coke oven can be either by-product recovery type which works under pressurized conditions or non-recovery type which works under negative pressure. As the fugitive emissions are less in non-recovery batteries, the non-recovery type is considered as Clean Technology.

For decades these problems were taken up very lightly till UNO declared Coke Making as the most eco-hostile industry on the Planet which must be saved. This was followed by Kyoto Protocol and new environmental standards were set for coke ovens. Many nations including India worked out stringent parameters and set acts and laws to address the issue.

2.1 Indian Initiatives and Role of Central Pollution Control Board (CPCB)

In March 1995, CPCB first assigned MECON, the Consultancy wing of Ministry of Steel, Govt. of India, to take up the preparation of report intending to:

i) develop national standards for control of emissions in coke oven plants in integrated steel plants.

ii) describe clean technology for coke making after assessment of pollution potential from each source in coke ovens and the latest developments that have taken place in developed countries.

MoEF&CC, on the advice of CPCB, notified the emission and effluent standards and guidelines for by-product recovery coke oven plants vide Notification no. G.S.R. 631 (E) dated October 31,1997. These rules have come into force on the date of publication (i.e., 31.10.1997) in the official Gazette. The subject was, thereafter, discussed at various levels and as desired by CPCB, a comprehensive report detailing the feasibility and constraints of retrofitting new facilities in the existing batteries for achieving standards was submitted in July 1999 for perusal.

Ministry of Environment and Forests through its gazette notification dated 31st March 2012, have issued rules further to amend the Environment (Protection rules) 1986 as given in Table X.1 and X.2.

TABLE X.1: COKE OVENS (BY-PRODUCT TYPE), NEW DELHI, THE 31ST of MARCH 2012, SL. NO. 24. INTEGRATED IRON AND STEEL PLANT:

B. EMISSION STANDARDS

PARAMETER	NEW BATTERIES AT GREENFIELD SITE	RE-BUILT BATTERIES	EXISTING BATTERIES
Fugitive Visible Emission			
1. Leakage from door	5 (PLD)*	10 (PLD)*	10 (PLD)*
2. Leakage from charging lids	1 (PLL)**	1 (PLL)**	1 (PLL)**
3. Leakage from AP Covers	4 (PLO)***	4 (PLO)***	4 (PLO)***
4. Charging Emission (sec/charge)	16 (with HPLA) ****	50 (HPLA)****	75
* PLD-% leaking doors **PLL-% leaking leads ***PLO-% leaking off-takes ****HPLA-high pressure liquor aspiration through gooseneck			
Stack Emission Standards			
1.SO_2 (mg/Nm^3)	800	800	800
2.NOx (mg/Nm^3)	500	500	500
3. Particulate matter (mg/Nm^3)	50	50	50
4. PM during charging in stamp Charging batteries (mg/Nm^3)	25	25	25
5. Sulphur in coke oven gas Used for heating (mg/Nm^3)	800	800	
Fugitive Emission Benzo(a) Pyrene (BaP)			
1. Battery area (top of Battery) ($\mu g/m^3$)	5	5	5
2. Other units of coke plant ($\mu g/m^3$)	2	2	2

Note:

1. For measuring leakages in %, refer to US EPA document "Coke Oven Emission, Visible Observations of Coke Oven Door Leaks" 1979.
2. Refer 25 Pa. Code § 123.44. Limitations of visible fugitive air contaminants from the operation of any coke oven battery.

In India, before the 1997 notification, there were only the following limits for coke ovens emissions (as per Environment Protection Rules 1986).

- Carbon monoxide emissions from coke ovens: 3 kg /tonne of coke
- Combustion Stack Particulate matter emissions: 50 mg/Nm3.

TABLE X.2: COKE OVENS (BY-PRODUCT TYPE), NEW DELHI, THE 31ST of MARCH 2012, SL. NO. 24. INTEGRATED IRON AND STEEL PLANT.

A. EFFLUENT STANDARDS

PARAMETER	LIMITING CONCENTRATION IN MG/I, EXCEPT FOR PH
pH	6.0-8.50
Suspended solids	100
BOD, 3 days at 27°C	30
COD	250
Oil and grease	10
Ammoniacal nitrogen as N	~ 50
Cyanide (as CN)	0.2
Phenol	1.0

2.2 US-EPA Standards for Recovery Type Coke Oven Batteries

It is interesting to see below how the emission limits of recovery type coke oven batteries were gradually reduced by US-EPA based on the Clean Air Act.

The Clean Air Act (CAA) of the United States of America was passed in the year 1990. This act of law describes standards for air quality, which exert a very strong influence on the requirements which have to be fulfilled for obtaining the permit to run an industrial plant. The so-called Residual Risk Standard (RRS) should provide an ample margin of safety to protect public health and to reduce the risk to cause cancer to a minimum.

In the case of coking plants, amongst others, standards are set for the allowed number of visible emissions (leaking rates as %) from battery operation to reach this goal, as described by the US EPA (US-EPA, 1993a, 2005). For the construction of new coke plants at the green site, the CAA calls for zero visible emissions from battery operation. *That means in practice, that in the USA, the non-recovery technology is the only one, which is allowed by the USEPA for new greenfield plants because of the prevailing negative pressure and consequently of the prevention of leaks at the oven. (ref. Environmental Control and Emission Reduction for Coking Plants by Michael Hein and Manfred Kaiser Submitted:*

November 17th 2011Reviewed: April 18th 2012Published: August 22nd, 2012, DOI: 10.5772/48275)

For existing conventional coking plants, the Residual Risk Standard, which is still open, has to be reached from 2020. It is to assume that the relevant legal demands will be very ambitious. During the recent 20 years, the US coke oven plant operators had the chance to approach this target on different tracks, which specify different compliance timetables (Ailor, 2003; US-EPA, 1993a). While the MACT-track (Maximum Achievable Control Technology) allows less stringent standards for a long period to fulfil the highest level of emission standards already in 2005, operators who have chosen the LAER-track (Lowest Achievable Emissions Rate) got an extension to reach this standard only in the year 2010.

The relevant standards for the allowed visible emissions are shown in Table X.3. Estimates of visible emissions should be based on the results of daily visible emission inspections using EPA Method 303 (US-EPA, 1993b).

TABLE X.3: TIMETABLE TO COMPLY WITH THE LEGAL DEMANDS OF THE US CLEAN AIR ACT

SOURCE	MACT (MAXIMUM ACHIEVABLE CONTROL TECHNOLOGY) FROM 01.01.2003	LAER (LOWEST ACHIEVABLE EMISSIONS RATE) FROM 01.01.2010	REMARK
Doors	5.5 %	4 %	≥ 6 m
Doors	5.0 %	4 %	foundry coke
Doors	5.0 %	3.3 %	< 6 m
Lids	0.6 %	0.4 %	all plants
Offtakes	3.0 %	2.5 %	all plants
Charging secs per charge	12	12	all plants

Thus, the Indian Standards are to gradually shift to Risk-Based Standards.

2.3 Emissions And Controls

2.3.1 Emission Sources and Factors

For the purposes of presenting emission factors for coke oven charging, door leaks, lid leaks, and offtake leaks, emission control levels Table X.1 indicates Indian Standards and Table X.3, US EPA standards.

Emissions from coke ovens include conventional pollutants (particulate matter [PM], sulfur dioxide [SO2], nitrogen oxides [NOx], etc.) and numerous organic compounds, including polycyclic organic matters (POM), volatile organic compounds

(VOCs), and others. Emissions originate from several operations at the coke plant and by-product recovery plant. The following paragraphs describe emissions and control characteristics of by-product coke production, non-recovery coke production, and by-product recovery plant.

2.3.1.1 By-Product Recovery Coke Production

As at present, this is the most common practice in India. At the coke plant, PM is emitted from raw coal unloading, storage, and handling; mixing, crushing, and screening; blending; charging; leaks from doors, lids, and off takes during coking; soaking; pushing coke from the oven; hot coke quenching; combustion stacks; and coke crushing, sizing, screening, handling, and storage. (*Figure X.2*)

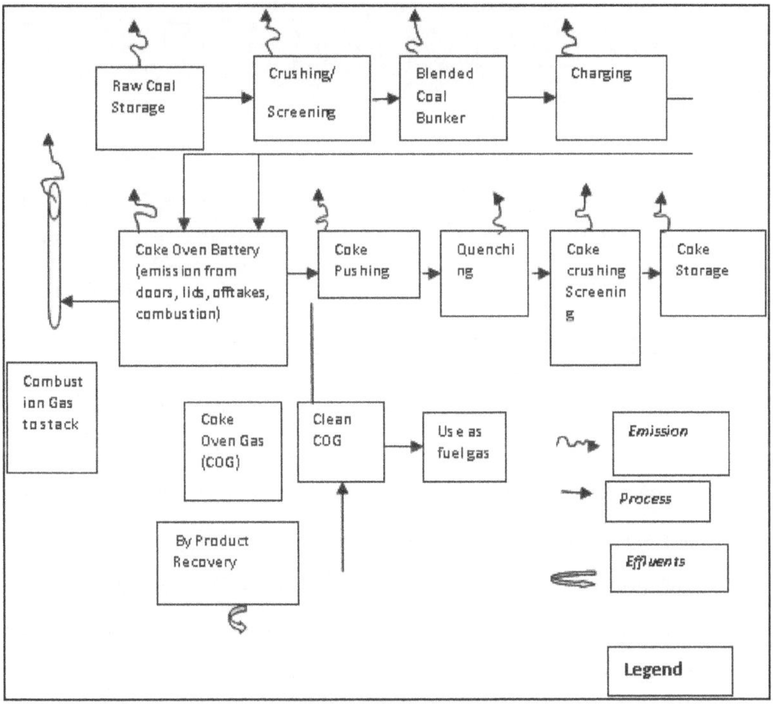

FIG. X.2: TYPICAL PROCESS DETAILS WITH EMISSION AND EFFLUENT FOR RECOVERY TYPE COKE OVEN BATTERY

Emissions from the by-product recovery plant are primarily organic vapours such as benzene and other light aromatics, POM, cyanides, phenols, and light oils (*Figure X.3*). These emissions occur from the separation processes, process vents, and transfer operations for recovered intermediates or products. These emissions also occur from wastewater that has contacted either the coke-oven gas or is generated from

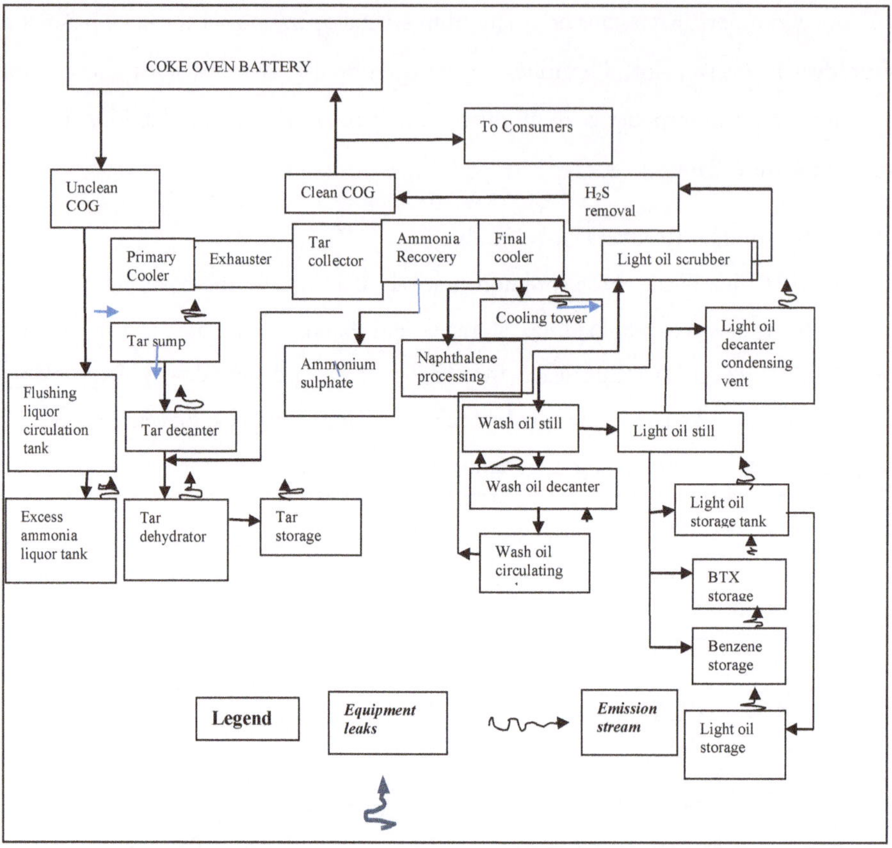

FIG. X.3: TYPICAL PROCESS DETAILS WITH EMISSION AND EFFLUENT FOR RECOVERY TYPE COKE OVEN BATTERY (BY PRODUCT RECOVERY SIDE)

separation processes when the water is handled in open wastewater treatment systems. Many plants control these emissions using gas blanketing or vapour balance/recovery techniques that collect the organic vapours and contain them within the gas handling system where they are eventually recovered. Some plants use carbon adsorbers to capture emissions, such as those from vents on storage tanks. VOCs are emitted from coke oven leaks, coke pushing, and coke quenching. Sulfur dioxide, nitrogen oxides and carbon monoxide are also emitted from coke oven leaks. Organic compounds soluble in benzene are the major constituents of the PM emissions and are also included as VOCs. Among the hazardous air pollutants (HAPs) included in the organic emissions are benzene, toluene, xylenes, cyanide compounds, naphthalene, phenol and polycyclic organic matter (POM), all of which are contained in coke oven gas.

2.3.2 Control Techniques to Reduce Emissions from Battery – Present and Future

2.3.2.1 Leakage Control

Because the oven is maintained under positive pressure, these leaks occur from small openings, such as gaps where metal seals mate against some other part of the oven as shown below.

Small gaps seal by the condensation of tar. Door leaks on most batteries are controlled by repairing and maintaining doors, door seals and jambs to prevent large gaps between the metal seal and the jamb. The manual application of a supplemental sealant such as sodium silicate is used at some plants to further reduce door leaks. A few batteries control door leaks by the external application of a luting material to provide a seal (called hand-luted doors). Lid leaks and offtake leaks are controlled by applying luting material around sealing edges to stop leaks and re-luting when leaks are observed. The control of leaks requires continuous monitoring by opacity experts and a diligent work practice program that includes locating leaks and then identifying and correcting their cause.

2.3.2.2 Land-Based Pushing Emission Control System

Pushing red hot and fuming coke into the quench car is a significant source of PM emissions. The smoke and fumes produced during the pushing of red-hot coke contain a huge amount of coke dust (estimated at 11% of the total pollution in the coke oven). Land-based pushing emission control systems mitigate this pollution. New steel plants in India are installing this system as technology is now available. It consists of three parts: (1) a large gas suction hood fixed on the coke guide car and moving with the coke guide, sending fumes to the coke side dust collecting duct; (2) the dust collection duct; and (3) the final equipment for smoke purification on the ground (ground piping, accumulator cooler, pulse bag dust collector, silencer, ventilation unit, stack, etc.). The large amount of paroxysmal high-temperature smoke produced during coke discharging is collected under the hot float fan into the large gas suction hood installed in the coke guide car and enters the dust collection duct through the other equipment.

The air is dissipated into the atmosphere after purification by the pulse jet bag filter and after being cooled by the accumulator cooler. The total de-dusting system is controlled by PLC. Fig X.4 shows a typical Coke Side Enclosure with moving and fixed ducts going to the Land-Based Collection System.

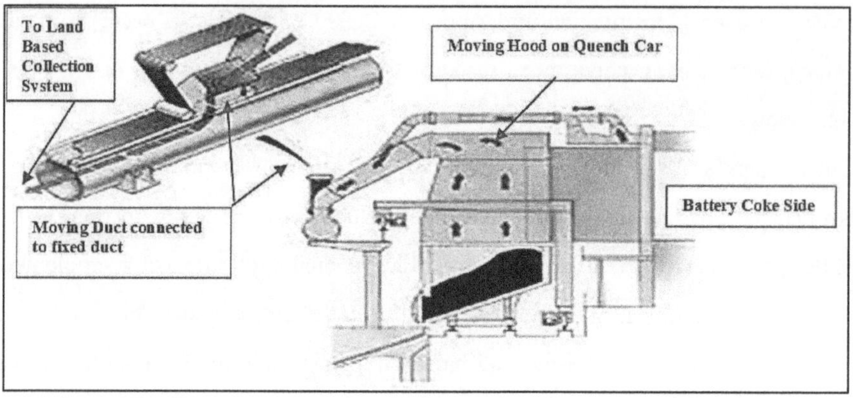

FIG. X.4: COKE SIDE ENCLOSURE CONNECTED TO LAND-BASED CONTROL SYSTEM

If the coke has been fully coked, emissions of benzene-soluble organics (BSO) and other organics are not expected to be significant; however, when coke that is still not fully coked (called "green" coke) is pushed, the various types of organic compounds found in coke oven gas are emitted.

2.3.2.3 Modified Coke Quenching

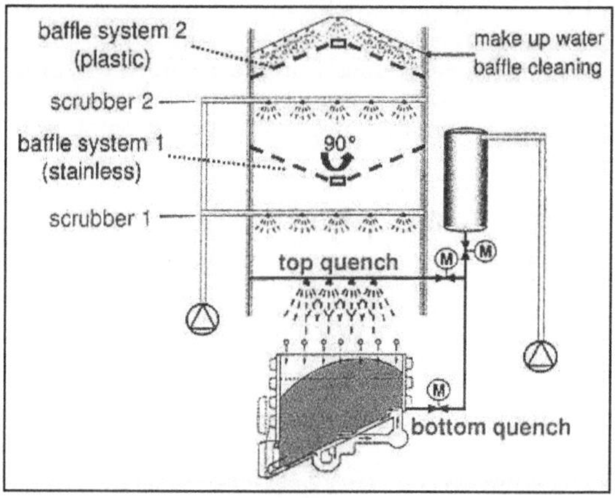

FIG. X.5: MODIFIED WET QUENCHING

Coke Wet Quenching (CWQ) entrains PM from the breakup of the red-hot coke when it is hit with water. The PM is carried up the quench tower by the velocity of the steam plume. In addition, suspended and dissolved solids from the quench water may become entrained in the steam plume rising from the tower. Trace organic compounds may also be present. As with pushing, other organic compounds may be released during the quenching of green coke. The typical control is to install baffles in the quench tower to reduce these emissions. Quenching can be controlled by installing baffles in the quench tower to impede PM flow, known as Modified Wet Quenching **(Fig. X.5)**

2.3.2.4 Modified Wet Quenching

Another "control" is to use clean water (recycled water that does not include process water) instead of "dirty" water (i. e., water high in solids or other pollutants like untreated phenolic water from by-product plants) to quench the coke.

2.3.2.5 Coke Dry-Quenching

Coke Dry Quenching (CDQ) is a heat recovery system to quench red hot coke from a coke oven to a temperature appropriate for transportation. In CDQ, red-hot coke is quenched in a closed cooling gas circulation. CDQ remarkably reduces the dust to be released into the atmosphere as compared with the CWQ. It is an energy-saving system in which, during the quenching process, the sensible heat of the red-hot coke is recovered and utilized. Recovered heat by CDQ is used to generate steam or electric power, which leads to a reduction of power need sand a baseline fossil fuel usage at the power plant. Thus, it can reduce CO2 emissions in the power plants supplying equivalent power to the plant and is one of the excellent countermeasures to energy recovery and global warming.

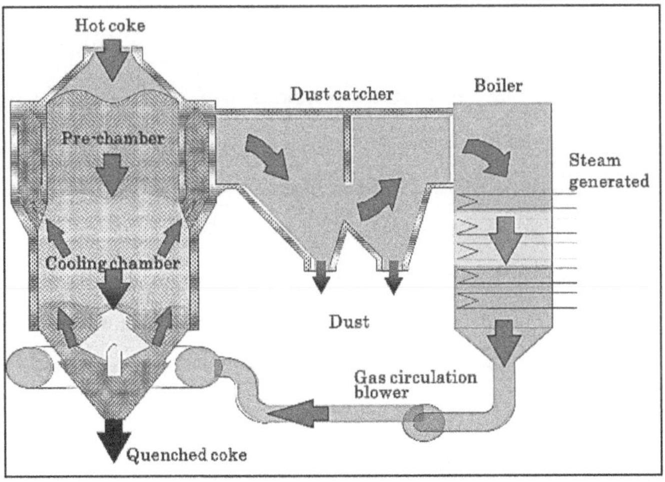

FIG. X.6: SCHEME OF CDQ

Coke Dry Quenching, in which the coke is cooled by a re-circulating inert gas, is recommended as clean technology to reduce emissions and energy conservation as the hot inert gas can be used to generate steam through a waste heat boiler. (Fig. X.6). CDQ is mandatory for new greenfield coke oven plants. It is also installed in a good number of existing Indian plants. However, existing plants hesitate to install it due to a shortage of space and technological issues.

2.3.2.6 Coke Oven Gas Cleaning

Coke Oven gas is a good fuel and the surplus gas after meeting the needs of the battery is used for reheating and other furnaces. Sulfur dioxide emissions may also occur if the coke oven gas is not desulfurized to remove hydrogen sulphide. Even after desulfurization, substantial amounts of sulfur dioxides are emitted if coke oven gas leaks through damaged oven walls and mix with the combustion gases to increase emissions.

2.3.2.7 Battery Stack Control

Battery stack emissions are usually controlled by maintaining oven walls to avoid leakage and by maintaining good combustion conditions. Conventional gas cleaning equipment, including ESPs and fabric filters, may be installed on battery combustion stacks.

2.3.2.8 Fugitive Particulate Emissions Control

These are associated with material handling operations consisting of unloading, storing, grinding, and sizing of coals, as well as the screening, crushing, and storing of coke and can be controlled by wet suppression techniques or installing bag filters.

2.3.3 Control Techniques to Reduce Emissions from By-Product Plant

Emissions from the by-product recovery plant are primarily organic vapours such as benzene and other light aromatics, POM, cyanides, phenols, and light oils. These emissions occur from the separation processes, process vents, and transfer operations for recovered intermediates or products. These emissions also occur from wastewater that has contacted either the coke-oven gas or is generated from separation processes when the water is handled in open wastewater treatment systems. Although not a criteria pollutant or HAP, ammonia (a particulate precursor) also is emitted from the excess ammonium liquor tank, tar decanter, and flushing liquor tank.

2.4 Effluents From By-Product Recovery Plant and Their Control

When coal is heated in the absence of air, it becomes buoyant, and all the volatile matter break down to yield gases, liquid and solid organic compounds of lower molecular weight and the non-volatile carbonaceous residue known as coke. Then, the coke oven gas is cooled and washed by water to recover by-products and clean the gas to be used as a fuel (Fig. X.7).

The resulting wastewater contains a sizable amount of ammonia salts and toxic compounds such as phenols, PAHs, SCN- and CN-. Much of the chemical oxygen demand (COD) occurs from phenols, which is a carbon source for acclimatized microorganisms, but also a toxic inhibitory substrate for micro-organisms. The metabolism of phenols and cyanides can easily be accomplished with biological treatment, such as an active sludge process with a surplus of oxygen, a pH of 7-9, adequate amount of nitrogen and phosphor. If the amount of phenol is high, i.e., > 2000 mg/l it requires to be diluted, to gain a non-volatile working metabolization (Anox).

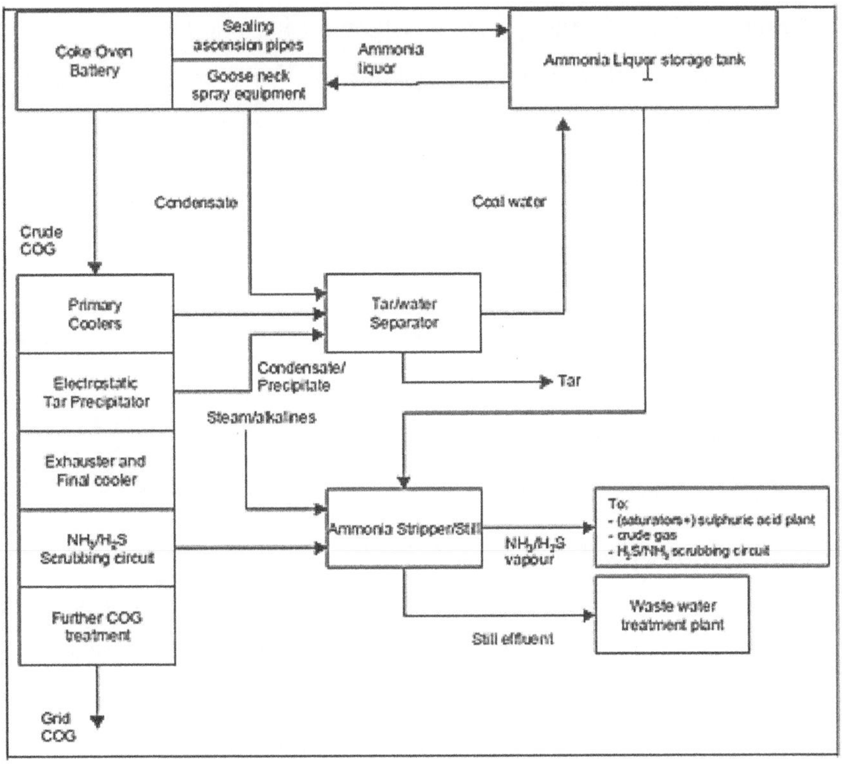

FIG. X.7: BY PRODUCT RECOVERY AND WASTEWATER

In a typical treatment plant, the wastewater is directed through a series of treatment steps with specific waste load reduction tasks, such as:

- Pre-treatment Physical and/or chemical
- Primary treatment Physical
- Secondary treatment Biological
- Advanced treatment Physical and/or chemical and/or biological

2.4.1 Biological Treatment

SBR, MBBR and MBR technologies are found to be more efficient than conventional treatment generally being followed in BOD plants in India.

2.4.2 Activated Sludge Process

The activated sludge process is the most widely used form of wastewater treatment for industrial wastewater. However, research has indicated that the technique *Sequential Batch Reactor (SBR)* is more efficient in reducing ammonia, phenol, COD, SS and BOD5 concentration than an active sludge process. This is probably a result that the SBR activated sludge microorganism proved to be more resistant to the variation of influent phenol content than the corresponding activated sludge process, i.e., having a higher survival ratio.

2.4.3 Sequencing Batch Reactor (SBR)

An SBR is a complete mix activated sludge system, with suspended growth and without a secondary clarifier. Aeration and clarification are carried out in one tank and within the single aeration basin.

There are five stages in the treatment process: Fill –React-Settle-Decant-Idle

2.4.4 Moving Bed Biofilm Reactor (MBBR)

MBBR is based on the process of fixed films **(Fig. X.8)**. The advantages of MBBR are a continuous process without any risks for clogs and needs for back flash, with low pressure and high accessible specific surface. This is achieved by letting biofilm grow on a numerous small plastic carrier that moves along with the water in the reactor. It also remains suspended through aeration. The carriers vary in size and form, they normally look like small cylinders and are made by polyethylene or polypropylene with a density close to water.

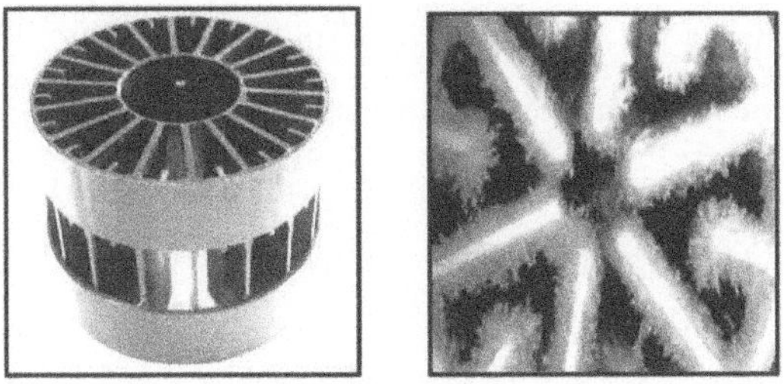

Fig. X.8: MBBR

The left picture shows a typical MBBR carrier. The right picture shows the fixed film layer on the carrier.

The microorganisms are well protected which make the process strong towards variations, disturbance and extreme strains. The process is easy-manageable, and the amount of active biomass is self-regulated and depends on the income strains. The carrier is continuous in movement due to oxygen from a bottom air system, which makes the process insensitive towards suspended material in the influent water. The effluent leaves the reactor through grating or strainers, which keeps the carrier behind in the process. The surplus sludge that continuously repeals from the carrier in a natural process, transports with the effluent through the grating next to a post-treatment step.

The MBBR process is built after the request of emission standards plus it is very efficient and would decrease the emission of phenol to less than 10 mg/L. The nitrogen level is going to reduce due to assimilation. Phenol is an easy degradable substance and COD contain between 60-80 % of it. It is uncertain what the latter 20-40 % of COD material is, but PAHs are probably one of them. PAHs are not easy biodegradable but treatment with MBBR would probably remove a great amount of them. MBBR is both easy to maintain and also good to withstand strains. Therefore, it is highly recommended to use MBBR as a post-treatment method[4].

2.4.5 Membrane Bioreactor (MBR)

MBR is the combination of a membrane process like microfiltration or ultrafiltration with a suspended growth bioreactor and is now widely used for municipal and industrial wastewater treatment with plant sizes up to 48 million litres per day. Fig. X.9 indicates a comparison between conventional biological treatment and Membrane Reactor.

Appropriate equalization and control of process conditions were identified as essential for the stable performance of the system.

A very detailed and exhaustive report on this subject as **TGM (Technical Guidance Manual for Coke Oven Plants)** Document has been prepared for the Ministry of Environment and Forests (published by IL&FS, Ecosmart Limited, Hyderabad) in 2010.

Other Control Measures:

- CMC (Coal moisture control)
- HPLA (High-pressure liquor aspiration system)
- Modern leak proof oven doors.

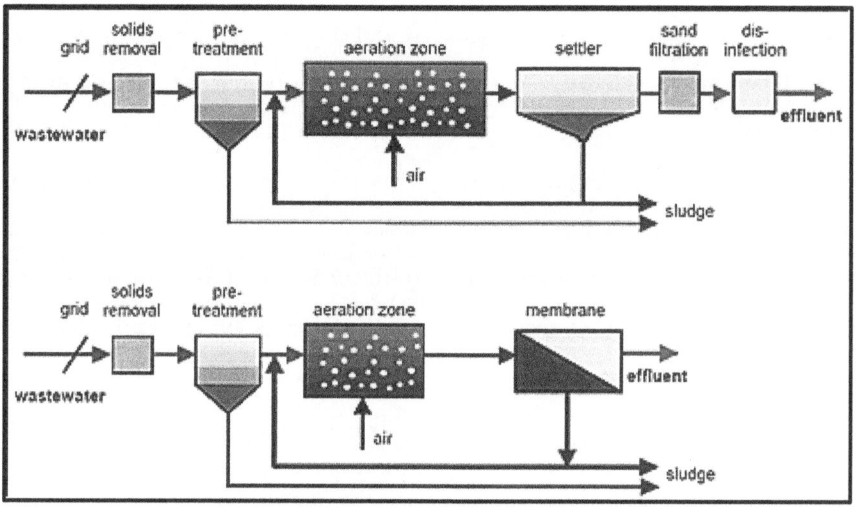

FIG. X.9: SCHEMATIC-DIAGRAM-OF-CONVENTIONAL-ACTIVATED-SLUDGE-PROCESS-(TOP)-AND-EXTERNAL-(SIDESTREAM)-MEMBRANE-BIOREACTOR(BOTTOM)

3. Health And Hazard Risks

Any unscientific operation of a coke-making plant invites considerable health hazards and risks. The process risks are from inhalation of particulate matter and carcinogens. A few real-time photographs (Fig. X.10 and X.11) were added for visual impacts.

If we want to save the Coke Oven Industry, we have no other way but to go for a Clean and Green Coke Oven as imagined by an Artist (Fig. X.12).

Fig. X.10: The Hostile Environment in some poorly managed Indian Coke plant.

Fig. X.11: The Changing Scenario elsewhere: Respect Nature and better Living. (The Living Condition of Women and Children of Coke Oven workers Then and No

FIG. X.12: AN ARTIST'S IMPRESSION - CLEAN AND GREEN COKE-MAKING PLANT.

4. Bibliography And References

1. 25 PA Code § 123.44. Limitations of visible fugitive air contaminants from the operation of any coke oven battery.

2. § 63.302 Standards for by-product coke oven batteries.

3. § 63.309 Performance tests and procedures.

4. Background Report, AP-42 Section 12.2, Coke Production

5. February 28, 2003, Fact Sheet, Final Rule to Reduce Toxic Air Emissions from Coke Ovens

6. The Environment Protection Rules, 1969, item 79, Coke Oven Plant, By-product Recovery Type and its amendments.

7. Environmental Groups Win New Coke Oven Regulation, Julie Grant, July 10, 2020

8. US EPA, Coke Oven Emissions, Visible Observations of Coke Oven Door Leaks, November 1979.

9. The Environmental Challenges for Indian Coke Making (Jointly prepared in collaboration with Mr KD Choudhury ex G.M (Environment). MECON under the guidance of Dr B Sengupta ex Member Secretary, Central Pollution Control Board)

10. US EPA § 63.303 Standards for Non-Recovery Coke Oven Batteries.

11. Method 22—Visual Determination of Fugitive Emissions from Material Sources and Smoke Emissions from Flares.

12. Atmospheric Pollution Control in Coke Ovens, N. K. Ghosh and L. Parthasarathy, R&D Centre, SAIL, Ranchi — 834 002

13. https://www.vizagsteel.com/code/energy/more_design.asp

14. Environmental Control and Emission Reduction for Coking Plants by Michael Hein and Manfred Kaiser Submitted: November 17th 2011Reviewed: April 18th 2012Published: August 22nd, 2012, DOI: 10.5772/48275

15. TGM Coke Ovens, MoEF&CC, November 2010, Resource Person K.D. Choudhury.

16. Environmental Protection Agency, 40 CFR Part 63, [OAR–2003–0051; FRL–7895–8], RIN 2060–AJ 96, National Emission Standards for Coke Oven Batteries, Environmental Protection Agency (EPA), Action: Final Rule; Amendments.

About the Author

K.D. Choudhury is a graduate from Mechanical Engineering, MIT, Muzaffarpur, 1970; MBA Industrial Mgt., BIT Mesra, 1983; Postgraduate, Environmental Law, National Law School of India University, Bangalore, 2003. Trained by GIPROMEZ, Moscow in Industrial Ventilation, Air Pollution Control and Dispersion Modelling.

TUV Rhineland, Cologne, Germany, Air Pollution Control, Monitoring and Dispersion Modelling. Presently working with Greencindia Consulting Pvt. Ltd., (GCPL), Ghaziabad; Worked with Grass Roots Research and Creations India (P) Ltd., (GRC), NOIDA till 2016; Superannuated from MECON Limited, A Government of India Undertaking, under the Ministry of Steel, as General Manager, Technical Services (Environment) in March 2009; Worked with Bokaro Steel Limited since 1971.

He has about 49 years of post-qualification experience in Steel Plants and Engineering Consultancy Organizations in various aspects of Environmental Engineering; setting up of Environmental Laboratory; developing environmental norms and guidelines for statutory bodies like CPCB and MoEF&CC; a Team Leader in the World Bank project "Environmental Issues in the Power Sector – India"; Guest Faculty for institutes like Management Training Institute (MTI), SAIL, Ranchi; Indian Institute of Coal Management (IICM), CCL, Ranchi; Engineering Staff Collage of India, Hyderabad; Headed Core Group on Clean Technology and Climate of MECON; Based on technology developed by Steelplantech, Japan, designed a Model Project to generate 20.6 MW power from waste heat of sinter cooler at RINL, Vizag, working satisfactorily; Represented Steel Task Force of Ministry of Steel in 'Asia Pacific Cooperation in Clean Technology and Climate"; Member in the Committee of Technical Experts constituted by the Hon'ble Supreme Court for submitting report on Ship Breaking Activities in India; Resource Person constituted by the MoEF&CC, Govt. of India, to develop Technical Environmental Impact Assessment Guidance Manuals on Metallurgical Plants, Coke Ovens, Electric Arc Furnace, Induction Furnace & Cupola and Expert Peer Committee Member on developing Technical Environmental Impact Assessment Guidance Manuals for Ship Breaking. Participated in capacity building for environmental engineers in Indonesia and Myanmar. Fellow of the Institution of Engineers (India); Life Member, Indian Society of Heating, Refrigeration and Air-conditioning Engineers (ISHRAE); Ex. Chairperson, Worldwide Fund for Nature (India), Jharkhand & Bihar State Centre; Ex. Member, Joint Committee on Safety in Steel Industry, Ministry of Steel. 11 Papers Published and 53 articles in Seminars/Workshops.

VI.2. Steps Involved in Environmental Clearance Of Industrial Projects

Dr Jayanta K Moitra

1. Introduction

The Ministry of Environment, Forests and Climate Change, Govt of India (MOEFCC) vide Gazette of India Notification S.O. 1533 (E) dated 14-9-2006 directed that construction of new projects or activities or the expansion or modernization of existing projects or activities entailing capacity addition with change in process and or technology and listed in the Schedule of the Notification shall be undertaken in any part of India only after obtaining the prior Environmental Clearance (EC) from the Central Government, or as the case may be, by the State Environment Impact Assessment Authority (SEIAA).

The Schedule listed 8 types of projects and 38 different activities for which prior EC is required. Two categories of projects are listed based on the threshold potential environmental impacts: Category-A and Category-B: The projects or activities that require prior EC, are referred to the MOEFCC for matters falling under Category 'A' and to the SEIAA for matters falling under Category 'B'.

General Condition of the Schedule mentions the condition when Category B project/activity will be treated as Category A project/activity. Specific Condition particularly dispenses with the requirement of prior EC for industries proposed to be located inside Industrial Areas/Estates/ Parks/ SEZ, etc. that have already obtained prior EC.

Coke Oven plant is listed under 4b of the Schedule. Metallurgical industry, coal washery and mining are listed under 3a, 2a and 1a, respectively.

2. Steps For Obtaining EC

An application seeking prior EC shall be made in the prescribed Form 1 (as given in Appendix II of the Notification), after the identification of the prospective site(s) by the applicant. Application for Category A and Category B is submitted in a single window at the Parivesh portal of MOEFCC (www.parivesh.nic.in). Applicant should register in the portal, provide the email id and phone number and obtain the user-id and password (always keep the user-id and password at safe place) The applicant shall furnish the pre-feasibility project report prepared as per guidelines of MOEFCC, Latitudes and Longitudes of the site boundary in the form of KML file, Survey of India toposheet of the project site along with the Form 1.

The EC process comprises of maximum four stages:

Stage-1: Screening is done only for Category B project/activity to establish the sub-category B1 [projects that require Environmental Impact Assessment (EIA) study] and B2 [projects that do not require EIA Study].

Stage-2: Scoping is the process where the project proponent makes a presentation before the Expert Appraisal Committee (EAC) for approving Terms of Reference for conducting the EIA study. The project proponent describes the project technology, project site and environmental sensitivity of the surrounding 10 km area,

Stage-3: Public Consultation is the process by which the concerns of local affected persons and others who have a plausible stake in the environmental impacts of the project are obtained. It is conducted by the State Pollution Control Board (SPCB) in consultation with the District Administration. The public consultation is done 'AT or CLOSE' to the project site. The District Magistrate or Additional District Magistrate presides over the public consultation. The proceedings of the public consultation and video recordings are forwarded to MOEFCC/SEIAA, as per the case of the SPCB.

Stage-4: Appraisal is the process where Final EIA Reports, Detailed Project Report/Feasibility Report, Engineering Drawings, Risk Assessment Report, the outcome of public consultation and Proceedings of the Public Hearing are scrutinized in detail by the Expert Appraisal Committee for grant of EC.

The steps involved in the EC process and documentation required is shown in the attached figure.

3. Preparation Of Environmental Impact Assessment (EIA) Report

The project proponent engages an accredited EIA Consulting Firm to prepare the EIA report. The accreditation is given by National Accreditation Board for Education & Training (NABET), a constituent body of the Quality Council of India (QCI). Accreditation is granted to EIA consulting firms for project-specific sectors. The list of the accredited consulting firms is available at the website of MOEFCC (www.parivesh.nic.in) and OCI (www.qcin.org). The generic structure of the EIA report along with the contents are given in Annexure III of the Notification. The Chapters of Summary EIA is mentioned in Annexure IIIA of the Notification. Procedure for conduct of Public Hearing is given in Annexure IV of Notification. The procedure prescribed for Appraisal of Application and EIA documents by the Expert Appraisal Committee is given in Annexure V of the Notification.

4. Validity Of Environmental Clearance

The prior EC granted for an industrial project is valid for a period of seven years, extendable to another three years for specific cases (maximum 10 years).

Project management must submit half-yearly compliance reports in respect of the stipulated prior EC terms and conditions, on 1st June and 1st December of each calendar year. A prior EC granted for a specific project or activity to an applicant may be transferred during its validity to another legal person entitled to undertake the project or activity on application by the transferor or transferee.

Forms and Formats: MOEFCC has prescribed standard forms and formats for online submission, which are scrutinized by the EAC during the appraisal process. The list of Forms is given below:

Form 1: Main application for prior EC

Form 2: Information Format for EC appraisal

Form 3: Application for amendment in TOR

Form 4: Application for extension of validity of TOR

Form 5: Application for extension of validity of EC

Form 6: Application for transfer of EC

Form 6: Application for transfer of TOR

5. Parivesh Portal Of MOEFCC

This portal can be accessed by the project proponent for official communication after online registration by using the user-id and password. The general public can also access the portal, any time for general information related to the composition of EAC, Agenda and meeting schedules, Minutes of the EAC Meetings, and several other circulars, office memorandums, official documents and notifications. Few important reference documents that are available in Parivesh Portal are

1. Compendium of Office Memorandum and Circulars related to EIA Notification,

2. Gazette Notifications and all amendments related to EC process,
3. Sector-Specific EIA Guidance Manual for all Project Categories,
4. Pollution Control Acts, Rules and Notifications issued thereunder (CPCB-2021),
5. OM-Sector Specific Guidelines to be Prescribed in Critically Polluted Areas (43 industrial clusters) and Seriously Polluted Areas (32 industrial clusters) (12-6-2019),
6. OM-Consideration of Proposals for ex-post-facto EC to the Project/ Activity involving Violation of the EIA Notification (28-12-2017, 30-5-2018),
7. OM-Consideration of Projects to be Treated within 10 km of National Park and Wildlife Sanctuary seeking EC (30-3-2015, 1-5-2015, 8-8-2019)
8. OM-Corporate Environment Responsibility (1-5-2018, 19-6-2018, 30-9-2018)
9. OM-Standardization of EC Conditions (19-6-2018, 19-8-2018, 4-1-2019, 8-1-2019)

6. Important Points

No construction work or preparation of land should be started on the project or activity without obtaining prior EC. Only structures necessary for securing the land is permitted. Deliberate concealment and/or submission of false or misleading information or data which is material to screening or scoping or appraisal or decision on the application shall make the application liable for rejection, and cancellation of prior EC granted on that basis.

About the Author

Dr Jayanta Kumar Moitra obtained his M. Sc (1986 - Chemistry) and PhD (1992-Environmental Chemistry. He has more than 35 years of experience in Environmental Impact Assessment Study, Carrying Capacity Study, Social Impact Assessment Study, Human Health Risk Assessment Study, Air Pollution Source Apportionment Study, Monitoring and Management of Air Pollution, Meteorology and Air Quality Impact Predictions and Technical Evaluation of Air and Water Pollution Control Devices. He has published/presented more than 100 research papers.

He started his career with the Central Pollution Control Board, as a Scientist. He has a good understanding of environmental legislation and the development of national standards. He also worked in Larsen & Toubro Limited as Manager, responsible for obtaining Environmental Clearances and implementation of Environmental Management Plan of infrastructure and industrial projects.

Dr JK Moitra is presently working as Director in EMTRC Consultants Private Limited (Environmental Consulting Firm; website www.emtrc.in). Dr Moitra is designated as Government Analyst by the MOEFCC. Dr Moitra is Accredited by NABET-QCI as EIA Coordinator of Metallurgy, Power, Cement, Coal Washery and Mining projects. He is a subject expert for Air Pollution Control, Meteorology, Dispersion Modelling, Water Pollution Control, Solid Wastes Management and Risk & Hazard Assessment. He has completed more than 150 EIA of large projects at various locations in India. He coordinated 8 Human Health Risk Assessment Study, 7 Air Pollution Source Apportionment Study and two Carrying Capacity Study. Dr JK Moitra was engaged by the Asian Development Bank in preparing an Action Plan for Improving the Air Quality of Ahmedabad City. Dr Moitra was engaged by the Consultancy Development Centre for conducting the Annual Audit of the CFC and CTC Phase-out Plan in India. He worked for Environment Defence Fund (EDF-USA) on Emission Inventory for air pollution management. Dr Moitra was invited by the London based Publishing House - Kogan Page – to write the Chapter on Environment for their Book "Doing Business with India". Dr Moitra is the Vice President of the Indian Association for Air Pollution Control and General Secretary of CPCB Alumni Association. Currently, he is working for the UNIDO Project on Sound Environmental Management of Biomedical Wastes.

VI.3. "Smokeless Charging" In 7.6m Tall Top Charge Battery in Bhushan Steel Limited

Susanta Kumar Halder

1. Introduction

Target: 100 % Smokeless Charging of Ovens to achieve visible emission < 16 sec during the oven charging. The target was set as a challenge during the project Conception stage itself.

2. Historical Background

In the early sixties above was the scenario of top charging. It was off main charging allowing 100% escaping of charging emission to the atmosphere. In the late nineties under the pressure of Government regulations on pollution, the concept of On-Main Charging started getting retrofitted in the top charge batteries across the country. Charging Cars with Screw Feeders for controlled charging came into the picture with High-Pressure Liquor Aspiration in Goose Neck. But these were merely introductory exercises with not much of an in-depth understanding of the complete process. Process of Sequential Charging, Step Charging, Jumper Pipes for Gas Transfer to next oven etc could only result in partial success which was miles away from the ultimate target of smokeless charging.

The above was the level of achievement, a partial success story.

The main reasons for not achieving the desired results are as under:

1. Improper sealing of Telescopic Chute of the Charging Car with the charge hole frame resulting in gassing through 3rd & 4th charge hole due to less suction effect on the 3rd and 4th charge hole.
2. Coal blockage in 2nd / 3rd Charge hole at the end of Charging leading to profuse gassing from coke side doors due to pressurization.

3. Suction in the charging oven by HPLA resulting in an increase in GCM pressure thereby increase of pressure in other ovens leading to gassing from other ovens
4. Fixed type jumper pipe for gas transfer to next oven also did not yield the desired result due to failure of jumper pipes.
5. A long-time of oven levelling led to a reduction in suction inside the oven due to air ingression from the leveller door under the effect of HPLA in the goose neck. This resulted in pressure increase inside the oven and gassing through weak points.

With the introduction of Stamp Charging, the scenario changed. In place of fugitive emission during charging from oven top unlike top charge ovens the emission source became the pusher side door. Pollution Control systems like "Suction Hood in Pusher Car, Suction Hood above the Pusher side door frame connected to ID fan and bag filter, GCT Car with a land-based bag filter / GCT Car with burning chamber and cyclone separators are being used with partial success.

Following Factors which are responsible for fugitive emissions during top charging of ovens were studied in depth also reviewed the working of the screw feeders, HPLA etc.

- Charging Time
- Chocking of Charge Hole
- Levelling stroke and time
- Design of Telescopes
- Charging sequence and methodology
- Charging and levelling synchronization between the Pusher and Charging Machine
- Raw gas evacuation from the oven during charging.

All the above negative factors are to be dealt with to bring them under a common controllable umbrella to get Smokeless charging in a top charge battery with a single GC Main.

Following targets were set for each and every factor mentioned above to eliminate the slippages.

- Total Gross Charging time < 240 sec. Net Charging time without levelling 100 sec

- Peakless Charging to avoid the phenomenon of Charge hole Choking
- Minimum Levelling (1 / 2 stroke)
- Double sleeve Telescope
- Simultaneous charging through all 4 holes in a controlled mode.
- PLC controlled synchronization between Pusher and Charging Car for oven charging.
- Individual oven pressure control system for gas evacuation during charging
- HPLA optional
- Provision of Suction Measurement at the 4th charging hole during the oven charging.
- Leveller Muff with Air Curtain

Design evaluation of a few major and reputed coke oven machine manufacturers were done, Plants, where such types of machinery are working, were visited. Series of Technical discussions held with various machine manufacturers. It was decided to procure machines from M/s Schalke of Germany which could fulfil the requirements. Schalke made coke oven machinery working in the world tallest coke oven of 8. 25m high ovens in Swelgran Plant found giving 100 % smokeless charging with Individual oven pressure control system.

Following 2 new concepts to achieve 100% Smokeless Charging are planned to be introduced.

2.1 Individual Oven Pressure Control System

At the end of the 20th Century, DMT and Thyssen Stahl AG Jointly developed this concept experimentally where the GC main of the battery shall be working at around -30mmwc all time. A special variable depth liquor flexi cup will be there to act as a barrier between the oven and the GC main which will ensure positive pressure inside the oven even at high suction at GC main. This will eliminate the use of HPLA and its bad effects. The system will work on auto mode based on the measurement of pressure at the AP base.

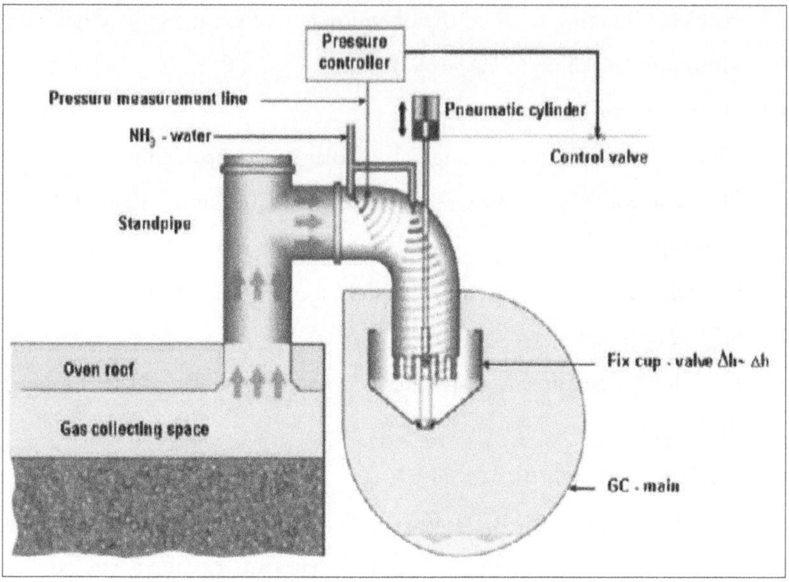

DMT & THYSSEN DESIGN

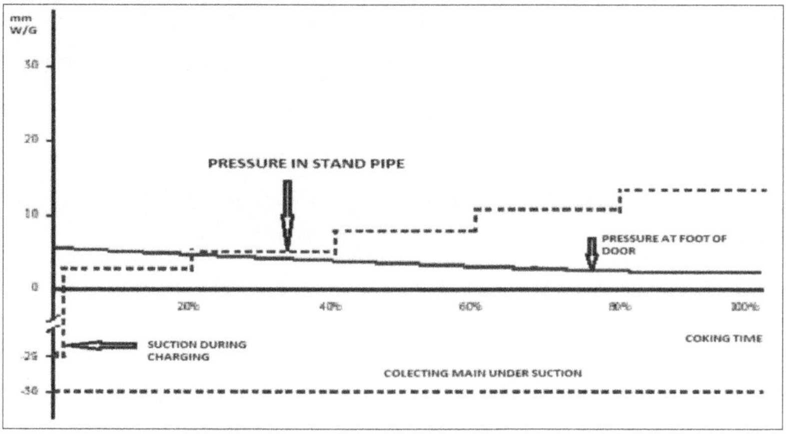

PRESSURE VARIATION AT DIFFERENT POINTS OF THE OVEN AND THE GC MAIN WITH INDIVIDUAL OVEN PRESSURE CONTROL SYSTEM

2.2 Peakless Charging

Filling of the oven chamber at a very high speed up to 95% of the desired quantity simultaneously without causing blockage of any charge hole and still leaving free space at the top for easy evacuation of the gas. Then one level. Rest 5% charging at slow speed with 1/2 level.

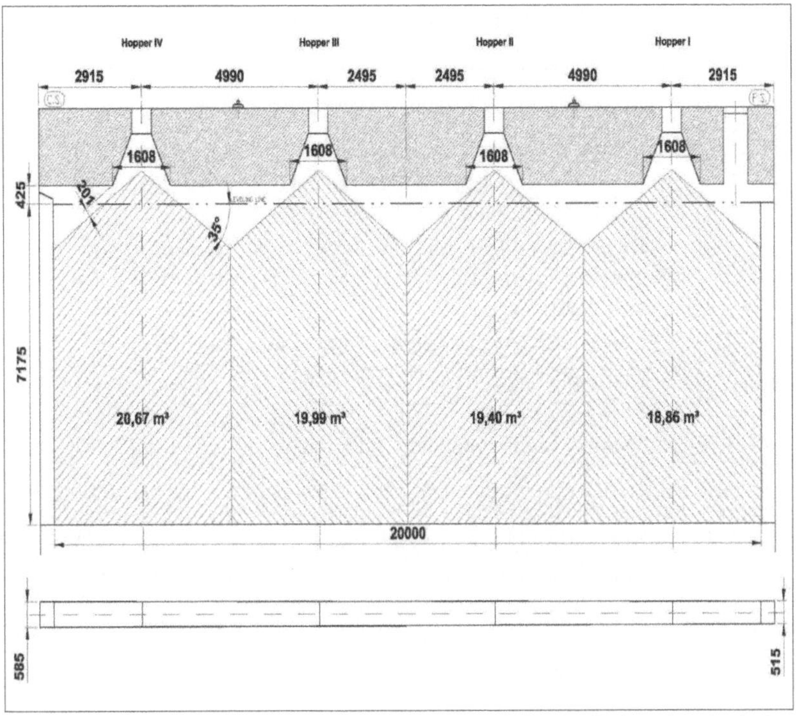

Battery Design was taken from M/s Paul Wurth, Italy who had the same technology of Individual Oven Pressure Control system. In this system in place of the flexi cup, there will be a special valve box with a semi spherical disc valve in between the Oven and the GC Main. Percentage opening of the steel plate based on the online measurement of pressure at AP base will be done through the PLC using special control software.

PWIT DESIGN

Advantages of Individual Oven Pressure control system over

- **Huge driving force for evacuation of Raw Gas from the oven .**
- **HPLA Not Required**
- **No Pressure surge in the GC Main and the battery in initial stage resulting to no emissions from doors in other ovens.**
- **Very small pressure surge in 4th charge hole at the end of 1st stage (fast charging) of oven charging (95 %) having no visible emissions.**
- **Visible emissions at the end of 2nd stage (slow Charging) reduced to the acceptable norm of CPCB (16 sec).**

Anticipating the main would be a practical Challenge of filling 78. 19 m^3 of useful oven volume in peakless mode with 2 stage charging (95% charging in fast mode and 5% in slow mode within 250 seconds including levelling time), a computer simulated model of oven filling was developed with M/s Schalke to arrive at the final settings of Hopper volume, Screw feeder rpm and total revolution etc. Provisions were kept for adjustment in Screw feeder rpm and total revolution based on coal moisture while in use.

2.3 Results Of Field Experiments

The battery was commissioned with HPLA. Following is the graphical representation of the pressure variations at "AP Base, GC Main and at 4th Charging Hole during charging.

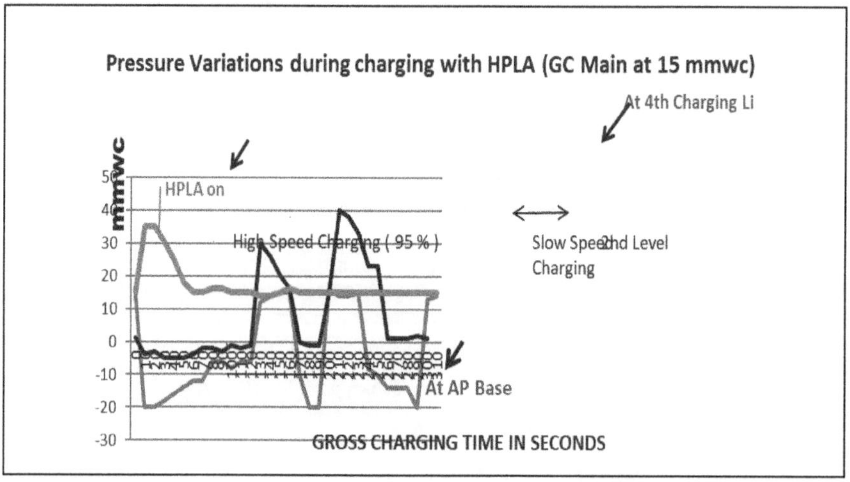

The pressure variations at these points of measurement clearly prove that even with peakless charging use of HPLA and GCM pressure at 15mmwc, visible Charging emissions at the end of 1st stage of charging and at the end of 2nd of charging together is > 16 sec. Gassing through oven doors during GCM pressurization was also observed.

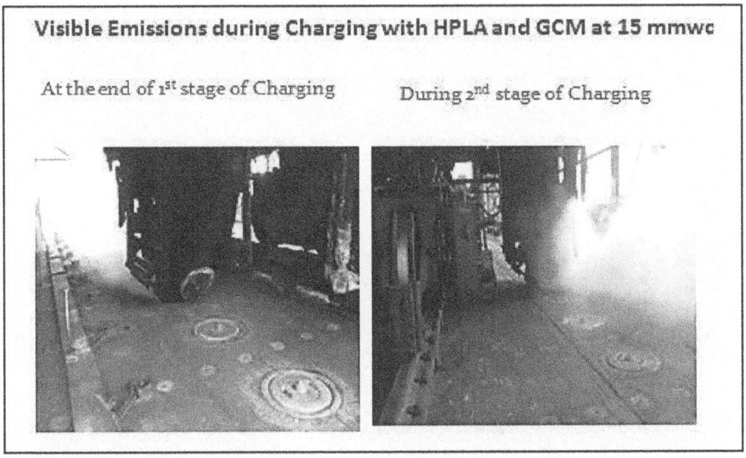

Later on, during a trial run of the Individual oven pressure control system pressure measurements were done with visual observations of Charging emissions and the type and total time of visible emissions were noted.

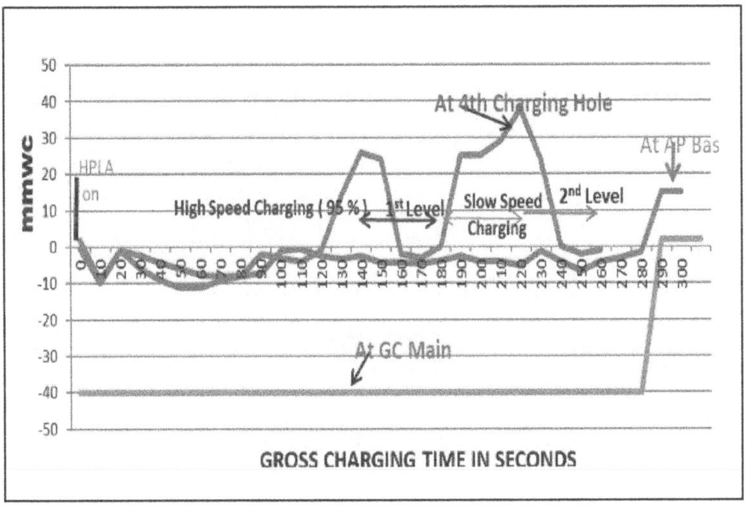

The above pressure measurement graphs show that:

- No surge or fluctuation in GCM Pressure
- AP Base pressure also remain at near 0 levels without fluctuation

- The extent of pressure surge at the 4th charge hole after fast charging and at the end of slow charging remains within control leading to visible emission within 12 to 14 seconds during charging.
- No gassing through doors in absence of any pressure surge in GCM

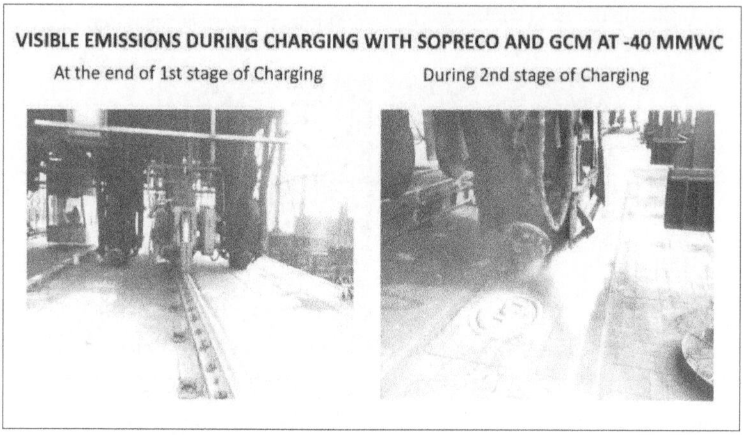

Only 12 to 14 sec of visible emission through the 4th charge hole was observed at the end of Charging. which can be put to more fine-tuning.

COKE OVEN 2, BHUSHAN STEEL

About the Author

Susanta Kumar Halder is B. Tech (Metallurgy) from I.I.T Kanpur.

Ex SAIL & BHUSHAN-TATA STEEL

VI.4. State Of the Art Wastewater Treatment Plant Through Bod Route

Susanta Kumar Halder

1. Introduction

To Treat the Wastewater generated in the BPP Plant, a State of the Art, fully automatic PLC controlled BOD Plant of 130 Cu.m/hr of treatment capacity has been installed in the Coke Oven # 2 complex of erstwhile Bhushan Steel Limited in 2013/2014.

Technology Supplier: MULTI Umwelt Technologie AG, Germany

Bhushan Steel Ltd. planned and built a biological wastewater treatment plant to have a robust, functional effluent water treatment facility that can treat non-recoverable effluent streams to a level of purity meeting the CPCB Norms at which the cleaned effluent can be re-used in the production facility.

Influent wastewater specification Considered for Design of the Plant:

Wastewater normal flow rate:	≤ 110	m^3/h
Temperature:	≤ 40	°C
pH value:	9.0 - 9.5	
Phenol content (as C6H5OH):	180 – 220	mg/l
Cyanides (as CN):	5 – 10	mg/l
NH3 –N free:	≤ 50	mg/l
NH3 –N total:	≤ 200	mg/l
N total:	≤ 300	mg/l
Sulphides:	≤ 50	mg/l
Tar and oil:	≤ 200	mg/l
Suspended solids:	≤ 500	mg/l
BOD5:	800 – 1000	mg/l
COD:	1800 – 2500	mg/l

During maximum conditions, the water flow rate is assumed to be 130 m^3/h, considering its increase due to rainwater which is mainly slightly contaminated and collected in the safety basins of storage tanks, tar separators, etc.

The total chemical load of this stream will be considered the same as the flow under normal conditions due to the dilution effect; except for tar, oil and suspended solids that are to be considered with the same concentration as und normal flow conditions.

Effluent wastewater quality after treatment:

pH value:	6-8	
Suspended solids:	< 100	mg/l
Phenol:	≤ 0.7	mg/l
Cyanide:	≤ 0.2	mg/l
BOD5:	≤ 30	mg/l
COD:	≤ 250	mg/l
Ammonia N:	≤ 50	mg/l
Oil and grease:	≤ 10	mg/l
PAH:	≤ 1	mg/l

2. Process Description of The Biological Treatment Plant

The 130 m3/hr biological treatment plant is designed for treating Wastewater coming from Coke Oven Battery complex no.2. The main source is the Ammonia Stripping section.

The water enters BOD PLANT through the Gutter channel. Here wastewater quality is checked and flocculent is added. From here water goes to Pre clarifier. In Pre-clarifier, a scrapper unit is installed to remove Tar particles from the feed stream as a floating layer. From the Pre-clarifier the wastewater enters Buffer Tank-1 where a mixer is installed for homogenisation of water to avoid deposition & sedimentation of sludge.

After that, the Tar free wastewater enters Equalisation Tank-1 or 2. Here also, mixers are provided to ensure homogenisation and achieve feed quality compatible with the biological treatment process. An emergency tank is also provided to take care of excess generation or lower off-take. The temperature of wastewater is reduced to 34 deg C in 3 nos. of Heat Exchangers operating in parallel before going to the Denitrification tank.

In the denitrification process, H3PO4 is added as food for biomass. Here the organic nitrogen is reduced to Ammonium compounds in 3 steps: - a) Ammonium compounds breaks down to Nitrate, b) Nitrate to Nitrites c) liberation of free Nitrogen.

Then wastewater enters the C.O.D aeration tank, where remaining carbonaceous compounds like Cyanide, Thio-cyanide, Sulphides gets oxidised & then gets reduced by Biomass growth. Then wastewater enters Degassing Tank-1 to remove air & Oxygen bubbles. Here Ferric Chloride is dosed as a Flocculent and polyacrylamides as the coagulant. After that, the wastewater enters to Intermediate Clarifier to separate biological sludge from water, & subsequently, the excess sludge is pumped to the Return Sludge tank.

The sludge from the Return sludge tank enters to Bio-sludge thickener & the water part from Intermediate Clarifier enters to Nitrification unit for growth of microorganisms in presence of Oxygen & to degrade the compounds (like Cyanide, Phenol, thio-cyanide etc.) present in wastewater. NaOH and antifoaming agents are added here to maintain pH. Bio Chips are envisaged in this stage to enhance and sustain Biomass growth.

Finally, the water stream comes to Final Clarifier after passing through Degassing-2, Post De-nitrification, post aeration and degassing -3 for achieving the desired quality of treated water.

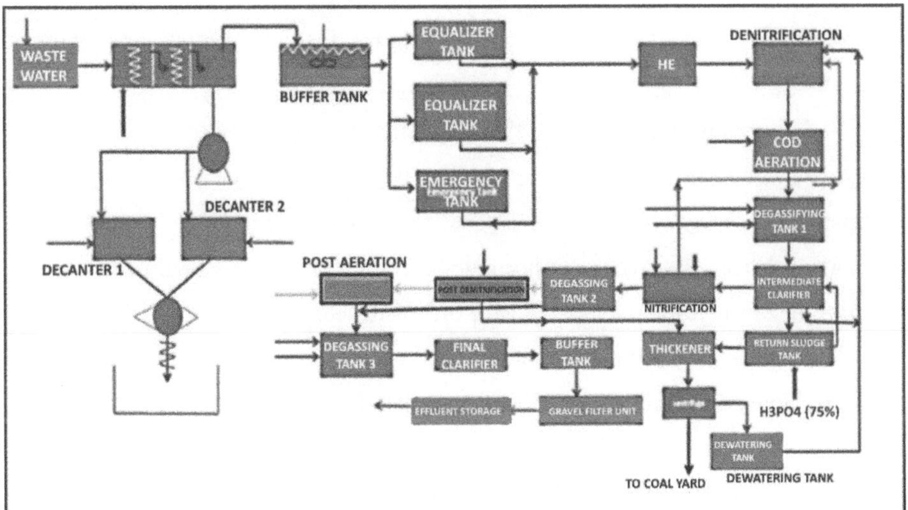

FLOW CHART OF BOD (MUMWELT TECHNOLOGIES) 110 m3/hr

For better efficiency of solid removal, additionally, three numbers of Gravel Filters are installed to separate suspended solids & the outlet water flows to Effluent Storage Tank. From here this treated water undergoes UV disinfection and is then recycled to coke oven complex and utilized for wet Quenching of Coke / BOF Slag Cooling etc.

3. Conventional Single Step Biological Treatment Plant

In a conventional Biological Treatment Plant for wastewater, For the simultaneous elimination of organic and nitrogenous pollutants, the bioreactors are designed for the rate-limiting step, which is nitrification. It requires very low organic loadings to achieve the necessary high sludge retention time (SRT), which means correspondingly large reactor volumes.

Moreover, Cyanide is highly toxic, particularly for nitrifying bacteria, and can only be accepted at low influent concentrations (< 50 mg/l). Other pollutants such as poly-nuclear aromatics or tar, which are adsorbed by the biomass, are also inhibitory. The recovery period of a nitrifying activated sludge population from a toxic shock usually takes several weeks

Due to the high concentration of pollutants and the low solubility of oxygen in the warm water, a high rate of airflow through the aeration devices are required, leading to correspondingly high waste air flows, stripping out substances that can create odour and corrosion problems in the vicinity. Excessive foaming can also be observed frequently.

As an improvement over the conventional single-step process, the plant supplied by MUTAG, Germany is designed on the concept of two-step treatment. The treatment for Heterotrophe (COD-oxidizers) and the Chemolithoautotrophe (ammonia-oxidizers) bacteria has been divided into a two-step treatment plant to optimize the biochemical potential of each organism category.

Separation of carbonaceous removal from nitrification allows the design of the respective bioreactors according to their determined function:

A two-step biological treatment process, consist of an activated sludge unit, subdivided into an anoxic denitrification zone and an oxygen aerated zone, followed by a carrier-based nitrification reactor for the treatment of industrial wastewaters containing both high organic and nitrogenous pollution loads, such as coke oven effluents, without requiring excessive reactor volume compared to conventional one-step treatment

The first activated sludge unit does not have to cope with the constraints valid for nitrification. It can be designed for a higher specific loading, i.e., with a significantly lower reactor volume. Toxic and inhibitory pollutants are retained or degraded by the less sensitive heterotrophic activated sludge of the first step and can no longer inhibit the sensitive nitrification reaction.

Application of pure oxygen aeration in place of Air avoids stripping and foaming, as gas throughput and waste gas flow are very low compared to aeration with air.

The nitrification reactor as the second step receives pre-treated, i.e., organically low loaded wastewater. This automatically leads to low organic loadings or a high sludge age (SRT) even at a comparatively low reactor volume, as the slow-growing autotrophic nitrifier can no longer be outgrown by the fast-growing heterotrophic bacteria.

Nitrates recycle from the nitrification step to the anoxic zone of the first reactor reduces the toxicity risk there due to an internal dilution effect with water of high purity.

Toxic and inhibitory substances (e.g., phenols, cyanide, thiocyanate, sulphide) are eliminated or hydrolysed in the robust first step, thus preventing negative effects on the more sensitive nitrification taking place in the second step. The nitrate return cycle from the second to the first step, necessary for denitrification, has the effect of an internal dilution of critical pollutants and therefore stabilizes overall biodegradation.

3.1 A Two-Step Biological Process

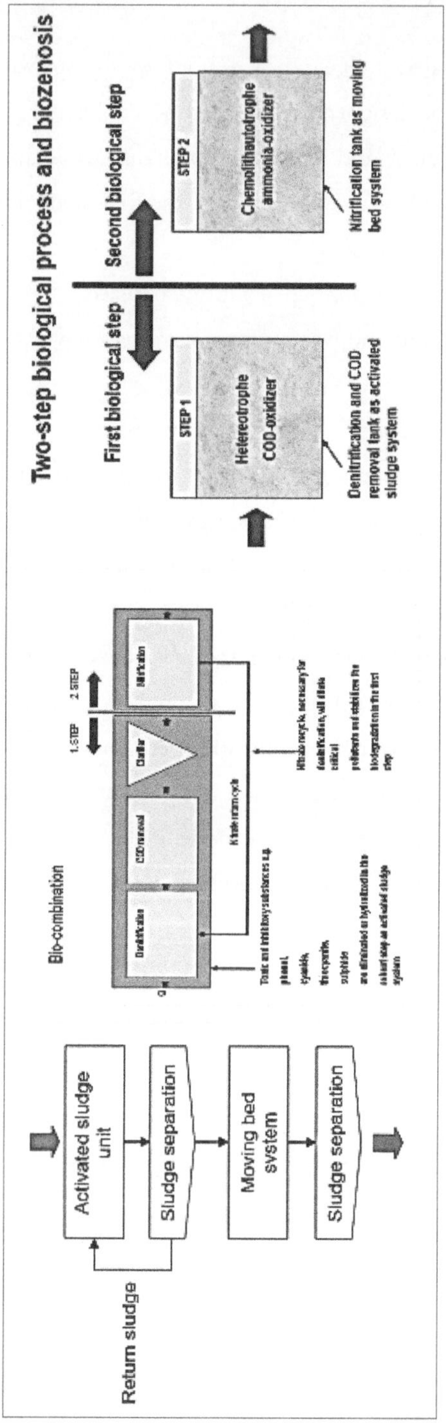

3.2 Pre-Treatment And First Biological Activated Sludge Step

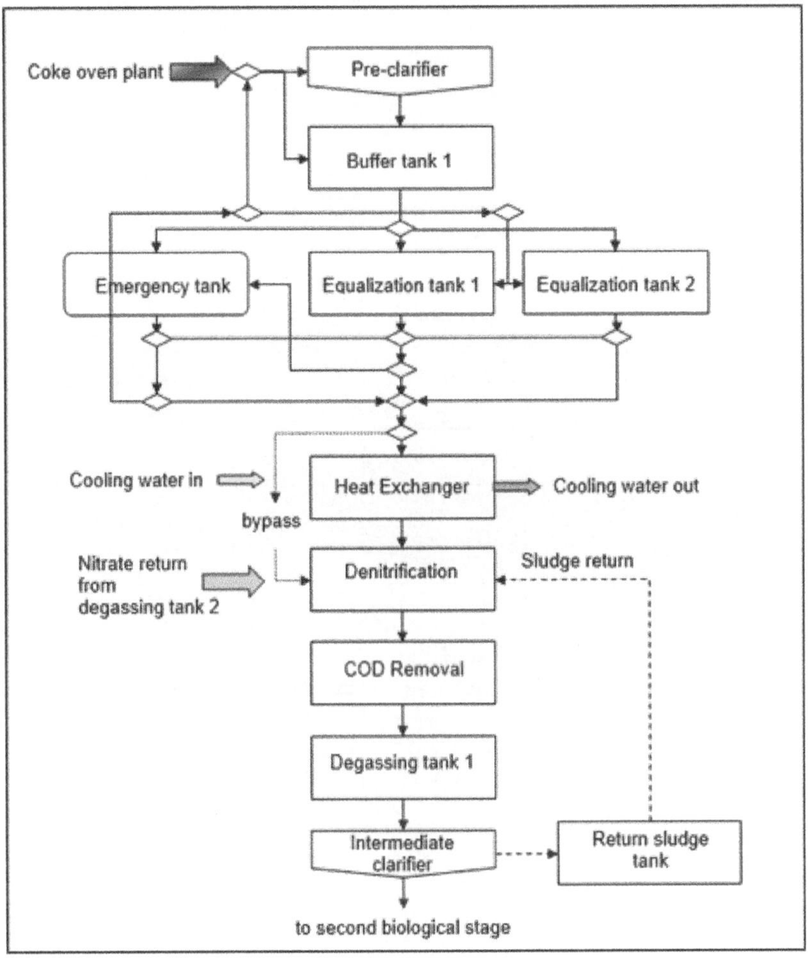

3.3 Pre-Treatment And First Biological Activated Sludge Step

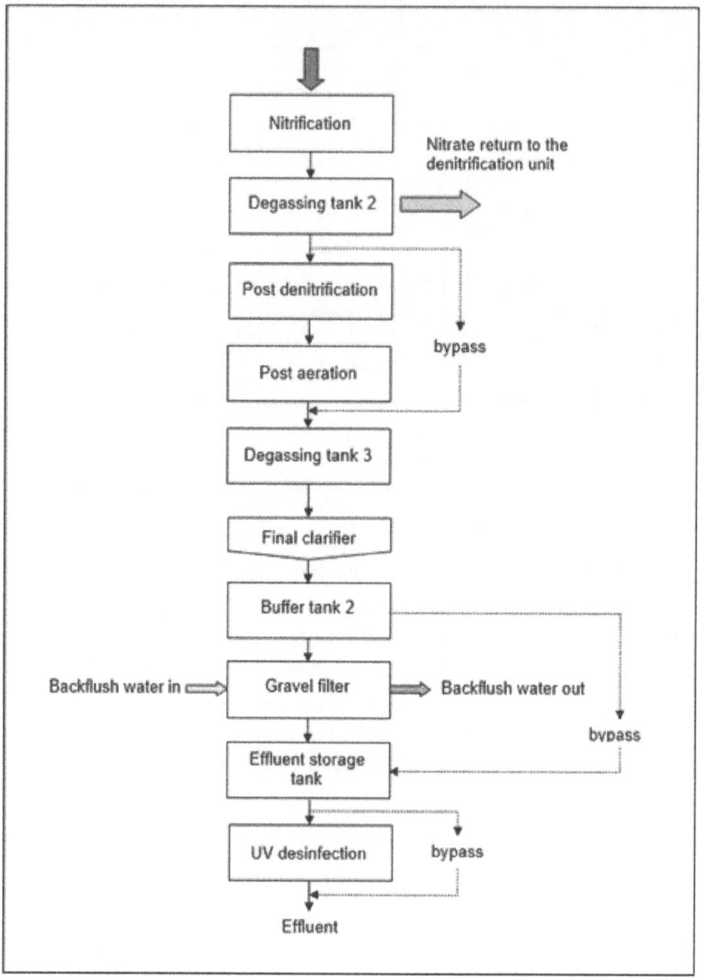

3.4 Special Design Features

Use of specially developed Mutag Bio Chips with a large surface area used for faster and safe growth of Bacteria.

In Place of Aeration by Air, Pure Oxygen is used for Aeration.

3.5 Plant Layout

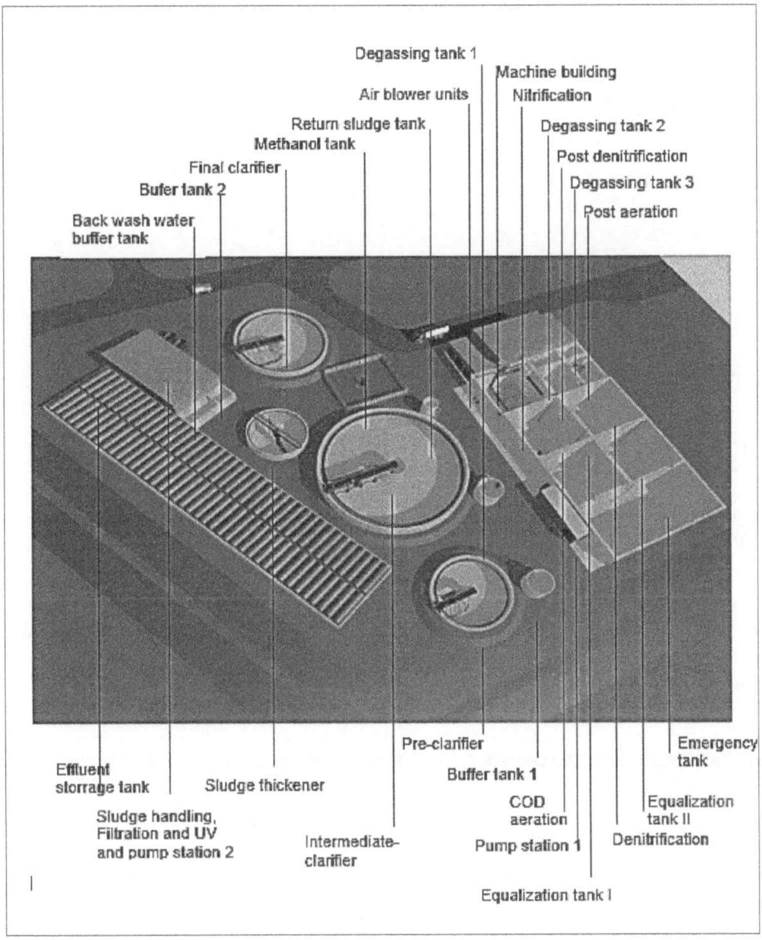

About the Author

Susanta Kumar Halder is B. Tech (Metallurgy) from I.I.T Kanpur.

Ex SAIL & BHUSHAN-TATA STEEL

VII. SHARING OF PLANT EXPERIENCES FROM COMMISSIONING TO OPERATION AND DISASTER MANAGEMENT BY THE TOP EXPERTS OF INDIA.

VII.1. *Challenges To Coke Makers and Some Non-Negotiable Issues To Achieve Best Results*

Sujit Kumar Haldar

1. Abstracts

Coke Makers all over the globe are focusing on improving coke quality with optimum cost in a cost-competitive environment in the steel market. To lower carbon footprint in the steel industry coal consumption per ton of Hot Metal Production become the biggest lever. To facilitate this improvement in coke quality, especially the hot strength of coke (CSR) becomes a dominant factor. The other aspect of the steel business is to enhance the life of a highly capital-intensive coke oven Battery. Hence, making a balance between higher battery life with high productivity and producing good quality coke from low-cost coal blend, is a challenge. Steps that need to be focused on from concept to commissioning and later adopting the right operation and maintenance practices sometimes gets deviated for various reasons due to prevailing circumstances. This paper will delve into such issues and some remedial measures and good practices in the coke industry.

2. Key Words

Suga Index, Anchorage, swelling of coal, Coal Blend, Internal Gas Pressure, push force, Battery expansion, Repair Strategy, Ceramic welding, MICOWAL, preventive maintenance.

3. Introduction

Recovery batteries are designed to produce good quality coke consistently and maintain the good health of the battery. Since the coke oven battery is a capital-intensive installation, optimizing cost and providing maximum flexibility in operation and maintenance is the most important focus to the designers. Structural stability of coke oven battery is important today as there is a tendency for the higher capacity battery to reduce operating cost (OPEX) and improve productivity is the other challenge from an overall business perspective. Despite robustness in the design, there are factors related to operation and maintenance that dominates in achieving the desired life of the battery. Coke Makers now has a wide variety of repair techniques to choose from, like Ceramic

Welding, Gunning, Dusting, patch repair, end-flue rebuilds, full wall rebuilds, etc. These repair methods have been presented and discussed at various Technical Seminars and Conferences. However, to arrive at the right strategy for battery life extension, it is important to consider each aspect of battery operation, such as heating, structural integrity, operational practices, etc., in conjunction with refractory repair strategies. The first step in determining the best repair practice for coke batteries is to adopt a strong system of inspection and prediction of residual life of each area, especially the oven wall refractory. Once the standard is set based on the prevailing condition and practices of the plant it is always quite easy to determine the right strategy and timing of repair. This is termed condition monitoring, and the need could range from urgent repair to quickly get back into production or to satisfy an emission problem or even a major rebuild to restore oven refractory structural integrity.

This Strategic approach should be driven by two key directions in the selection of repair options – whether to take the" Preventive Maintenance" approach based on prediction or to opt for a "Remedial Repair" based on need or urgency. These are the two core strategies for prolonging battery life. The purpose of such strategies is not simply refractory refurbishment but to restore the battery to proper operational, thermodynamic and environmental performance.

In the absence of such a strategy, a random repair will lead to aimless use of resources and continuous "firefighting" and the goal of battery life extension will ultimately not be achieved.

Further, if the repair performance is not monitored then no benchmark can be established to predict future battery life or deterioration trends.

4. Battery Design Consideration

SUGA Index is one such parameter normally ignored by operators while selecting a coal or coal blend. Of course, the designers while designing the battery considers a range of coal blends that can exert maximum pressure to the designed level during the normal carbonization period. As a rule of thumb, it is 1 psi with a certain factor of safety. To achieve this, the design of oven bracings and loading on walls are calculated. This total Battery anchorage system is made in a very precise through a pre-stressing on oven buck-stay through a number of springs at a different level of the wall and cross tie rods. The oven walls are subjected to various kinds of load, both static and dynamic arising out of

Swelling Pressure of coal blend, heat stress during coal charging and coke pushing process, the load of oven roof, during oven door opening and closing especially the modern doors with the spring-loaded closing system, high push force due to poor heating profile or coal blend issue, dynamic load for movement of charging car or gas transfer car on the oven top are the main concerns. Higher capacity batteries are designed with a high SUGA Index of more than 12 Kpa. However, the higher capital cost in achieving higher SUGA is one of the factors to optimize this number. The SUGA of earlier designed coke oven batteries was limited to 7.5-9.0Kpa.

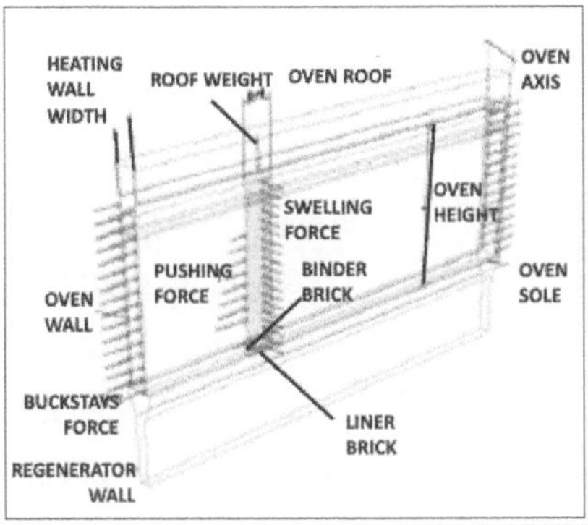

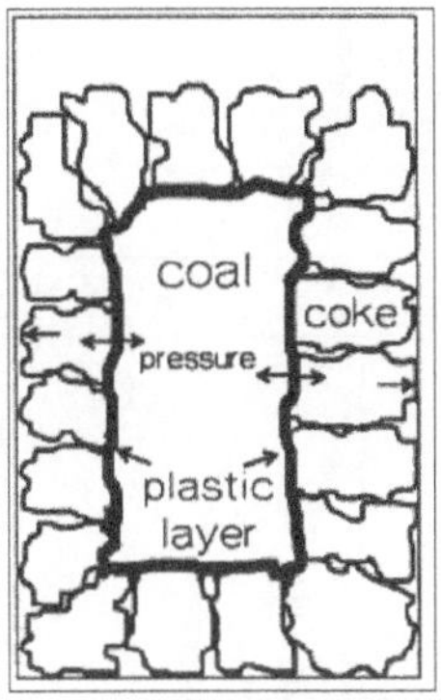

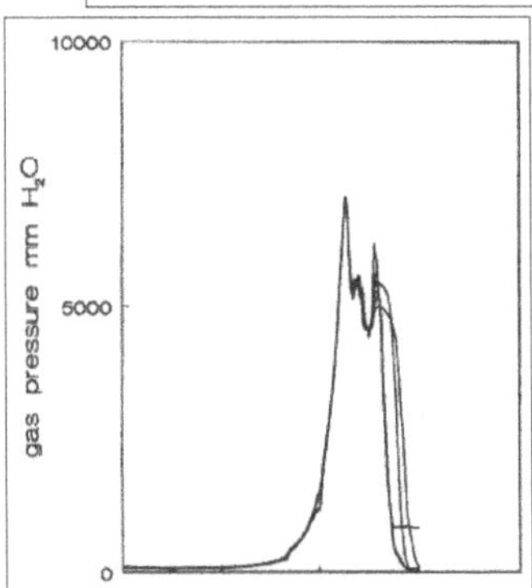

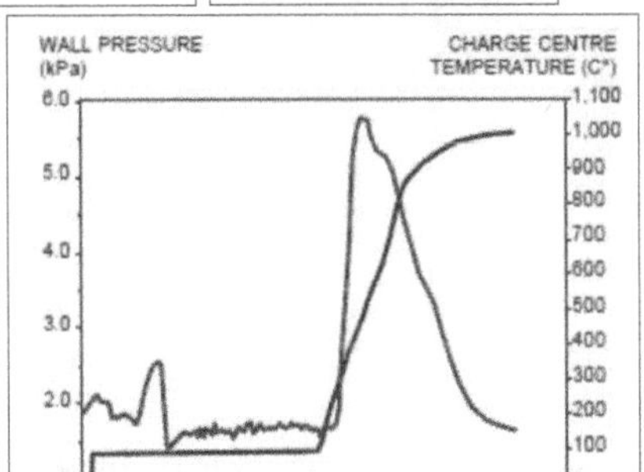

Factors influencing SUGA Index.

- Internal Gas Pressure/Swelling Pressure
- Wall Pressure

During heating up and commissioning of batteries the control in the rate of heating w.r.t silica material used for construction and the regulation of battery anchorage very meticulously is of paramount importance. Once ovens are commissioned the adjustment of anchorage in the operating range is another important aspect for good health and longer life of the battery.

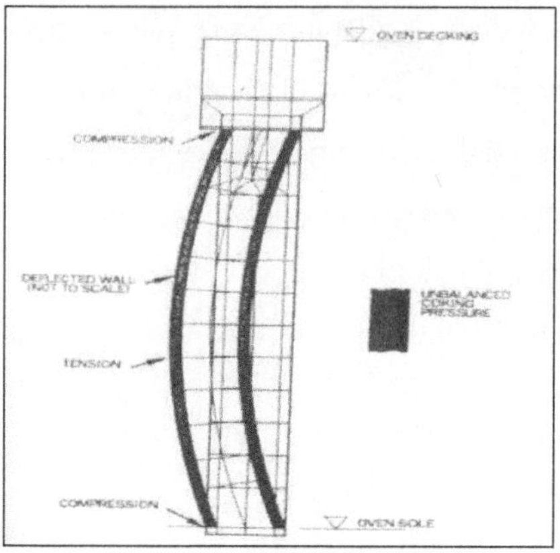

VERTICAL BENDING OF WALL

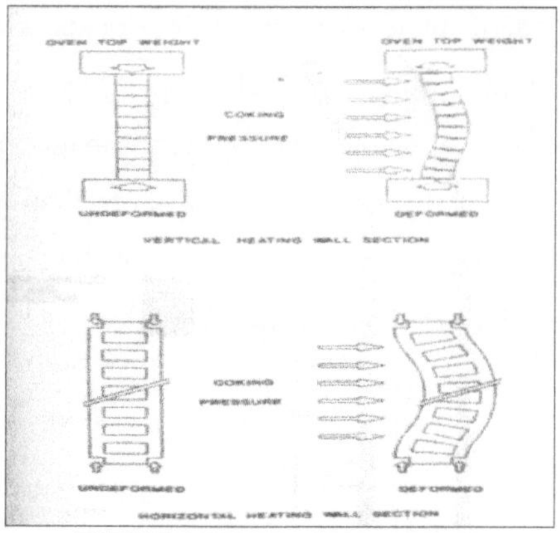

EFFECT ON REFRACTORY JOINTS

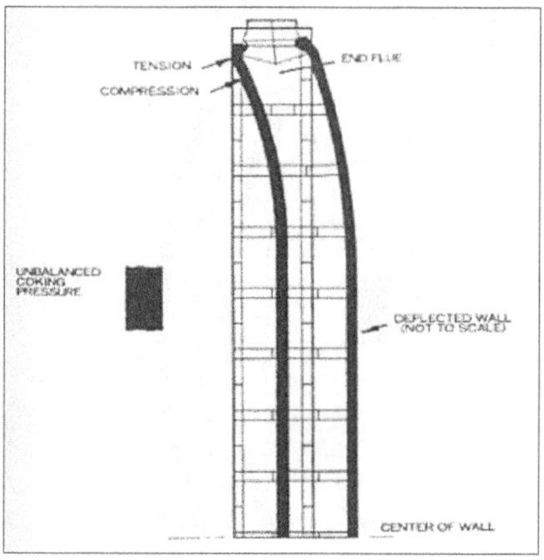

HORIZONTAL BENDING OF WALL

Regulating battery anchorage at a regular interval, especially checking the movement of buck-stay and all the loads on buck-stay as per schedule cannot be ignored or neglected. As the battery grows old, checking and replacing springs that are close to higher temp or flame also non-negotiable. At this juncture, regular checking of cross tie rod canal temp and inspection of tie rod cover in the proximity of Ascension pipe and charging hole should be done. In no case, buck-stays should come under flame for a long period. Sometimes it happens that hot coke mass falling from oven pushing is kept at the base of buck-stay in the service platform level. This resulted in bowing of buck-stay and loosening of anchorage. The dynamic load on the wall during carbonization is making the walls subjected to bending moment and the longer the wall or higher is the height of the wall the bending moment will be more. To counter all these stresses the pre-stressing of backstays is meticulously designed and must be maintained although the lifespan.

5. Operation And Operational Discipline

Service life, both economical and environmental of a coke oven battery is fully dependent on operation discipline and routine refractory and anchorage maintenance of the battery.

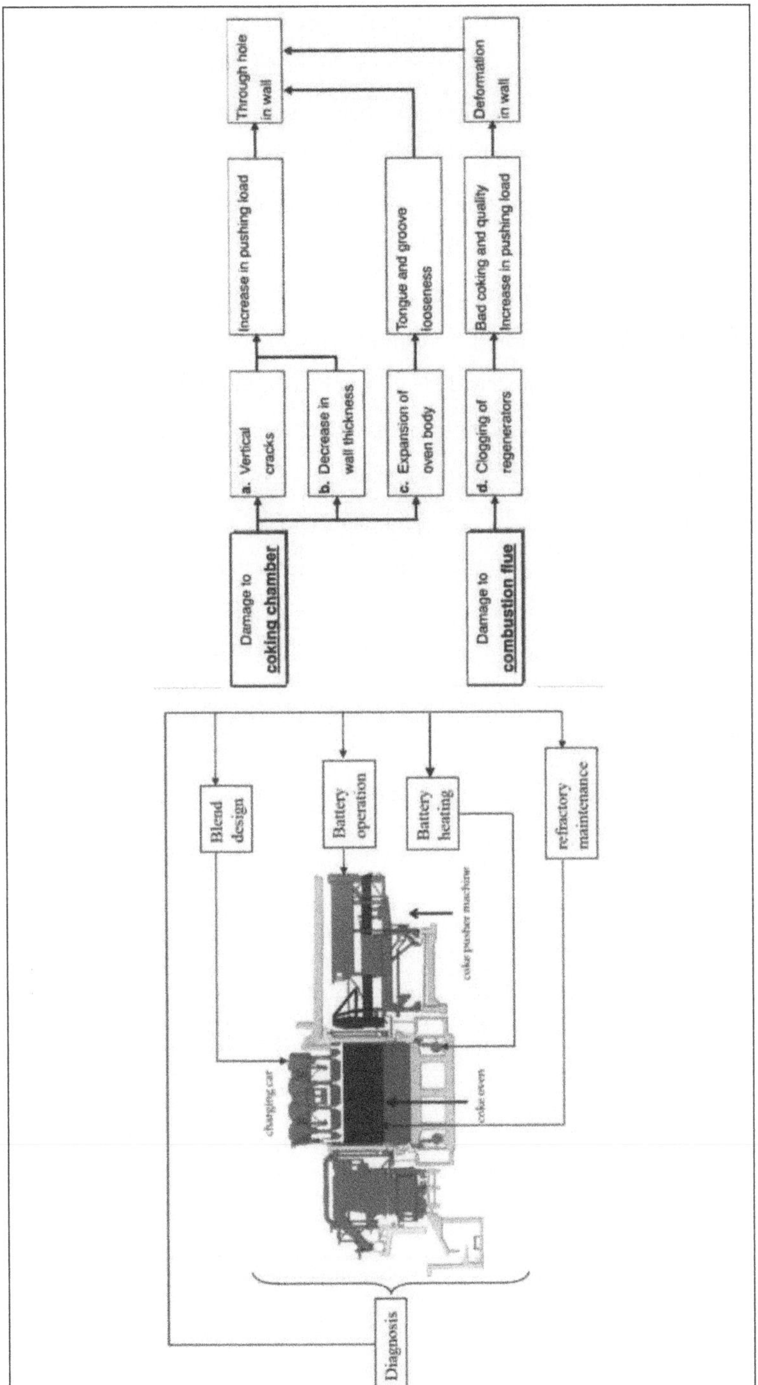
Diagnosis
Blend design
Battery operation
Battery heating
refractory maintenance
charging car
coke oven
coke pusher machine
Damage to coking chamber
Damage to combustion flue
a. Vertical cracks
b. Decrease in wall thickness
c. Expansion of oven body
d. Clogging of regenerators
Increase in pushing load
Tongue and groove looseness
Bad coking and quality Increase in pushing load
Through hole in wall
Deformation in wall

Once we talk about the consistent operation for consistent coke quality and good health of the battery, we must start with the coal and coal handling process. Optimizing coke cost is the largest lever in the hand of Coke Oven operators. At the same time demand for high-quality coke becomes the order of the day as the sizes of Blast Furnace increases and pulverized coal injection increases. Hence the tendency of going for a higher percentage of high-rank coal in the blend, depending on coke-making processes like Stamp Charge or Top Charge or Heat Recovery is gaining popularity. Another aspect of coal blending is to use Semi-Soft coal or non-coking coal-like PCI also gaining popularity for cost optimization or preservation of scarce prime hard coking coal. Balancing these two aspects depending on coke making process, the health of the battery and demand for coke quality is to be done very carefully.

Crushing fineness of coal especially + 3.15mm/3.0 mm and the micro-fines control must be done carefully. A higher size fraction will lead to a reduction in coke strength both cold and hot. Higher micro-fines lead to fouling of tar and upstream of by-product plant. Micro-fines also can contribute to roof carbon formation due to attrition and blockages of gas passages and obstruct the free movement of raw gas. In the top charge system micro-fines also reduces the bulk density of the charge and adversely affects the coke quality. Blockages gas passages lead to oven pressurization and oven burning and subsequent damage to oven structure.

6. Charging Of Oven

The charging of the oven always should be full. In the case of a stamp charge battery, the coal cake breakage should be minimum. In the case of top charge battery uniform levelling all through the length by proper charging and levelling method is the key to get the result.

Less charging due to any reason will lead to roof and wall carbon formation and gas passage blockages and the consequences.

7. Hydraulic Regime of Battery

Balancing Exhauster inlet pressure and Gas collecting main pressure and the controller in the cross over main is important to battery operation and oven health. Regular servicing of the control valves to prevent jamming from coal tar and sludge is non-negotiable. Maintaining Gas collecting main pressure depending on the height of the oven will prevent oven pressure from going negative at the sole level at the last phase of carbonization. Controlling oxygen levels in the inlet of Exhauster is non-negotiable.

8. Thermal Regime

The thermal profile of the entire battery for a consistent coke quality and good health of the battery is the basic requirement. Temp. the gradient on both vertical and horizontal direction for a smooth pushing operation and coke quality is important. Any major deviation can impact on pushing force and roof carbon formation and adverse impact on oven health.

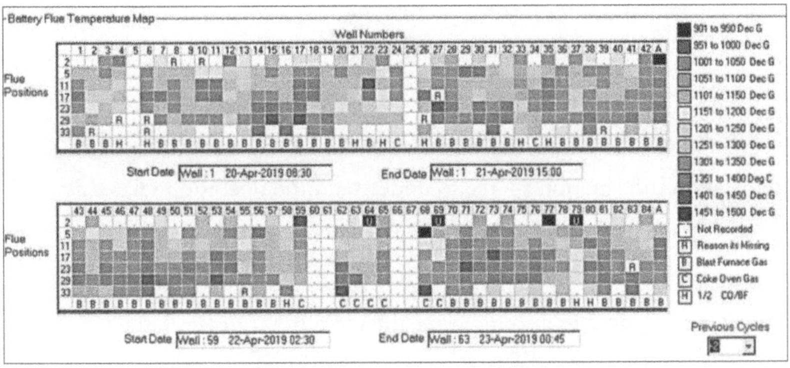

THERMAL MAP OF OVEN WALLS

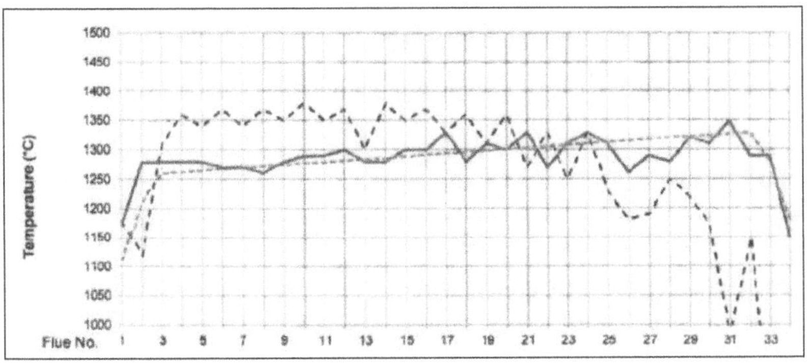

TEMP. PROFILE OF OVEN WALL

9. Battery Operation

One of the most important parameters for good operation and health of the oven is regularity. Some plants measure it and give it a lot of importance. Pushing Regularity Index (PRI) is one such measure. This ensures not only uniform heat distribution, but also take care of uniform swelling or coking pressure on all walls. The pushing sequence and its adherence all the time is another non-negotiable operation discipline.

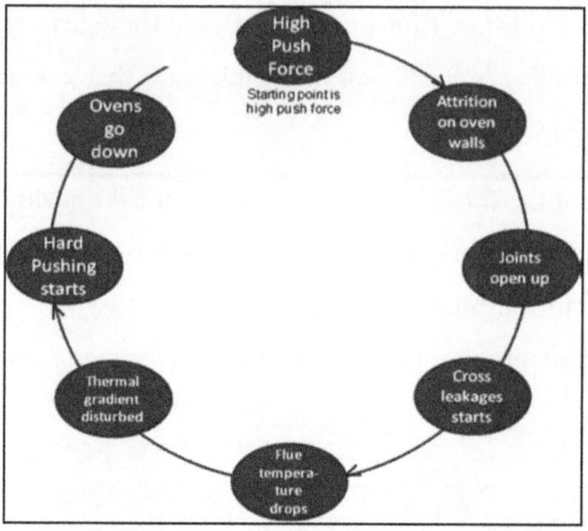

THE VICIOUS CIRCLE OF PUSH FORCE EFFECTS

Push force measurement system in the pusher car and its recording is an incredibly good information source to understand the operation related issues like oven temp. temp gradient, coal blend behaviour, blend moisture, coal crushing fineness and coal charge bulk density (BD) and at times the functioning of pusher beam operation. Sometimes this parameter also indicates the condition of oven refractories or oven sole refractories and pusher track level. Even it can indicate the issue of coke guide alignment. By looking at the trend of push force for a particular oven operator can decide for action based on detailed analysis. Some plants adopted it fully or partially and extracting the benefit. This parameter also needs to be monitored by all coke makers and a non-negotiable parameter for good operation and good health of the oven.

10. Battery Anchorage and Refractory Maintenance

Coke oven battery is a huge stand-alone refractory structure, anchored (supported) by mechanical structural, meticulously designed to put thrust at a different point on refractory structure in all situations so that the refractory brick joints remain intact. Refractories are strong under compression but very weak under tension. Once the anchoring pressure on the refractory wall gets reduced due to various reasons, the first manifestation will be the opening of refractory mortar joints, as joints are weaker in tension. Hence to maintain the good health of the battery specially to prevent itfrom cross leakages and chimney smoke it is important to maintain the tightness of the battery anchorage in totality. This single parameter is the key to the success of refractory maintenance of the battery.

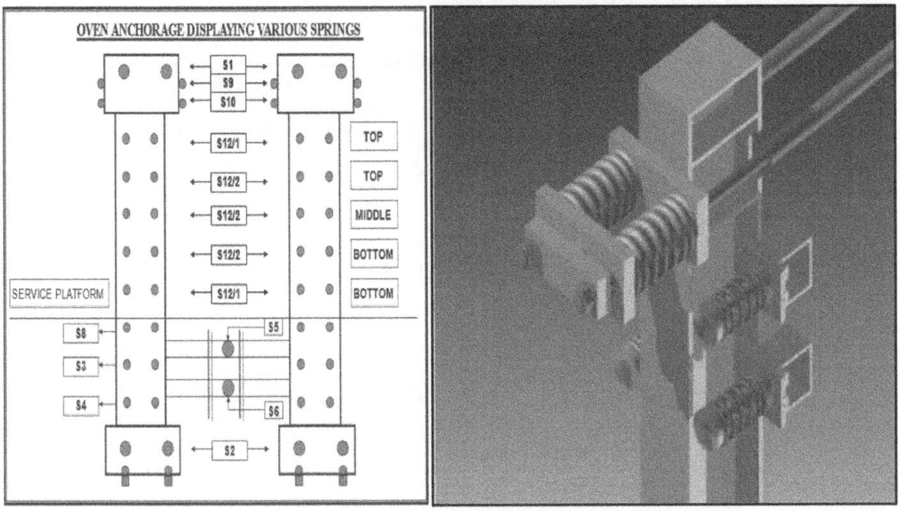

OVEN ANCHORAGE

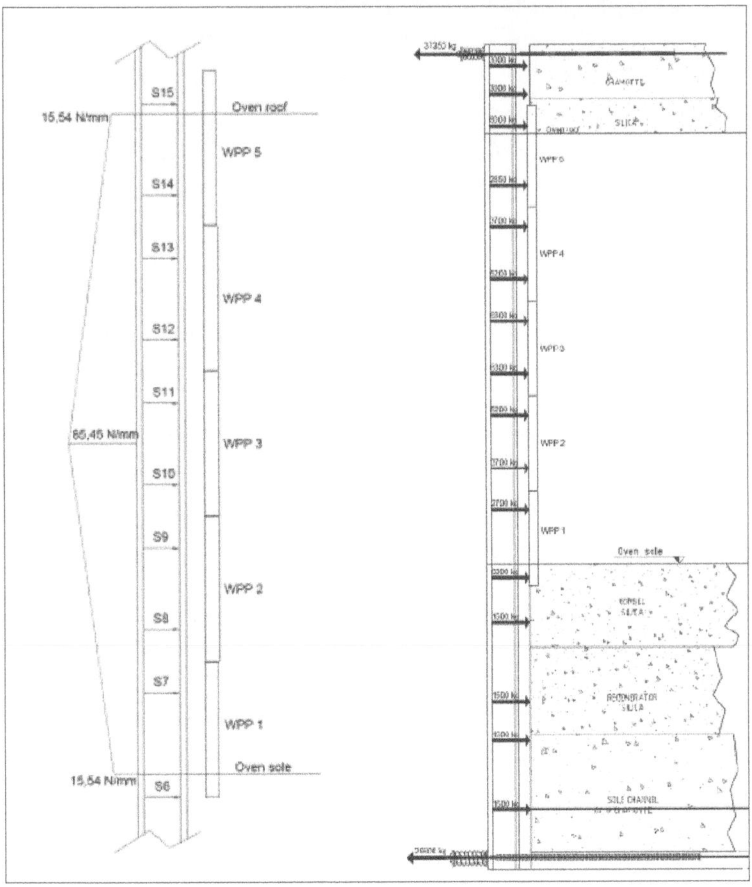

11. Battery Life Extension Strategy

As discussed earlier all Coke oven batteries are designed for a desired length of life and as a rule of thumb, it is 16000 pushing per oven for a well-designed battery. This number speaks several aspects of operation and maintenance like level of operation (operating ratio) and thermal and hydraulic regime, coal blend designed, Pushing regularity Index (PRI) and Anchorage and Refractory maintenance practices.

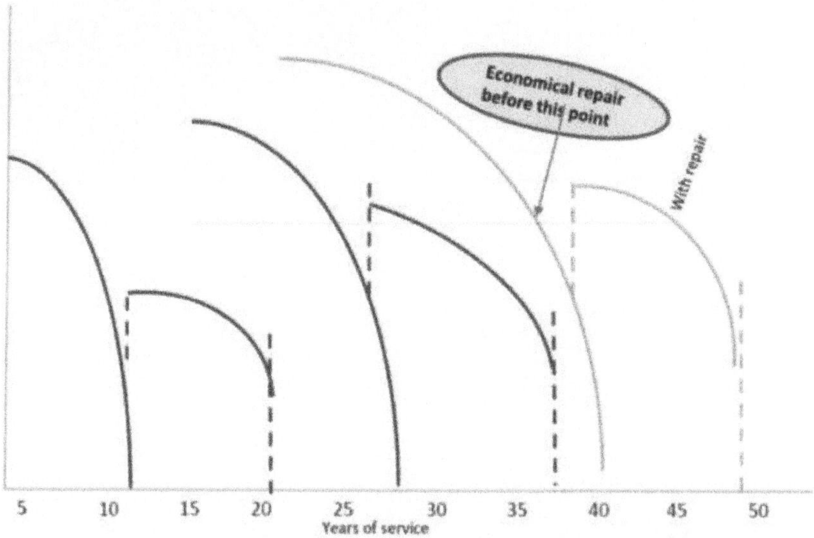

OVEN LIFE IS A FUNCTION OF COAL CARBONIZED & OPERATION PRACTICES

Without strict adherence to all these, the lives of the battery slowly deteriorate, and the rate of deterioration will vary depending on design, heating and commissioning operation and post-commissioning regulation and operation and maintenances practices. Fig. below explains how battery life can be extended what strategy one should try to adopt simultaneously.

12. Some Of the Common Refractory Maintenance

As discussed earlier all Coke oven batteries are designed for a desired length of life and as a rule of thumb, it is 16000 pushing per oven for a well-designed battery. This number speaks several aspects of operation and maintenance like level of operation (operating ratio) and thermal and hydraulic regime, coal blend designed, Pushing regularity Index (PRI) and Anchorage and Refractory maintenance practices.

Without strict adherence to all these, the lives of the battery slowly deteriorate, and the rate of deterioration will vary depending on design, heating and commissioning operation and post-commissioning regulation and operation and maintenances practices. Fig. below explains how battery life can be extended what strategy one should try to adopt simultaneously.

12.1. Semi-Dry Gunniting

As discussed earlier all Coke oven batteries are designed for a desired length of life and as a rule of thumb, it is 16000 pushing per oven for a well-designed battery. This number speaks several aspects of operation and maintenance like level of operation (operating ratio) and thermal and hydraulic regime, coal blend designed, Pushing regularity Index (PRI) and Anchorage and Refractory maintenance practices.

- Limited Short-term control measure as and when required and mostly in ovenends.
- Quicker Process but short Life (typically 2 to 3 months)
- Repeated Thermal Shock leads to Progressive refractory deterioration.

12.2. Dusting

Good for sealing Micro-Cracks to prevent chimney smoke.

12.3. Ceramic Welding

Ceramic welding is a widely recognized method of refractory repair. Ceramic welding technology provides ample opportunities for repair comparatively faster than other methods to increase the service life of batteries. This technology was developed in

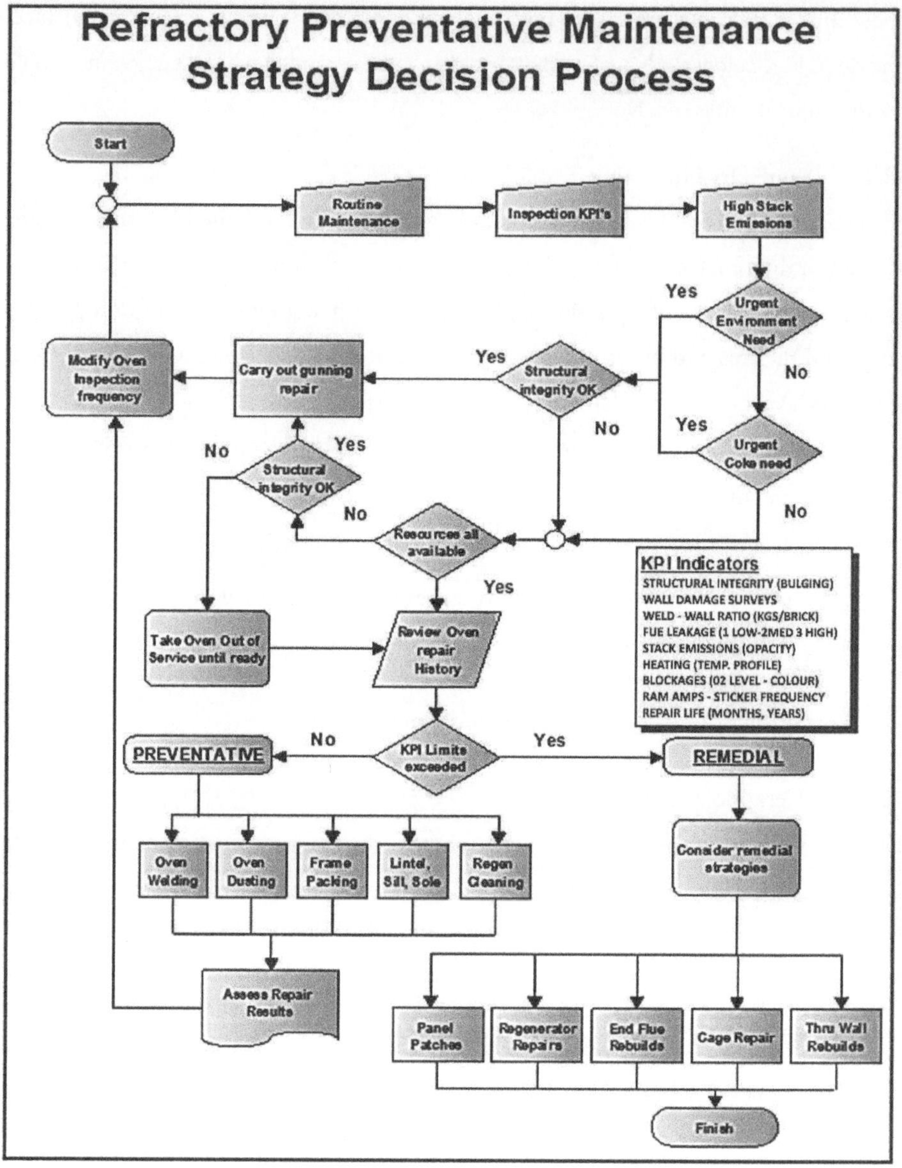

BATTERY LIFE EXTENSION STRATEGY

Belgium for application in glass furnaces and was later adapted to coke ovens for carrying out the hot repair of the damaged refractories. The technology provides maximum monolithic bondage with a minimal negative impact on the brickwork.

The ceramic welding process was first applied in the 1970s. This hot repair process is used to repair the coke oven refractory at operating temperature with minimal disruption to production. The repair material is ceramic bonded to the virgin surface of the damaged refractory, utilizing a fusion process that generates an exothermic reaction in excess of 2200oC. The ceramic bond strength of welding allows for a superior repair that restores structural integrity to the original brickwork.

Before the ceramic welding, a pneumatic blasting system is used to clean the refractory. This cleaning process is to be carried out with the objective of producing minimal trauma to the surrounding refractory.

Ceramic welding is accomplished by conveying a powder mixture, rich in silica and some metal through a small pipe lance via an oxygen stream. The powdered silica mixture as soon as comes out of the end of the lance and touches the hot surface, ignites against the hot wall of the coke oven. The heat of the reaction penetrates the silica brick creating a plastic zone of silica. The reaction also turns the powder silica mixture into a molten mass, which bonds to the plastic zone on the silica brick. The appearance is like that of a welding joint. Also, the effects are similar, as new, equally strong material is deposited onto existing worn and cracked silica brick. The following aspects are important for ceramic welding.

The ceramic welding method can not only weld cracks and chips but also large area defects, such as cavities. Furthermore, the technology allows for reliable welding of joints of old and new brickwork laying.

Professionally installed, ceramic welding material has provided twenty years of service. However, the average ceramic weld will last approximately 10 years.

Ceramic welding is best utilized as a preventive maintenance tool, but can also address large scale catastrophic repairs, thus providing an economic alternative to extend battery life and prevent unscheduled downtime.

Few major concerns faced by most of the Coke Oven Operators are:

Wall cracks in the end flues and minor spall between silica and alumina refractories. Joints open mostly in the end flues.

Developing undercuts in case of Stamp Charge operation.

Deep undercut in the wall in long run for mainly stamp charge battery. Silica wall hot corrosion/ chemical corrosion from coal ash fusion.

Small holes in the walls or wall spalling due to leakages of liquor through AP holes.

The advantages of Ceramic Welding are it is a dry process and minimum thermal shock tosilica refractory, Flexibility in repair anywhere in the walls and longer life due to crystalline bonding.

The main deliverables are:

Early Prevention of oven Refractories fabrics from minor or big damages to operate the Battery trouble-free for long life.

Restoring the health of the oven from critical damage quickly and economically. Extending the residual life of a Battery by adopting a composite repairing technique. Extending the environmental life of a Battery by controlling the stack emission.

The final sermon for Ceramic welding is that it is not a total solution to repair the refractory wall.

It is a total solution to wall undercut repair and a specific solution to a specific problem. Deferment of end flue repair and residual life extension to several years is possible.

Hence, it is the best solution in achieving long battery life in a most cost (overall) effective manner.

Extending the Residual life of the Battery is purely based on economic feasibility.

It is a supplement to traditional repair techniques and not a total replacement.

Ceramic Welding

Wall Hole

Joints open/undercut

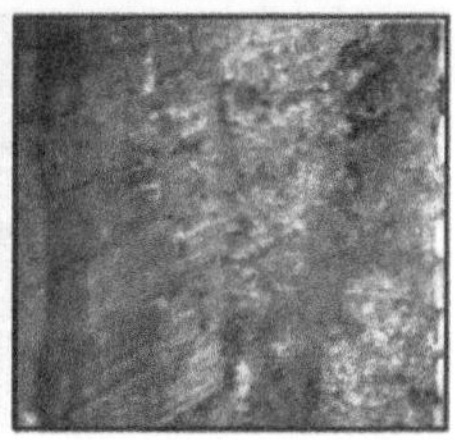

Hot Corros

13. Refractory Repairs

Refractory Repairs are of two types generally, Hot Repair and Cold Repair.

Hot repairs consist of Ceramic Welding, Semidry Gunning, Patch repair of hole in Hot Coke Oven Walls, Oven sole repair in running the oven, Regenerator Checker replacement, replacement of AP, Elbow, ISO Valve, Buck Stay, Tie Rod etc

Cold Repairs consist of Through/Full Wall Repair (Single or Multiple), Partial Wall Repair (flue). One-off or Systematic from one end, Complete overhauling of the full block from oven deck.

13.1Partial Wall or Flue Repair

Inspection and fix number of flues (normally even number) to be renewed, preparing a bill of material (BOM) for total job, oven down including two additional side ovens as a buffer for temp. control on the repaired chamber. Job execution including dismantling, relining and mechanical work. Repair and replacement of regenerators checker bricks. Heating up and commissioning.

Partial wall Repair

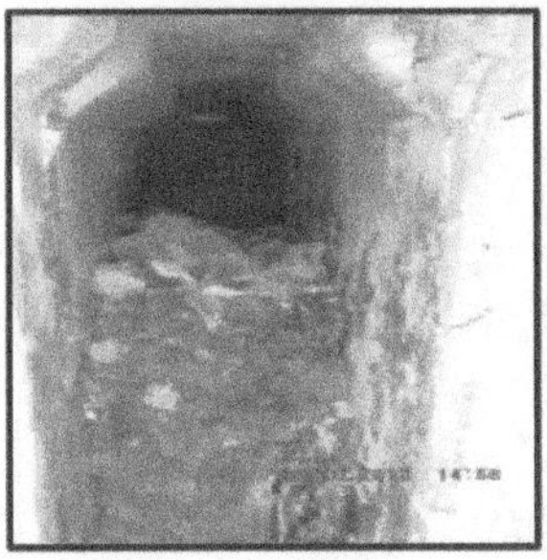

Damaged Regenerator

Refilling of Checker

13.2Through Wall/Full Wall Repair

The process is as follows: Finalize the number of walls to be renewed at a time, prepare the bill of material (BOM) including some under sole refractory and mechanical spares. Ovens down including two additional side ovens as a buffer for temp. control on the repaired chamber. Job execution including dismantling, relining and mechanical work. Repair of regenerators and replacement of checker bricks Inspection, bracing support, oven closure, Heating up and commissioning.

13.3 Innovations In Full Wall Repair

Instead of using normal silica shapes to build the wall nowadays fused silica-based castables are used to pre-fabricate two to three flues together and installed with the help of lifting gadgets. This is making the wall erection extremely fast and heating up time also gets reduced extensively as it is zero expansion. This makes the wall much more stable and stronger.

Full Wall Repair

Full Wall repair with pre-cast block

13.4 Some Developments in Coke Oven Repair

Use of high-temperature camera for inspection at remote access and decision on repair and other actions.

High Temperature Camera

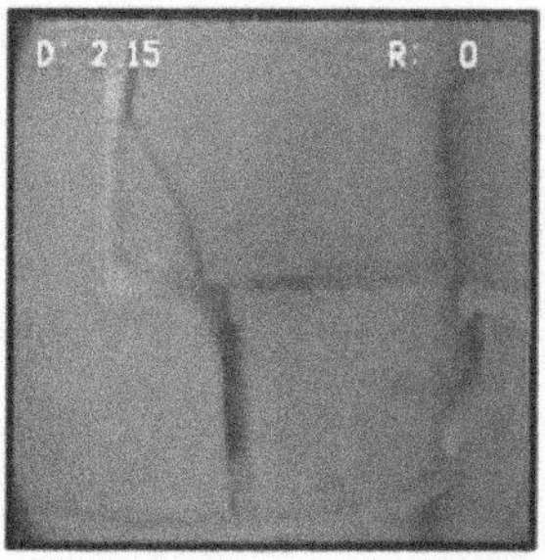 

Endoscopic inspection

13.5 Micowall Repair

This is a modular repair technique. Execution is extremely fast. Few numbers of fused silica shapes are used instead of a huge number of shapes. Reduces the number of mortars joints and makes the wall stronger. Less time is required to heat the repaired wall and it matches perfectly with the existing wall as it is zero expansion. Less inventory carrying cost as the number of shapes are exceptionally low.

 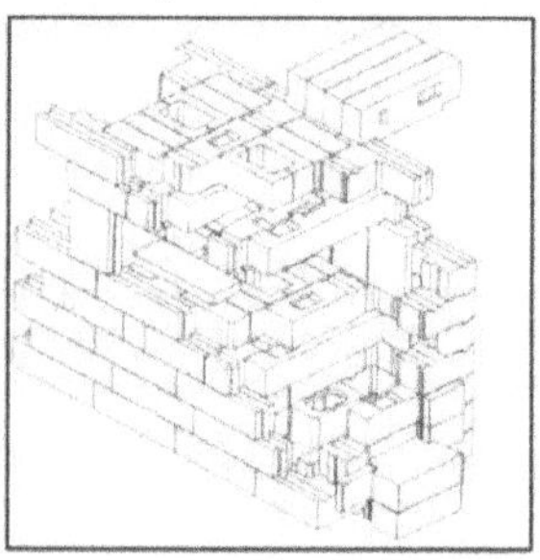 

14. Conclusion

Many efforts are made in the design of a coke oven battery to make the battery more stable for higher productivity and longer life. Efforts are made to improve the residual life of all the existing batteries by adopting various methods in operating and maintenance practices.

These are safe coal blend, correct operating ratio, high pushing regularity index, regulating pushing force within the limit, regulating battery anchorage system at pre-determined intervals, quantifying the residual life of the battery by surveying battery on its physical movement and proactive maintenance to get back the original shape and last but not the least the skill development of people operating the battery including maintenance. Some new techniques have been developed in battery inspection and maintenance. Few plants adopted those techniques. The rest of the plants may like to adopt depending on their requirement for achieving quick results towards the long life of the battery.

However necessary care during construction, heating up- Commissioning and usages of superior quality of materials, especially Refractories cannot be undervalued.

About the Author

Sujit Kumar Haldar is a Consultant at Aqua Terra Logistics Pvt. Ltd. He was the Ex-Chief of Coke plants, Tata steel. His learning carrier started in IISCO, SAIL Coke Ovens in 1980. After 16 years, switched over to Tata Steel. Continued learning in the next 21 years in the Coke Making field in various capacities. Coverage was wide and starting from Coal selection, coal blending, Battery Operation and Battery Maintenance, By Product Plant operation to Battery Construction as Project In-Charge. Spearheaded the Green Field Heat Recovery 1.6 mil tonne capacity Coke Making Project. Visited many Coke Plants for technical discussion and troubleshooting around the Globe. Visited Coal mines in Australia and the USA. Provided support to Port Talbot Plant UK and Ijmuiden Plant Netherland in improving their operating and safety performances in my capacity as Consultant of Tata Consulting Engineers, post-retirement. Independently given consultancy through BCG to JSW, Tornagalu and Dolvi Plant in improving Coke Oven performances.

VII.2. My Journey with LAM Coke Making

Susanta Kumar Halder

1. First Leg of The Journey

It all started in Oct 1980 when I got my second appointment letter from SAIL after the first one from Mazagaon Dock Limited. Being attached to the family of SAIL, it took few minutes for me and my parents to decide in favour of SAIL. Joined Bhilai Steel Plant the best of the SAIL lot as an MTT.

Industrial life started with a single room in the hostel, a bicycle and 10 to 5 in the Training Institute with occasional plant visits. Cool Life with a lot of dreams for the future and also some fear of the unknown.

Then came the D Day when the entire batch got a dramatic and horrific entry to the work environment in coke oven due to a sudden strike by the CITU led workforce. We were put in actual jobs starting from operators to the most dangerous job of Doorman and Oven top man. Few of us left the plant and went home out of fear. After about 8 days normalcy was restored and we were back at the training centre. Within the next few days, all of us were given permanent postings. My destiny brought me back to Coke Oven with placement in Battery operation.

In an environment of Gas, Dust, Fire, Heat and working in shifts, it took me some time to adjust. Probably the 8 days of traumatic and shocking exposer to the very harsh environment of the coke oven gave the mental strength to survive. Of course, there was a silver lining too and that is "Highest Incentive Earning Department."

Eight side fired top charge batteries of 65 ovens each with no pollution control system. Manual cleaning of Oven Doors & Goose Neck. The manual backfiring of hot coke spillage into the oven. Very Crude but very rugged battery machinery. The riskiest work was gravity charging of ovens. Gassing through doors, burning through oven doors & AP Cap, Undercharging of Ovens, Cleaning of Doors and DF, Failure of charging lid covers, Failure of Tie Rods were some of the major problems with the design. All these batteries gave good life due to the very robust Battery Anchorage.

In 1987/1988 got a chance to be associated with the commissioning of a 7.0 m tall battery with additional facilities of Pollution Control systems like Water Sealed AP Cap, Zero Leak Oven Doors, Controlled charging of Ovens with screw feeders in Charging

Car, HPLA, Pushing Emission Control etc. Barring proper control on charging emission, pollution due to emission through oven doors, charging lids & AP Cap improved a lot.

2. Second Leg of The Journey

In 1989 got transferred to Durgapur Steel Plant. Came to Durgapur and Joined DSP Coke Oven in Sept 1989. 5.5 batteries with 78 ovens in each battery in 2 blocks. Underjet type of German and British Design. All top charge batteries with single GC Main. A too bad condition of Batteries. Very Weak design of Battery Machines. No pollution control. A very challenging task for operation and maintenance. Oven End and Anchorage failure was a frequent phenomenon. Gassing, Burning, Gas main chocking etc were regular features.

In 1992, after the visit to Linz Plant, Austria charging emission control by the installation of gas transfer through 5th hole to next oven was installed in Battery # 3 during rebuilding with HPLA system based on our recommendation for the first time in SAIL. Could not give the desired result due to gravity feed uncontrolled oven charging but some improvement was observed in the overall scenario. Then in 1995, a Greenfield battery of OTTO India Design with single GC Main and Gas Transfer through 5th hole was commissioned. No proper pushing emission control. Challenge was to achieve visible emission below 16 sec during oven charging. Could achieve partial success. Peak less charging could not be ensured by the charging car. Availability of clean coke oven gas was a problem due to poor efficiency of Primary Coolers, Non-availability of chilled water to have better cooling in the Primary Coolers, Non-working of ETP etc.

Battery life overall was not good due to the failure-prone Anchorage system.

3. Third Leg of The Journey

In 2006 took VR from SAIL and Joined Electrosteel. Posted in Kalahasti DI Pipe Plant Heat Recovery Coke Oven. The first exposer to Heat Recovery Coke Oven. 100000 tpa top charge modified Beehive oven of Indian Design without any waste base Power Plant. The oven Condition was bad.

The block wise repair was carried out with the following modifications in the battery proper to make the battery suitable for stamp charging with the installation of Waste gas base Power generation.

"Double Layer in Oven sole with 62% Alumina Brick, Modification in Oven Raft Cooling system, HRC concrete stopper in coke side, Installation of Water-cooled dampers in flue tunnel, Making of Cross Concrete beam between Pusher Tracks, Ceramic Module lining of the Waste Gas Tunnel, Installation of stamping station and a new pusher machine, Modification of the Coal Preparation Plant, Dove Tail modification of the stationary pusher and installation of a 12 MWH waste gas-based power plant with boiler and auxiliary facilities". A suitable adjustment was done in the coal blend to ensure good quality coke at a reasonable cost. A real success story.

Carried out the same modifications in the Haldia Coke Plant of EIL with the following additional major modifications

"Complete design modification in the cooling channel of the battery raft, use of better Refractories, Replacement of Waste gas underground tunnel from battery end with overground duct to avoid tunnel bulging ".

Took up the responsibility of the 1.0 Mt green field heat recovery coke plant of EIL DI Pipe Plant in Bokaro in 2009. Based on the detailed analysis of Technology, Space Requirement, CAPEX & OPEX decision was taken to install 0.5 MT Vertical slot type heat recovery coke oven for the first time in India and 0.5mt Horizontal type heat recovery type coke oven.

Challenge was to gain a proper understanding of the VCO technology, process control etc and to do the engineering accordingly. The technology supplier was SDM, China. Except for Electrical all supply was in the scope of supply of SDM. Engineering and ordering for major Chinese equipment and Refractories were completed in time. Inspections of various items, machinery and Refractories were carried out.

Unfortunately, due to some local problems left EIL during the construction stage.

4. Fourth Leg of The Journey

Started in April 2010 with Bhushan Steel Plant at Angul, Odisah as Head of Coke Ovens. 0.85mt tpa coke oven consisting of 2 batteries of 65 ovens each were under construction. A huge project delay was there. One of the main contractors TPL backed out in Sept 2010. Eleven Small contractors were put in service along with L&T to complete the pending jobs. In March 2011 Battery #1 was commissioned after heating with LPG. The battery was put on under firing with its own generated COG which was a challenging task.

The main problem faced during the initial period of commissioning was frequent coal cake breakage and Charging Emission Control. The Chinese made Stamping machine was replaced with a German make stamping machine which solved the problem of coal cake breakage to a great extent. Battery # 2 was commissioned in Dec 2011. In 2017 both the Gas transfer Cars were replaced with new gas transfer cars. Land-based gas cleaning system was converted to wet from dry type to avoid incidences of internal fires in the baghouse.

Usage of many new verities of Coking Coal was introduced during the O&M. Optimum blend ratios were adopted with an eye on cost and quality based on the typical spec sheet of the imported coal in absence of a pilot oven. No Sticker, No hard push so far. Coke quality was good. Usage of non-coking coal % was gradually increased to 25 %. In 2012 both the battery was retrofitted with a radar base positioning system supplied by Lotus Wireless with visual monitoring from a remote place.

From 2012 onwards also got involved in the project engineering of the 7.6m tall Green Field, top charge Chemical Recovery Battery with state-of-the-art facilities and automation. This was the largest single battery complex in entire Asia and the second largest in the world after Schwelgern Coke Plant in Duisburg, Germany.

In 2013 end came a lifetime best opportunity to head the new 7.6m tall complex. Project Engineering, Project Execution, Heating and Commissioning, O&M. 74 ovens of 1.3 MTPA capacity. The technology supplier was Paul Wurth.

4.1 Salient Features

- Single GC Main
- Water Sealed AP Cap
- HPLA system in Goose Neck and HPL flushing in GC Main
- Individual Oven Pressure Control system (First time in India)
- Zero-Leak Oven Doors
- Land-based pushing emission control system
- Pusher Car based Charging & Pushing pollution control system on pusher side.
- Battery Machines from Schallke Germany capable of running in fully auto mode
- Auto Positioning system for all machines

- Automatic under firing gas reversing system operated by 10 hydraulic cylinders.
- Campa Block Heat Exchangers in BPP
- Claus Plant
- Most Modern and state of the art BOD Plant from Muttag, Germany.
- 3 stage air entry to vertical flues to minimize NOx in the exhaust gas.

The battery was Commissioned on 22nd June 2014 after completion of Heating & Drying using COG with auto controlled burners supplied by PWIT. Ovens were charged without any wooden log/Wooden plank charging into the oven before coal charging as suggested by Paul Wurth against stiff opposition by PWIT. Initial Charging Blend was decided by me bypassing the PWIT suggested blend. It was a smooth sail. Within one month of O&M, the percentage usage of non-coking coal was increased to 10 % to reduce the cost of coke. Within a year of Operation, the usage percentage of non-coking coal was increased to 15%.

After initial teething problems, Smoke-Free Charging was achieved with HPLA and Peakless Charging ensured by the working of Charging Car and Pusher Car in auto mode. Visible emission during charging was below 16 sec (CPCB Norm). Combination of high speed and low speed charging with 2 stage levelling. A Computer simulated model for oven charging was developed in peakless mode by M/s Schalke, Germany and BSL Jointly to ensure smokeless charging and levelling.

After 1.5 years of operation, an individual Oven Pressure control system was commissioned where the GC main works under negative suction and at the same time maintain positive pressure inside the oven. A special PLC operated section valve was installed in the Goose Neck before the Isolation Valve. No HPLA system is required to get smokeless charging. Due to some financial issue with the supplier, the system could not be operated for long. The system was under revival till I left TATA BSL on 31st March 2019. Excellent level of pollution control was achieved in the area of Oven Pushing, Oven Charging, Chimney Emission etc.

But there were certain weaknesses in PWIT Design as mentioned below where BSL did major modifications barring few areas where the same could not be done in the running plant.

- The use of Butterfly valves as isolation valves in the under-firing system does not ensure complete isolation. Could not replace with Plug Valve due to insufficient riser pipe height.

- The door Repair rack was very flimsy and not user friendly. Modified in house
- No Tar Chasing Hole in the Cross Over main up to the separator. Installed by BSL during construction.
- No Bypass for Ammonia Liquor across the isolation valve in the GC Main. Installed by BSL during construction.
- No "U" seal on the Flare stack. Designed and installed by BSL.
- No "U" seals in COG and BFG Inlet line to battery. Designed and installed by BSL.
- Gas mixing station for mixed gas under firing not proper due to faulty design. Remade with an in-house design by BSL.
- No Preheater in COG inlet line for under firing. Installed with an in-house design modification by BSL.
- Usage of Butterfly valves in Main COG raw gas line from Primary Coolerer inlet to final exit. Valves getting replaced one by one with Leakage class 6 triple eccentric butterfly valves.
- The problem of high Tar sludge in the Tar due to poor tar decanting system. Installed a boat type Tar Decanter with all necessary modifications.
- Failure of the Tar Centrifuge. Bypassed.
- Chocking of Compa block Heat Exchangers. Developed an in-house system of CIP.
- Frequent failure of Insulators in the ETP of BETH from Germany. Electrode Earthling problem. Replaced the Insulators with 70% Alumina Insulators. electrode fixing system was modified.

4.2 Silver Lining

- Flawless, Supper efficient, Self-sustaining Battery machines capable of running in auto mode. Being operated in semi-auto mode.
- Very reliable Auto Positioning System.
- A well designed, efficient, reliable BOD plant. Within one year of operation, the plant could meet all the norms of CPCB for wastewater. It is running very efficiently giving rise to zero outside discharge from the coke oven plant on account of reuse. Treated water is being used for Coal spray, Road washing, Gardening and SMS slag cooling in the slag yard.

During 2014 to 2015 completed the engineering for 2 CDQ plant with Nippon Steel (NSENGI). 180 tph single chamber with 95 tph WHRB tall battery and 120 tph single chamber with 65 tph WHRB for stamp Charge Battery Complex. 180 tph CDQ got commissioned in May 2018. Within few days of operation, the operation got stabilized. Loading was taken to 90% level within a month. One of the most challenging tasks was to make the operation of the Quenching Car compatible with CDQ in auto mode making use of the existing auto-positioning system. This was done with in-house effort successfully. The plant is working since then without any major incidence of breakdown, fire or explosion. HP steam generation of 0.60 tons per ton of coke cooled was achieved which is more than the design capacity of 0.55 tons per ton of coke cooled. Marked Improvement in Coke Parameters observed. Coke Moisture < 1.0 %, CSR improved by 3.0 %, all other strength parameters improved using the same blend. That resulted in the use of 15% non-coking coal in the blend without sacrificing the coke quality.

120 tph CDQ is under installation.

5. Fifth Leg of The Journey

After retirement on 31st March 2019 from a permanent job, 5th phase of my journey with LAM Coke started in the role of a consultant. The first one started in Sept 2019 in Kirloskar Ferrous Industries Limited for their 200000 tpa Heat Recovery Coke Oven in Koppal District, Karnataka supplied by MINMETAL (SCCDI DESIGN) from China. Horizontal type ovens. The battery was under construction. Heating was done with Coal and Commissioned in May 2020.

Premature Failures like "failure of hot gas valves, failure of top air distribution plugs, Oven Door lining failures, hot gas piping lining failure, Arch brick falling, Coke Transfer Chute lining failure, failure of coke screen PU mesh etc "were observed within 1 year of operation. Heating Regime in the ovens is maintained with Primary and Secondary Air Inlet control only. the oven draft is fixed. Very difficult to achieve fine control of the process. Burning loss is to the tune of > 5 %. Coke Quality is good.

The brick lining of oven doors was replaced with castable lining. Primary Air inlet distribution plugs were replaced with an in house developed plug. Coke Screen Mesh being replaced with hard-faced steel mesh. Different material like fused silica is being used for making of hot gas valves. A different process control regime was developed and put in place. Coke Moisture brought down to 5.0 % from 8.0 % by modifying the impulse quenching time cycle and modification in the quenching wagon.

Presently Providing consultancy to WCL to set up a greenfield 200000 tpa Heat Recovery coke Oven as an integrated part of a Mini Steel Plant consisting of BF, Sinter and DIP in Gujrat.

O&M consultancy to Uttam Galva is in the offing and so on in some other plants.

My Journey with LAM COKE Continues…

About the Author

Susanta Kumar Halder is B. Tech (Metallurgy) from I.I.T Kanpur.

Ex SAIL & BHUSHAN-TATA STEEL

VII.3. How The Four Jumbo Coke Oven Batteries Were Saved During the Super Cyclone Hudhud In 2014 At Vizag Steel

Rathinasabapathy Nagarajan

1. Super-Cyclone Hudhud and Its Devastation Story

A cyclone is an area of closed, circular fluid motion rotating in the same direction as the earth. The cyclone can be in any one of the forms: Depression (D), Deep Depression (DD)Cyclonic Storm (CS), Severe Cyclonic Storm (SCS), Very Severe Cyclonic Storm (VSCS), Extremely Severe Cyclonic Storm (ESCS), Super Cyclonic Storm (SuCS). The cyclone 'Hudhud' has been named after the national bird of Israel, the Hoop hoe bird. Hudhud was termed by IMD as the strongest and devastating tropical Very Severe Cyclone Storm (VSCS) of the Bay of Bengal that ever hit any Indian city. It was formed in the North Indian Ocean and hit the land at Visakhapatnam on 12th October 2014. The winds were blowing at a peak of 220 km/hr with avg speed of 180 km/hr. According to IMD reports, after 1985, Hudhud was the first cyclone that crossed Visakhapatnam Coast in October 2014. The eye of the storm was particularly concentrated around the town of Visakhapatnam and actually passed through the city took the major brunt of the cyclone. Luckily, the loss of human life was considerably less but there was a huge loss of public and private property as expected. The city was completely devastated. Hundreds of huge trees were uprooted, impeding mobility on roads. Power and water supply were totally disrupted for the next ten days.

2. The Coke Oven and Coal Chemical Plant of VSP

The Coke Oven Plant of Vizag Steel have four numbers of 7M tall Top Charge Batteries of (capacity around 0.9 MTPY each of Met-Coke) with 100% CDQ plants with 16 Chambers and large Coal Chemical Plant Complex and the Status of the Plant before the cyclone was:

- Batteries 1&2 with CO gas heating
- Batteries 3&4 with Mixed gas heating.
- Oven pushing consistently @ 321 ovens/day,
- CO gas handling of 1,45,000 cum /hr

- Other Coal Chemical products are Benzol products,
- Tar products & Ammonium Sulphate etc.

3. Hudhud and Its Devastation Story

A cyclone is an area of closed, circular fluid motion rotating in the same direction as the earth. The cyclone can be in any one of the forms: Depression (D), Deep Depression (DD)Cyclonic Storm (CS), Severe Cyclonic Storm (SCS), Very Severe Cyclonic Storm (VSCS), Extremely Severe Cyclonic Storm (ESCS), Super Cyclonic Storm (SuCS).

The cyclone 'Hudhud' has been named after the national bird of Israel, the Hoop hoe bird. Hudhud was termed by IMD as the strongest and devastating tropical Very Severe Cyclone Storm (VSCS) of the Bay of Bengal that ever hit any Indian city. It was formed in the North Indian Ocean and hit the land at Visakhapatnam on 12th October 2014. The winds were blowing at a peak of 220 km/hr with avg speed of 180 km/hr. According to IMD reports, after 1985, Hudhud was the first cyclone that crossed Visakhapatnam Coast in October 2014. The eye of the storm was particularly concentrated around the town of Visakhapatnam and actually passed through the city took the major brunt of the cyclone. Luckily, the loss of human life was considerably less but there was a huge loss of public and private property as expected. The city was completely devastated. Hundreds of huge trees were uprooted, impeding mobility on roads. Power and water supply were totally disrupted for the next ten days.

3.1 Predicted Vs. unpredicted consequences

The magnitude of the cyclone effect could not be predicted exactly. We could not imagine the impact of winds blowing at a speed of 200Km/hr. The impact of the cyclone was a unique experience in the history of the steel plant. The VSP management along with employees struggled for ten days to revive production that came to a standstill due to zero power situations. We had to make decisions under extreme uncertainties. This was a unique situation that was never experienced by any steel plant across the world. It was later experienced during the landfall of the cyclone that the precautionary measures taken by the plant authorities and the employees concerned could not withstand the magnitude of the cyclone and wind speeds crossing 200Km/hr. The prediction was far away from what had actually happened. The effects of the cyclone led to unpredicted consequences, for which there was no remedial solution that could be planned.

Management of Visakhapatnam Steel Plant has stopped all the non-critical units in advance but some of the critical units could not be outrightly stopped.

3.1.1 Operating Parameters before the start of the Cyclone

Parameters	Battery 1	Battery 2	Battery 3	Battery 4
Operating Control Vertical Temperature	1256 ^{0}C 1271 ^{0}C	1284 ^{0}C 1301 ^{0}C	1223 ^{0}C 1252 ^{0}C	1236 ^{0}C 1238 ^{0}C
Coking Period Hrs	19:42	19:03	18:38	18:38
Heating Walls Conditions	Less than average	Average	Less than average	Good
Tar properties: QI	1.8			
Tar properties: Sp. Gr.	1.17			
Gas Make	145000NM3/Hr			
Pushing Amperage	Normal			
Coke Properties: M10 &M40	M10: 7.3 & M40: 79			
Coke Properties: CSR & CRI	CSR: 66.5& CRI: 24			
Oven Stickers Record	No Sticker or hard Pushing reported			

3.1.2 A chronological description of Events

Day 1: 10th October 2014

Day 2: On the 11th of October, Saturday

There were no climatic changes but strategies to minimise loss due to the cyclone continued

1. Both the coal towers were filled to the brim by the end of the "C" shift to ensure the availability of raw material.
2. The normal pushing from 107 ovens was reduced to 92 pushing on 11th October in the "A" shift.

Day 3: On the 12th of October, Sunday

The atmosphere became very turbulent with strong gales and high tides in the sea from early morning and the district authorities indicated that the cyclone would strike the city at 11.30 am.

Instructions were passed to all the shifts that pushing should be stopped and no charging be done when there is heavy gale and torrential rains.

3.1.2.1 Responsive Actions Taken at Coke Oven Were:

1. Immediately the entire gas stop cocks to the Coke Oven gas heating system were closed.
2. As the Electrically driven Flushing liquor pumps tripped, diesel pump was started to pump ammonia liquor for cooling of Coke Oven Gas 250-300 cum per battery was ensured and all the Coke Oven gas bleeders' water seal was maintained
3. All Chimney dampers were closed as far as possible to retain the heat in the batteries
4. In order to keep the Gas Collection Mains (GCMs) in positive pressure, Suction Main Gate valves were closed in all the batteries.
5. At CDCP's,
    i) One hot bucket got held up.
    ii) As there was no power and all CDCP control rooms became pitch dark, neither the mill fan nor reserve fan could be started to cool the hot coke in the working chambers.
    iii) The UPS systems also had gone out of operation.

The "A" shift personnel were stranded inside the plant till the night of 12th October 2014 and the "B" shift employees could not reach the plant as the roads were blocked due to uprooted trees and electrical poles. All the chimney pumps failed, and the diesel pumps brought in by CISF did not work. Flushing liquor diesel pump which was running since the morning ran out of diesel stock despite the reasonable stock arranged at Coke Oven & Coal Chemical plant and the pump tripped to the airlock. Bringing diesel was very difficult as the roads were blocked by fallen trees and it was pitch dark all around.

Day 4: On the 13th of October, Monday

The day aftermath "Hudhud" (13th October 2014) employees started reporting to the plant with great difficulty as roads were blocked not only in the plant but all over the city. Public transportation was not available. Drinking water lines dried up and though the groundwater level was high, there was no power to pump water. The pusher car which was stationed in Battery No. 4 was pushed away to Battery No.3 due to the high-speed gales. Plenty of roof sheets, corrugated sheets on the conveyor galleries, Ammonium Sulphate Godowns, Mechanical, Biological and Chemical (MBC) Plant roof etc were blown away.

Initially, it was planned to seal all the doors, ascension caps, charging lids and waste heat box discs. The entire air ingress area was covered by placing ceramic fibre blankets below the water heat box flaps.

Loaders were used to clear the road and diesel could be arranged. Flushing Liquor diesel pump air lock problem was sorted, and the pump was put into operation from 9.00 am. There was no flushing liquor supply for 8hrs (from 1.00 am to 9.00 am on 13th October 2014).

There was no gas from batteries and a lot of air ingress was felt in the gas network. As a consequence, Oxygen content inside the gas network was substantially high to avoid ingress of air in the cooling chamber, sealing off at lids and other openings were done with glass wool. It had to be done to avoid explosive mixture formation, clinker formation, burning out of coke and blowing devices inside the cooling chamber.

The attempt to cool the hot spots rising intermittently in one chamber or the other was inadequate due to the insufficiency of water with firefighting engines.

Average control vertical temperature (CVT) fell by around 2000C in all the batteries on both sides i.e., the pusher side as well as the coke side.

Average Control Vertical Temperature (CVT) on Day 4.

	Pusher Side in deg C	Coke Side in deg C
Battery No 1	1047	1031
Battery No 2	1020	1001
Battery No 3	986	992
Battery No 4	1034	1037

As initial CO gas generation will be less than 4000 Cubic meters, we need to regulate the gas outlet from Exhauster thru 300 dia pipe and not thru regular1500 dia pipe. The exiting bypass line was in damaged condition and had not been used since commissioning. Mechanical persons rose to the occasion prepared themselves to take shutdown. In the absence of power and Nitrogen, a shutdown could not be given. Accordingly, one group was kept ready to erect and weld the pipeline as soon as Power and Nitrogen is made available.

It was decided to start the coke oven and Air separation plant (ASP) after the restoration of power. So, the CO&CCP collective had to stand on their own strength as there was no choice left to the revival of batteries. We have to produce coke oven gas by pushing out the Coke from some ovens, charging these Ovens with coal and generate some amount of coke oven gas with the residual heat of the heating walls, start the Coke oven gas exhauster and put Coke Oven batteries under heating with the generated Coke Oven gas as fuel. As soon as the power supply is restored, this needs to be done immediately as Batteries temp are falling alarmingly. Immediately a handwritten protocol for option 2 was drafted at night with the help of torch lights and submitted to the Office of Director (Operations) on the same night.

Meanwhile, power was not yet restored, and employees were kept round the clock in all sections with a common control room in Battery Shift Managers Room for Battery related operations and another control room was set up in the recovery shift room for Coal Chemical Plant related operations.

3.1.2.2 The Typical Issues Involved in Any Disaster Management

1. Identification of risk and hazards
2. Assessment and prioritization
3. Organization and resources
4. Awareness and competence
5. Evacuation, relief and rehabilitation etc.

Having understood the Disaster management system, we fully prepared ourselves for the Unpredictability, Unfamiliarity, Urgency, Uncertainty, Speed and Threat we would be facing in Revival of the CO&CCP.

Day 5: On 14th October, Tuesday

As the power situation was getting delayed inordinately and battery chimneys were getting flooded, Coke Dry Cooling Plant chambers were literally burning, and the battery temperatures were dropping alarmingly and were below the measurable range of optical pyrometers. it was decided to plug the end flues of all batteries before taking up the heating. Accordingly, 1088 dummy plugs to plug 1, 2, 31 and 32 flues for all four batteries were fabricated using gas cylinders and were kept ready for welding once power restores.

All the senior and available officers of Coke Oven & Coal Chemical Plant met on the evening of 14th October on the terrace of the Area Shop Office and explained the risk and challenge ahead of us. After all the options were discussed in detail, it was agreed upon to start the batteries 2, 1 4 and 3 in order and as per the protocol.

Around 8 pm on 14th October, in the night people heard a big sound. Nothing could be accessed due to pitch darkness. The next day it was found that a 30 meters height civil staircase attached with the coke Bins of CSP collapsed and fell on the Gallery, two of the coke oven conveyor belts K-13 and K-14, causing heavy damage and the gallery sagged. These two conveyors convey BF Coke from CDCP1&2. K-14 conveyor belt was extensively damaged and with no alternative, we decided to take a risk and run K-13 for the revival of Batteries.

Day 6: On 15th October, Wednesday

To reduce the heat loss in the batteries, a small opening of 500mm X 500mm was created in the chimneys at preheating manhole areas to create less draft from the battery waste heat system. Due to excessive heat, the blowing devices of a few Chambers were in a precarious condition and these chambers were somehow cooled manually with the available water. Employees were inspecting during daytime and all abnormalities; surprises were recorded and an action plan to address the same was prepared by Sr Officers. Sealing of oven doors to the maximum height was done by employees

As a strategy, the battery revival strategies and the risks involved were discussed with the union members of Coke Oven &Coal Chemical Plant and their co-operation was sought which they appreciated and expressed confidence in the leadership and readily agreed to support.

Day 7: On 16th October, Thursday

Uncertainty in the power situation still existed but the conditions were becoming worse and hope for reviving the batteries and resuming production in the plant was becoming critical. Due to very less temperature, sinking had taken place in the freshly relined Heating walls of Battery No. 3. Temporary packing plates were inserted between the brickwork and charging car track so that the charging car does not sink, and the area was kept under total surveillance throughout day and night. Employees continued to do sealing of doors, oven top erupts, WHBs and minimize the air ingress. As the Power supply is getting delayed, we were thinking about the LPG option for pushing out the coke and reviving the Batteries.

Day 8: On 17th October, Friday

Finally, the waiting period came to an end and around 2.00 pm, Coke Oven & Coal Chemical Plant could get 7Mega Watts of power when Visakhapatnam Steel Plant as a whole could get 50 Mega Watts power through the old APTRANSCO transmission line. With the available power following activities could be started.

1. Power was given to start 3 numbers of Flushing Liquor pumps, Scraper Conveyors of Decanters, Starting Oil Pumps and STG Mechanism of Exhausters, 4 numbers of Service Water pumps, Mill fans, Pusher 2 and 5 for sealing top part of the doors, two locos, 2 DEs, CSP 1 & 2 one track each, Cellar of all Batteries, All Control Rooms of batteries, CDCPs, CPP and CSP. QATD, Area Shop Office and Canteen were also given power. Two charging cars were also given the power to restore the Catenary lines.

2. After power is restored in CDCP control rooms one by one, it was observed that coke temperatures were too high from top to bottom in the range of 600-1000 deg C compared to the normal temperatures of 150-200 deg C at the bottom area. As starting of Boiler is essential to start any activity in the Chamber area, there was no other alternative but to wait till Boiler start up.

3. In the wee hours of the night, shift clearance came to start mill fans. As per the situation, it was not advisable to start mill fans, as it could damage blowing devices due to the sucking of oxygen due to the absence of nitrogen. Then it was decided to start reserve fans one by one in all cooling chambers and the temperatures were kept under observation. After three hours it is observed that the temperatures were gradually coming down. Then all the reserve fans were continued in running condition to maintain the regime and allow coke to cool up to 200 deg c. Discharging could be started from CDCP chambers after coke was cooled to take out coke and save the Blowing devices and discharging devices.

4. While discharging coke from some of the chambers, it is observed that coke was not getting discharged and some lumps were found in the gate 3&4 area. Later after intense poking lumps could be cleared from three Chambers and the other three chambers i.e., 2 were observed to have metallic lumps and it was suspected that metallic parts of the blowing device might have got melted and lumps formed. Except for these three chambers remaining all running chambers were kept ready for taking hot coke whenever pushing starts.

5. Even after taking all possible measures to arrest air ingression, lower dividers and some portions of liner plates at blowing devices of some of the chambers were damaged due to prolonged exposure to heat for about 6 days without any cooling or inert circulating gas and. With the quenching of water from the DD, area damages could be avoided in the Rotary drum and DD area only and, there is no access to quench coke above gate 1&2 as chambers were full of coke and no manholes were in place to do anything.

6. Meanwhile, all the Lifters were tested with empty buckets for readiness to receive hot coke without any problem.

7. In the evening one big surprise was there. It was found that the Cold bypass line of the Exhauster to Battery 2 header was in jammed condition. Hence there was no option left but to revive the Battery 1first.

8. It was planned to push 16 ovens in Battery 1 and keep another 8 ovens ready in case gas generation is not sufficient to start exhauster. By measuring the cross-wall temperature, the ovens were selected. It was expected that Nitrogen will be made available by night so that purging of all lines can be started.

Average Control Vertical Temperature (CVT)

	Pusher Side in deg C	Coke Side in deg C
Battery No 1	853	882
Battery No 2	872	890
Battery No 3	905	905
Battery No 4	883	885

Day 9: On 18th October, Saturday - Execution of Plan Battery-1

Nitrogen was available early in the morning at 6kg pressure. The Nitrogen injection points to GCM which were of ½ inch dia. got jammed especially in Battery No 1 and could not supply the required quantity for purging resulted in a delay in preparation jobs. Finally back purging from CCP was started to flush the lines. Following was the sequence of events taking place.

1. As the gas generation was expected to be around 4000 cum /hr, it was decided to erect a 300 mm dia bypass line after exhauster header with a butterfly control valve. This job also could not be taken up earlier as nitrogen was not available. It was sure that explosive mixture has formed and hence all cutting, welding erection of pipeline was taken up after nitrogen was made available.

2. After getting nitrogen at CDCP and ensuring that the temperature of the coke mass had come down with the running of the reserve fan, Mill fans in CDCP 1, 2&3(2,3,6,7,8&9) were started along with nitrogen supply at 6.00 am. Coke discharge was resumed slowly, and enough space was created in the chambers to receive 16 ovens coke from Battery-1 at the shortest possible time i.e. two hours. Accordingly, CSP conveyors one stream was started and through dumpers, the coke was taken to the yard. A lot of care was taken to run the K13 conveyor as its gallery was in damaged condition.
3. The earlier DM water system of CDCP boilers was established before starting the reserve fan.
4. After the Oxygen content was found to be less than 1% at all batteries, Cellar endpoints, GCMs, Suction Mains, Battery 1 was ready for re-commissioning. Erection of Bypass line of Exhauster header of 300mm dia. The pipeline with Butterfly valve was made ready by 04.00 pm.
5. Based on the Charging plan and expecting initial gas generation of 4,000- 5,000 Nm3 the following was done for smooth and safe commissioning of CO&CCP.
6. The shortest route and the minimum number of equipment were used initially to maximize the return of gas to the Battery. Accordingly, the gas route and equipment were identified and isolated by closing the valves and filled with water.
7. It was decided CO gas from Battery # 1 would be carried to Exhauster and ultimately to Battery #1 through the identified route. The rest of the network and equipment will remain isolated with nitrogen purging till the need comes to use them or adequate gas is available to put them under gas.
8. PGC inlet header was segmented, isolated Battery wise by closing all isolation valves and the header valves to batteries No 2,3 &4 were closed and filled running water hoses. Ideally, we should have blanked the headers, but we had to take a calculated risk. The vent valves, sampling cocks at PGC headers were not in service for the last 25 years, we had difficulty in removing the blanks, revisioning the valves and sample cocks. All these jobs were done after the nitrogen was made available.
9. PGC outlet header was kept common with the outlet valves of all PGCs in closed conditions.
10. All ESPs were under nitrogen purging with inlet and outlet valves in closed condition and filled with water.

11. Common by-pass of ESP was kept in open condition.
12. Only two PGC (# 1 & 2), one Saturator (# 1) and one FGC (# 2) were kept in operation initially.
13. Exhauster # 4 was identified for starting because it was fresh from capital repair. There was no steam available for purging the exhauster body and condensate lines also Even though TPP has resumed power generation, steam could not be made available to COCCP, and steam generated from CDCP was not sufficient to cater for the needs of CCP. It was decided to forgo this fundamental necessity and go ahead with the commissioning.
14. 1500 mm diameter gas valves on the delivery of Exhauster to ASP were closed. Gas valve 122/N with by-pass (300 mm dia.) line was restored with line modification for control release of gas after Exhauster with new 200mm dia Butterfly valve. Job started in the morning after N2 purging and line fabrication & erection was completed by 04.30 pm i.e., just before oven charging.
15. Isolating the gas network from the Consumer by closing the gas valves after Benzol Scrubbers was required to minimize the air ingress into the gas network.
16. While purging the GCM, we could see nitrogen coming out of C/s two bleeders and caused a lot of concern. There was an element of doubt of explosion that occurred earlier. We did not have time to repair/replace the bleeder, instead opened the side cover and stuffed a huge quantity of ceramic blanket and went ahead.
17. Following ovens were selected which were having better temperature and were not adjacent to each other and also to facilitate simultaneous pushing and charging uniformly in both the GCM's with sufficient gap in sequence for pushing: 111, 136, 103, 139, 107, 114, 116, 147, 121, 150, 125, 153, 129, 157, 133 and 161.
18. It was decided to use two Charging Cars for charging so that ovens could be charged quickly without further loss of temperature and gas. It was also decided to connect all the ovens to GCM at one go so that a sudden huge surge of gas could be dispatched to PGC and Exhauster.
19. First oven 111 was pushed at 02.40 pm and the balance 15 ovens were pushed by 04.25 pm, by two sets of machines and two groups of Officers/ operators. Enough preparations were done for quick removal of huge coke fall and to rake in the worst case of the sticker with two sets of machines. The arrangement was also made for

sole cleaning, wood charging from all sides. Charging commenced from 04.45 pm and was completed by 05.55 pm

20. It was decided to charge ovens full-on Pusher side and middle and 3/4th on Coke Side.
21. CO gas network valve was kept in closed condition after COCCP. Flare Stack set point was reduced from 800mm to 300 mm to take care of the safety of the Gas Network.
22. If everything goes well as per plan, it was also decided to utilize the available gas for heating only 50% of the H/Walls in batteries 1 & 2 and go for reviving Battery 4 if BF gas was not made available.
23. The gas generation was so low that it was almost impossible to push/bring the gas at the top of PGC. For half an hour effort was made to identify the gas at the vent on the top of PGC but no symptom was available including any rise in the temperature in the vent line except a slight increase in the line pressure, observed through connected U Tube manometer.
24. Considering the overall situation, it was decided to run the Exhauster after closing the purging vent. Accordingly, Exhauster # 4 was started at 18:40 hrs in the "B" shift itself.
25. The initial start of Exhauster was a really tricky and risky job. The main aim was to start the Exhauster with estimated 4,000 Nm3 gas and put it in continuous operation till Batteries was able to generate quality gas in adequate quantity. VSP Exhauster is a constant single-speed drive of 3900rpm, 72000Nm3/hour, hence a lot of problems was faced to maintain adequate suction, high temperature due to local circulation, load variation due to Nitrogen mixture and O2 content in CO gas. Despite load fluctuation, the continuous operation of Exhauster was managed without tripping for any reason. Initially, through Hot By-Pass and afterwards, through Cold By-Pass adequate minimum load to the Exhauster was maintained successfully. Though the Exhauster was in continuous operation, its load was fluctuating widely from 60 amps to 160 amps. It is mainly because of the high quantity of Nitrogen (91-92%) in the gas and the large quantity of gas was in continuous circulation through Cold-Bypass.

26. In addition to abnormal load fluctuation, a high percentage of N2 in the gas led to high body temperature (maximum 104°C) of Exhauster, which was managed by wet blanketing of Exhauster with continuous water spray on the body of Exhauster.
27. Adequate gas was not available to keep the exhauster in stable condition. So, it was decided to start some more ovens pushing and charging to generate more gas till 10 pm. Thus gradually 11 more ovens and a total of 27 ovens were charged by 10 pm.
28. Oxygen content became as high as 8.33% after exhauster and the nerve was kept as cool as possible. The effort was made to push and charge more and more ovens to generate gas and reduce the Oxygen content.
29. But as 100% isolation could not be guaranteed through the closing of valves, it created a lot of anxiety during abnormal Oxygen content in the handling gas, particularly after starting Exhauster. Operation and Mechanical worked very hard to ensure proper isolation and to prevent unwanted ingress of air into the system.
30. One hour after the start of Exhauster the analysis of gas showed 3.0% Oxygen and it went on increasing up to 8.33%. It was felt that the incident was a consequence of air ingression through vents on the suction side of the Exhauster and as a precautionary measure all the vents were blanked.
31. The initial gas generation from the 16 Ovens of Battery # 1 was much less than expected and estimated to be around 2,500 Nm3.
32. After running of Exhauster for about three to four hours, gas was not available to Battery # 1 for initiating under firing
33. When the exhauster delivery pressure increased to 700mmwc opening the 122N valve was started at 7.30 pm. Header pressure was hardly 50mm. exhauster load was varying between 65 to 120amps against the rated capacity of 135amps. The gas make was not known because the gas make was below the lower limit of the flow instrument.
34. Because of the higher percentage of Nitrogen in CO gas, and density difference between Nitrogen and Coke Oven gas had resulted in a huge variation in the exhauster load high temperature of the exhauster casing. Running water hose was continued on the exhauster body.
35. The Oxygen continued to increase up to 8.33% even after charging 30ovens at 11.30 pm. It was a testing time and some preferring to stop the exhauster. However, zero suction to -10mm at the PGC header could be maintained.

36. At around 9:00 pm, Batt-1 Reversing Winch was put in heating with all the stop cocks closed position. Initially, 111, 130, & 133 Heating Walls and corresponding waste heat boxes were connected to the system with the finger plate gap initially being 150mm and immediately flue inspection was done at the oven top for proper burning. But flames were not visible at the burner level due to the high nitrogen content in the gas.
37. As the gas network pressure was not building up the CO gas header valve was closed. There was an element of doubt that the generated gas was passing through the flare stack, and we could not get the adequate pressure in the header. So, the flare stack valve was also closed fully. The entire network pressure control was taken up by Battery people, manually through the cellar vent valve.
38. With this, there was an increase in the gas network pressure from 50mm to 100mm and then finally to 240mm.
39. The initial composition of gas after Exhauster was mainly Nitrogen & Oxygen (N_2 – 91.39%, O_2 -6.31%, H_2 – 0.74% & CO – 1.55%). Since oxygen content was alarming, we decided to continue pushing's and charging from 10.20 pm. We maintained our nerve cool and continued the tasks as it was in the expected line.
40. Again, Heating of flues was started around 02.00 am in the night, 111, 113, 103, 105, 133 one by one H/Wall up to 30 numbers of heating walls were connected by 6 am with stop cock 50% open in Battery 1. By 6 am we have pushed and charged 47ovens.
41. From midnight onwards there was an improvement in the CO gas composition showing a methane content of 12.38%. Oxygen content came down from 8.33% to 5.44%. Nitrogen content from 98% to 68.6%. From then onwards the situation was very much under control and oxygen content came less than 1% also by 4 am.
42. This was a very precarious, challenging and "do or die" situation which tested the planning, confidence, coordination skills, experience and courage of the entire CO&CCP collective. The threat before CO & CCP collective was converted to opportunity and finally to excellence. The stellar performance of the entire team was nothing but the culmination of all these traits and made the dream endeavour a reality.
43. It was decided to resume Battery-2 pushing series wise with End Flues in blanked condition and to hold further heating of Battery 1 so that the identified 4-Series

H/Walls in Battery No 2 could be put under heating for an hour before pushing. Accordingly, one group of persons started working in Battery -2 cellar and 4-series heating walls.

44. Pushing's done on this day:

Battery No	Pushing: "A" Shift	Pushing: "B" Shift	Pushing: "C" Shift
1	0	27	20

Average Control Vertical Temperature (CVT)

	Pusher Side in deg C	Coke Side in deg C
Battery No 1	860	880
Battery No 2	890	900
Battery No 3	887	876

Day 10: On 19th October, Sunday - Execution of the Plan in B-2

1. As planning was done to commission Battery No 2 with pushing a bunch of 13 ovens, CDCP Cooling Chambers No: 2, 3, 6, 7, 8&9 were kept ready with low coke levels.
2. Battery 2 pushing was started at 11.50 am after two reversals of heating of 4-series heating walls as per the following sequence: 204, 234, 209, 239, 214, 244, 219, 249, 224, 254, 229, 259 and 264. Pushing of all those ovens were completed by 01.30 pm. Charging was done with 2 Charging Cars between 12.02 pm to 01.45 pm.
3. The idea of heating the ovens before pushing has paid rich dividends of huge gas gushing from the standpipes. The ovens were connected simultaneously to GCM, and Gas quantity was good enough. Exhauster was connected quickly without much problem.
4. Then it was planned to heat adjacent heating walls in sequence i.e., 3-series 203, 208........... 258 & 263 before pushing. Further, it was extended to 2-series, 1-series and 5-series. This has helped in a systematic way of revival unlike Battery No 1.
5. As gas generation started showing improvement there was no problem in connecting all the heating walls in both the Batteries to the heating system.
6. While preparation was being done for changing Battery No 4 mixed gas heating system to CO gas heating system, it was informed that BF gas would be made available by 06.00 pm. Battery 4 heating started with the BF gas at 6.30 pm with all flues including end flues. All the flues including end flues started burning giving a lot of hope. Hence it was decided to keep the Battery 4 under heating for the day

and go for pushing and charging from 20/10/2014 onwards from the "A" shift. This break was necessary for the people to get rejuvenated for commissioning Battery No 3 & 4 on the next day.

7. Pushing schedules were made in a staggered way to ensure uniform gas make.
8. Pushings done on this day:

Battery No	Pushing: "A" Shift	Pushing: "B" Shift	Pushing: "C" Shift
1	9	11	Nil
2	13	25	25

Average Control Vertical Temperature (CVT)

	Pusher Side in Deg C	Coke Side in Deg C
Battery No 1	832	867
Battery No 2	863	877
Battery No 3	864	847
Battery No 4	931	938

Day 11: On 20th October, Monday - Execution of the Plan in B-3 & 4

It is slated for a very hectic day as it is planned to commission both Battery 3 and 4 on the same day.

1. As planning was being done to commission Battery No 4 by pushing a bunch of 16 ovens,
2. Mill fans in CDCP-4 were started and kept ready along with CDCP 2&3 cooling chambers in advance.
3. Continuous purging GCM and Suction mains and checking the Oxygen content pushing started in Battery No 4 from 08.00 am as per following sequence: 459,404, 434, 409, 439, 414, 444, 449, 419, 454, 424, 464, 429, 406, 411 and 416. Pushing of all above ovens was completed by 11.00 am. Charging was done from 11.00 am to 01.30 pm.
4. It was decided that 16 ovens charging will be required as Battery No4 GCM is a Km away from exhauster. Because of the heating done the previous night, gas was found to be bellowing out of standpipes after charging. Thereafter gas was connected to the exhauster at 1.30 pm.
5. Meanwhile, some amount of BF gas was diverted to Battery No 3 for heating from noon. Preparation was made for Battery-3 revival after lunch.
6. There were problems with Nitrogen availability, and this had delayed the commissioning activities of Battery 3. Nitrogen was purged back from PGC to

Battery-3 oven top. After purging GCM and Suction mains and checking the Oxygen content pushing started in Battery 3 from 03.12 pm as per the following sequence: 360, 310, 312, 314, 318, 322, 326, 330, 332, 344, 348, 352, 356, 364, 366 and 346. Pushing of all above ovens was completed by 05.15 pm. Charging has got delayed because of the low pressure of Nitrogen. The charging of 16 ovens was continued from 07.29 pm to 09.20 pm.

7. There was some hassle as Oxygen content was fluctuating and was on a higher side. Hence after ensuring Oxygen content is less than 1% the gas was finally connected by 10.00 pm.
8. It is to mention here that hot full wall repair job was going on two fronts (305 heating wall and 338 HW) and the heating in those areas were resumed later and successfully commissioned thereafter.
9. All the CDCP cooling chambers (12Nos) were commissioned as per normal operating procedure even though there were some anxious moments with Chamber No 4, 11, 12 and later on with Ch-3. The entire team worked hard to revive those chambers also and ensured the functioning of the remaining Chambers to fulfil the pushing targets. That marks the end of re-commissioning and revival of all Batteries
10. It is observed that after first pushing in Battery-1 coke stuck up in chamber-2 funnel and could be cleared with great difficulty. As the end flue coke was not ready it was decided to push coke only after ensuring coke mass readiness at the end flue.
11. In C- shift coke mass was ready but big lumps of coke were going to CDCP chambers and jamming was taking place which caused lots of anxious moments. Even though all were aware that a higher coking period would produce bigger size coke lumps and would create discharging problems at the rotary discharging device, we kept adequate manpower at strategic locations to clear the jamming. There was no other alternative. In this way, we could ensure the way for accommodating battery hot coke in cooling chambers of CDCP's.
12. Pushings done on this day:

Battery No	Pushing: "A" Shift	Pushing: "B" Shift	Pushing: "C" Shift
1	0	1 (Oven of following cycle)	10 (Oven of following cycle)
2	4	25	25
3	0	16	18
4	16	14	17

Average Control Vertical Temperature (CVT):

Battery No	Shifts	Pusher Side in Deg C	Coke Side in Deg C
Battery No 1	A	1060	1068
	B	1136	1157
	C	1191	1210
Battery No 2	A	972	1022
	B	1049	1068
	C	1090	1116
Battery No 3	A	NA	NA
	B	988	1013
	C	966	973
Battery No 4	A	1073	1077
	B	1052	1062
	C	1091	1093

Day 12: On 21st October, Tuesday

It is time to ensure that all the ovens of all the Batteries are pushed out without any sticker of high amperage; then only it can be affirmed that all the Batteries are revived. There are two blocks of ovens in repair areas of battery 3. The heating walls of the adjacent Heating walls could not be connected as the temperature was very low. 1. Pushings done on this day:

Battery No	Pushing: "A" Shift	Pushing: "B" Shift	Pushing: "C" Shift
1	11(Oven of following cycle)	14 (Oven of following cycle)	12 (Oven of following cycle)
2	4+6 (Oven of following cycle)	2+ 11 (Oven of following cycle)	0+15 (Oven of following cycle)
3	14	16	18
4	20	0	0

Average Vertical Control Temperature (CVT):

Battery No	Pusher Side in Deg C	Coke Side in Deg C
Battery No 1	1191	1211
Battery No 2	1193	1216
Battery No 3	1135	1155
Battery No 4	1152	1143

3.1.3 We Can Summarize That the First Cycles Of All Loaded Ovens Were Completed As Follows:

1. Battery 1 Start: 18/10/2014 14.40 hrs Ends: 19/10/2014 12.06 hrs
2. Battery 2 Start: 19/10/2014 11.50hrs Ends: 20/10/2014 08.20 hrs
3. Battery 3 Start: 20/10/2014 15.12 hrs Ends: 21/10/2014 17.17 hrs
4. Battery 4 Start: 20/10/2014 08.00 hrs Ends: 21/10/2014 13.49 hrs

3.1.4 First Cycles of ovens pushed after fresh charge after cyclones:

1. Battery 1 Start: 20/10/2014 20.12 hrs Ends: 22/10/2014 14.06 hrs
2. Battery 2 Start: 21/10/2014 12.03 hrs Ends: 22/10/2014 23.17 hrs
3. Battery 3 Start: 22/10/2014 07.32 hrs Ends: 23/10/2014 21.16 hrs
4. Battery 4 Start: 22/10/2014 06.26 hrs Ends: 23/10/2014 18.20 hrs

About the Author

Rathinasabapathy Nagarajan, is the former ED (Projects), RINL. He has started his carrier in RINL and has been actively involved since the commissioning of Coke oven Batteries and Coal Chemical Plant. He had successfully executed Hot Repair of 7-meter-tall Coke oven batteries. He also has vast experience in the Raw material handling system, traffic, Logistics and Projects. At present, he is with M/s YSR Steel Corporation Limited (An undertaking of the Government of Andhra Pradesh) as Director (Infrastructure) and M/s YSRSCL is the upcoming steel plant in YSR Kadapa District, Andhra Pradesh.

4. Appendix: The Chronological Sequence of Events in The Critical Areas Of Production In The Plant

10.10.2014		ALERT RECEIVED FROM THE DISTRICT ADMINISTRATION, emergency plans were put in place in all the departments, Diesel generator sets were checked for readiness, yard equipment was locked, dewatering pumps were kept ready, JCB's and Mobile lighting vans were put in readiness, control rooms in plant and township were manned with senior officers round the clock, CISF personnel were put on high alert.				
11.10.2014		CMD &directors periodically reviewed the preparations. It is reiterated that all measures planned shall be closely followed at all levels with a total focus on safety and proper communication with district administration and CISF.				
DATE	**TIME**	**UTILITIES DEPARTMENT**	**POWER PLANT**	**COKE OVENS & COAL CHEMICALS PLANT (CO&CCP)**	**BLAST FURNACE (BF)**	**STEEL MELT SHOP (SMS)**
11.10.2014					BF-1 was put down	
12.10.2014	0300hrs-0600hrs	High-speed winds were observed, some of the trees fell				
12.10.2014	0755hrs	Diesel Generators were started The liquid nitrogen pump started to supply N2. Action initiated for procurement of 1000 tonnes of N2 on an emergency basis Positive pressure maintained in LPG pipeline.	Incoming Line-1 tripped, PP cut-off from APTRANSCO; steam generation came down; Coke Oven gas tripped; boilers started tripping; turbo generators hand tripped; Alternate Line-2 (AL2) also failed due to tower collapsed; start-up power was not available for normalisation; Diesel generator (DG) supply started; emergency and other critical operations like heating of motor; maintenance of lube oil was done.	Power tripped. Entire operations came to halt; Maintaining of heat in the ovens; Maintaining positive pressure in the pipeline was of utmost importance; fuel supply was completely cut-off; The immediate requirement was to retain the heat in the battery. So all outlets for the gas was closed in order to maintain the gas collection mains(GCM) in the positive pressure; due to severity of cyclone and blockage of roads additional de-watering pumps could not be arranged; all possible steps were taken to plug leakages and prevent ingress of air	One BF was put down; one BF was required to be run for maintaining the gas network and to sustain BF gas to coke oven batteries for safe operations. However, due to power failure, the third furnace was stopped as is where in condition.	SMS was kept in operation to ensure off-take for hot metal production from BF; A sudden outage of power resulted in the total stoppage of the converter and one of the converters got badly jammed with metal;
15.10.2014			One of the old lines for power was revived by Power Grid Corporation India Ltd (PGCIL) & APTRANSCO			
16.10.2014	Early hrs		Power from grid restored. 50 Mega Watts of power only could be drawn from a single line.			
	2040hrs	The air compressor was started Air was supplied for running TPP.				

REVIVAL STATUS:

DATE	TIME	UTILITIES DEPARTMENT	POWER PLANT	COKE OVENS & COAL CHEMICALS PLANT	BLAST FURNACE	STEEL MELT SHOP
16.10.2014		Supply of process and instrumentation air for starting of the captive power plant and coke ovens, as well as N2 for purging all the gas lines, was ensured. Plant air compressor of compressor house-1 started. However, the damage to the cooling towers was huge and the revival operation was in				
17.10.2014	evening		Power generation from the captive unit Turbo Generator-3 could start			
18.10.2014	1500 hrs			The coke oven gas lines were purged and pushing from Battery-1 was commenced.		
19.10.2014	0700 hrs			Battery-2 Operations resumed	The criticality was to maintain the thermal regime so as to avoid solidification of the metal inside the ladles/ furnace resulting in damage to the furnace. The BF gas lines were purged and the third furnace operation resumed.	
20.10.2014	1505 hrs & 0700 hrs			The third and fourth batteries resumed operations	Blast Furnace -2 resumed operations	
21.10.2014					Blast Furnace -3 resumed operations	SMS is a vital unit for off-take and continuous running of the blast furnace Two converters of SMS-1 were brought to operation on the 21st and 22nd of October-2014. Operations starting activities at SMS-2 can be resumed only after the revival of power through the second line and restoration of
22.10.2014						

VII.4. Energy Conservation Initiatives in Steel Plant

Shailendra Prasad Bhargava

1. Introduction

India is on the path to becoming a developed country. In the recent past India has made a significant place in the World. India is creating World-class facilities.

To achieve this mission India is fully engaged in all-round work on infrastructure in almost all sectors like World-class Airports, new ports, waterways, dedicated Railway freight corridors, express highways, bridges, modern establishments, dams, waterways, river linking projects, defence corridor power plants and many more at state levels also. All these are being done with high-quality standards and are being completed in a time-bound schedule.

To make them realty Steel and cement are the main requirements have the main responsibility to develop infrastructure.

Therefore, our Steel industry has the main responsibility to provide quality steel in time to meet stringent targets of the projects.

In India, the steel plant journey started in ancient times. Iron and steel making developed and used for agriculture, armaments, and many other requirements. The best example is Loh Stambh in its premises is Qutub Minar. This was made in memory of Ashava Medha Yagya performed by great Samudra Gupt way back in 600. This has not rusted even after 1500 years. Somehow, we lost our homegrown technology and could not retrieve it.

In 1909 JRD Tata took the initiative to become a pioneer in implementing modern steel technology and TISCO Jamshedpur came into existence. Thanks to JRD Tata to do against heavy odds as Britishers discouraged the project.

Later Bhadravati, IISCO, and in the fifties, Bhilai steel plant, Rourkela steel plant, Durgapur steel plant and after some time Bokaro steel came in existence.

In these integrated steel plants, there was limited cost control and energy conservation initiative. This was the scenario in steel plants outside India also. Then came in crude oil crisis due to OPEC organization started increasing the price of crude oil way back in 1973. Suddenly petrol prices shot up from Rs1 toRs1.75 to Rs 3 per litre. This had affected the budget of all countries and of course of consumers.

At that time Japan took initiative and developed a tar injection in Blast Furnace. Later on, Russia and other countries went for Energy conservation initiatives in a big way.

In India, Visakhapatnam Steel Plant took initiative and became a pioneer since its inception to go for Energy conservation and Environment control facilities. Some features are Coke drying, Backpressure Turbine, Seven-meter-tall battery, high top pressure and big capacity Blast Furnace, Top pressure recovery turbine 100 per cent LD gas recovery in Steel Melt Sop 100 pc Continuous Casting and Most modern Energy Centre first time in India and many other features.

2. JSW Steel

In the late nineties, JSW Steel Bellary located at Torangagallu came in and present this as the largest Steel plant in the country producing 12 MT steel at one location. This is the only Steel Plant having all processes of Steelmaking. Following are the features which make it a Steel plant having taken initiative for Energy Conservation, cost control, water control and Environment control

1. They have developed a house iron ore beneficiation facility converting 45 pc iron ore up to 64pc. This process is utilizing the recycling of wastewater in a big way. Further, they have gone for Pellet making in a big way and are charging large quantities in Corex as well as in all Blast Furnace
2. In coke ovens they have gone for 100 pc stamp charging, mixed gas firing, Coke drying associated with Turbine to utilize waste heat of Coke drying.
3. In the Sinter plant, they have gone for efficient very low heat consumption Ignition furnaces. Sinter cooler heat is recovered to generate Steam to meet the humidification requirement of the Blast furnace. A recently developed project is part of waste heat of Sinter cooler is utilized for drying BF coke charged in Blast Furnace giving tremendous cost saving.
4. In blast furnaces, they have gone for large volume furnaces with top pressure up to 2.5 kg per square cm associated with Top pressure recovery turbine, consistent pulverized coal injection. The waste heat of Stove is utilized for Mixed gas and combustion air heating going to Stoves.

5. In steel melt shop oxygen consumption is lowest,100 pc LD gas recovery and best of Converter operation. There is a facility for in house granulation of LD slag
6. 100 pc Continuous casting and best Tundish management to ensure uninterrupted casting. There is hardly any Steel going to the emergency container.
7. In Hot strip rolling and for a long product more than 70 pc hot charging is achieved. Reheating furnaces are efficient with low heat consumption. Maintenance section changing role changing is carried out in lowest time.
8. In captive power plant JSW Steel is having the lowest heat rate in the country with multi-fuel firing mixed gas, LD gas Corex gas BF gas Boiler coal. Oil usage is zero in Power making. This is the first plant utilizing 100 pc BF gas to generate power.

JSW Steel Energy management is efficiently controlling the right Gas requirement of pressure flow and CV to ensure continuous operation of the plant EMD also controls the right flow pressure of oxygen-nitrogen argon LPG. EMD works along with all departments to control the heating regime, utility consumption and all other Energy parameters. EMD has installed the most modern Energy centre receiving real-time signals to control the regime of all departments. If we sum up JSW STEEL is an ideal Steel plant that continues on the path of improvement, innovation, always positive to modify for improvement and process improvement. JSW Steel developing new Steel and a variety of sections. All are done in house by Research and development. Continuously they are achieving the lowest energy consumption as well as lowest water consumption.

Our government may consider the following suggestions so that cost of Steelmaking is reduced and many Integrated Steel Plants and Medium and Mini Steel plants will become viable.

1. Steel scrap is considered to have zero energy and is used for Steelmaking and controlling the temperature of Steelmaking. Usage of scrap reduces the usage of hot metal to some extent having a high cost. In foreign countries Steel scrap management is ideal and there is continuous recycling of scrap to Steel plants. There Scrap is costing less than the cost of hot metal. This reduces the cost of Steelmaking. We are happy that Government is going for recycling automobile scrap, but it

should also consider collected from houses, industries, and all other establishments to make streamline the system. The government may consider a new entity like the Steel management Department for collecting and supplying Steel scrap at the right price. This will also be a step towards our prestigious Swatch Bharat mission.

2. Natural gas can be made available to Steel plants at the price being charged to Fertilizers plants and Power plants; this will further reduce the cost of steelmaking. This Natural Gas injection in the Blast furnace will make it more stable and increase its productivity.

About the Author

Shailendra Prasad Bhargava is the former General Manager of Visakhapatnam Steel Plant and former Consultant (Energy) JSW Steel. In the CII Energy Summit 2021, he acted as an honourable Judge for the Metal sector.

VII.5. The Journey of Electrical System In CO&CCP Of RINL-VSP

S N Singh

1. Charging Car Power System

The charging car of CO&CCP was provided with 3 phase electrical power system through trolley lines 3 Rails (90 lb size). Over time, when the Battery became old, this system became non-reliable. Rails that were placed on the insulators were disturbed due to heavy flames, smoke and tar. Power pad falling became a regular phenomenon. Since it was at a height of 9 meters, it was a grave concern to rectify the problem in case of power failure in the car.

To overcome this problem, the Catenary power system was commissioned with the help of Railways. There is a catenary wire made of copper in this system, which supports contact wire using droppers. Subsequently, this power collector system was modified and made, roller type. Six rollers (2 Rollers in each phase) were mounted at the top of each charging car. Rollers continuously rotate and touch the power contact wire, giving power to the charging car for its operation. Rollers and catenary systems have made the charging car power system very lightweight and maintenance-friendly with rare breakdowns.

2. DE Trolley Line System

DE car of coke oven also carried 90 lb Rails, which were placed on porcelain insulators. It was installed at the height of 12 meters from the ground. During the opening of the oven door for pushing, a heavy flame used to come out from the oven and sometimes it damaged the insulators. Because of this, the trolley line rail got twisted or even sometimes fell due to its weight. It was very unsafe and challenging to place the rail in its initial position again.

To make the rails lightweight, 125 mm size channels are welded back-to-back to match the base with the insulator for fixing. This method also gives more strength to the new system. One more additional advantage is an increase in top contact surface area with the power collector pads. Due to its lightweight, it's a rare phenomenon that rails fall from their position now. As a result, the maintenance has become more rugged, easy and reliable.

3. Skip System

The DOL operated skip upgraded to vvvf drive (Delta make) and Programmable Logic Controller-PLC(AB-1000) for the control circuit. PLC accommodated all the analogue timers needed for the skip operation in its programme, so it eliminated unnecessary components. For the end cutting of the bucket at the top and bottom, contactless-type magnetic limit switches are used. Also, one more very useful circuit modification is done. There are magnetic limit switches at the top and the bottom just 6 meters before the bunkers, when it operates drive operates at a slow speed of 18 Hz so that bucket does not hit the bunker at the full speed (50 Hz). After the up gradation of the skip, problems related to the skip system has come down drastically.

4. Reversing Winch

Reversing winch is the heart of a coke oven battery. In the beginning, the system was given with 24 hours timer made from a rotating set of gear pinion systems driven by a small motor called a K3P timer. It contained twelve contacts programmable at varied timing. Later, it was upgraded with a microprocessor-based programmable timer of twelve channels to control the application and thereafter the age of PLC took over for such applications.

Later on, it was updated to PLC (Allen Bradley make), so all the external analogue timers were accommodated in the Ladder logic itself, so now no timer related problem. Also, HMI is developed to monitor the status of all the necessary parameters on the computer screen. This made operation and troubleshooting easy and fast.

In the new Battery-5, reversing winch operation is by the hydraulic system.

5. LOCO (Hot Coke Transfer Car)

Loco is used for transferring 25 Tons of hot coke pushed in the bucket to the CDCP for cooling it. There are two DC series motors of 257 KW of capacity. There was an analogue controller system to control the firing angle of the thyristor so that armature and field currents could be adjusted according to the speed requirement. During the course of the time when this got old, it started malfunctioning. Since it was a running machine lot of vibration made it more prone to problems.

Recently locos are upgraded to the parker make DC drives. This is a digital drive so parameters are adjusted digitally. Now the field is controlled by the field converter and armature by DC drives. After this upgrade, the availability of the Locos improved significantly.

In the new Battery-5, to operate long travel of Locos, AC motors of 125 KW capacity is used which is operated by VVVF Drives.

6. Machine Interlocks During Pushing

In the Russian system, interlock was through HF (High frequency) system. This system started giving trouble and finding the spare parts were very difficult and costly.

This system of machines interlock is upgraded. In the new system, photosensors are used between DE and Loco. The transmitter of the photosensor is installed in DE and the receiver in Loco. Infrared lights are emitted in the transmitter and when it falls on the receiver it gives a contact which is used to give the signal to the loco operator. When the Loco operator acknowledges it using a Radiofrequency system, it gives a signal to the pusher operator and is also used for electrical interlock for pusher ram. This system is accurate up to 100 mm, whereas the old system was for 1000 mm.

Now, in Battery-5 for this interlock. Lotus make Radar system is commissioned, which gives the exact positioning of all the machines. Also, the machines which are involved in pushing are grouped so that there is no chance of wrong pushing. This system is accurate up to 5 mm.

7. Battery Machines (DE, Pusher Car, Charging Car)

From the time of starting of the battery, all the long travel (LT) motors of DE, Charging car and Pusher cars were slip ring induction motors. These LT motors were operated by the DOL method. All long travel motors are now changed to squirrel cage induction motors, which is controlled by VVVF drive. The pusher ram mechanism and Leveller bar mechanism also, are operated by VVVF drives individually.

All the relay logic are replaced with the PLC for the control circuit. In all the above machines wherever Drum controllers were used in the control circuit, it has been replaced with the non-contact type Magnetic limit switches.

In Battery-4&5 machines, other mechanisms are operated by the hydraulic system. All VVVF Drives in Battery-5 are VACON make, and PLC are SIEMENS make.

8. Current Monitoring of Pusher Car and Leveller Bar Mechanisms

To know the healthiness of the oven, it is necessary that the Pusher Ram current and Leveller bar current should not be abnormal. So, it is continuously monitored in the H&R control room. Earlier to send the pusher ram current, a 90 mm channel was laid in parallel to the LOCOTL, which was connected to the output of the CT from the pusher ram power cable. Subsequently in the coal tower from this channel, a transducer was connected which was input to a current transducer, and the output of the current transducer was connected to the display unit.

Now using the Radiofrequency system pusher ram current is directly transferred to the H&R control room without any physical contact. The same way leveller bar current is also monitored.

9. Electrics of Coke Dry Cooling Plant in VSP/RINL

9.1 Introduction

First Coke Dry Cooling Plant was installed in VSP/RINL in India.

There are four chambers & Lifters in each CDCP. Each chamber has its own Mill fan, Reserve mill fan, Circulation water pump, Rotary discharge with various valves and gates. Mill fan is used to circulate air into the Chamber and Boiler. The circulating air quenches hot coke from Battery and the heat recovered is transferred into the water of Boiler tubes to generate high-pressure steam to generate power. The quenched coke is discharged through Gates & Rotary Discharge on conveyor for transporting coke to coke sorting plant. The common mechanisms have Ventilation fans, intake fans, sump pumps, DE system etc.

Lifter: Hot coke pushed from Battery ovens is received in a special type of coke car. The coke-car with hot coke is brought to the coke dry cooling plant for its cooling. The Hot Coke bucket is lifted to the top of the cooling chamber with the help of a Lifter to discharge the coke into the chamber. The empty bucket is lowered and taken back into a coke car. The coke bucket lifter hoist mechanism is provided with four (4) numbers 160 KW Slip ring motors with associated external resistance banks.

9.2 Stages of CDCP Electrics

1989-90

All the Chambers at CDCP-1 & 2 were commissioned with open type electric panels, 6.6 KV HT Mill fans & Feedwater pumps. The individual mechanisms were controlled through hardwired physically operating switches on control desks.

All the Lifters at CDCP-1 & 2 were commissioned with a Slip-ring Induction motor with mammoth Resistance-bank panels. The hoisting operation of the hot coke bucket was controlled with shorting resistances of the connected rotor employing bar-type contactors. The lowering process was being controlled through the Dynamic-Braking technique by injecting DC current. The Lifter operation was controlled through a Lifter operator stationed at 40 metres height.

Both the two CDCPs were installed in collaboration with USSR technology.

1992

In line with CDCP-1 & 2; CDCP-3 was commissioned with closed type electric panels with L&T make switchgear in combination with L&T make PLCs.

1993

The lifter control post was divided into two segments as Lifter-Local-control-post and Lifter-Remote-Alignment-control-post. This modification got highlighted up to the national level QC project.

2004

Introduced continuous coke discharge (Rotary Discharge system) in place of Batch-wise Cooled coke discharge system with the help of micro-PLC and VFD.

2006

To avoid frequent breakdowns of one side grip of the bucket, the Alignment post was divided into two parts to ensure both sides proper hooks closing.

2007

To avoid frequent breakdowns of bucket lowering out of Loco-carriage, lowering of coke-bucket was controlled at 9 metres level.

2008

CDCP-1 Relay Logic Control of Chamber, Boiler & Common mechanism was upgraded to PLC control. The Operator-Desk-control system with hard-wired switches was converted to an HMI Screen-Control system.

2009

CDCP-4 was commissioned with an automated system like PLC control, HMIs, Intelligent MCCs, and VFDs. Lifter Hoist was commissioned with 600 KW single Induction Motor and Regenerative VFD, in place of four Nos. 160 KW slip ring induction motors as in CDCP-1, 2 & 3.

One-side-Grip Interlock was introduced to avoid frequent single-side lifting of coke-bucket. This was a breakthrough to reduce major damages to the coke bucket and relevant equipment.

2010

CDCP-2 Lifter Relay Logic was upgraded to PLC control with Lifter control in Auto-operation mode without Lifter operator.

2014-15

Introduced Auto-stop interlock in case of chain-snap in Continuous coke discharge system, which was subsequently presented in International QC conference at Seoul.

2017-20

A major up-gradation of electrics in CDCP- 2 & 3 was taken up as the Slip-ring Induction Motors, open type electric-panels, Dynamic braking system, DSL system, Relay Logic Control systems etc. have become obsolete.

The electrics of Chambers & Boilers were replaced with Intelligent MCCs with Closed type Panels, Latest PLCs & HMI controllers with Ethernet & Mod-bus communication.

At Lifters, Slip-Ring Induction motors, Resistance banks, DC injection Dynamic Braking system, DLS system were replaced with Inverter duty Motors, Regenerative VFDs and E-chain system. An air conditioning system was introduced as required for VFD rooms and Lifter MCCs.

With the upgrade, Lifter Auto-mode operation was commissioned; and all critical equipment historical trends are now readily available for troubleshooting & analysis.

Dec-2020

CDCP-5 was commissioned with all the latest electrics with the automated system. Air conditioning system introduced in VFD rooms and Lifter MCCs. Medium voltage (690 V) VFDs and Motors commissioned in chamber equipment fans.

9.3 Summary CDCP

SL. NO.	MECHANISM	PREVIOUS	PRESENT
1	Lifter technology	Dynamic Braking with Slip ring Induction Motor,	Re-generative VFD with VFD grade Induction motor.
2	Control System	Relay logic with Hardwired Control Desks	PLC with HMI controlled Screen
3	Power supply	DSL system	E-chain-system/Trailing Cable
4	Switchgear MCCs	Forced ventilation	Air conditioning system
5	Lifter Operation Mode	Manual mode with two operators.	Auto-mode with a single operator in all the Lifters.

About the Author

S N Singh is presently working as GM(CO&CCP) I/C & HOD at Vizag Steel, RINL. He did his Electrical Engineering from MIT Muzaffarpur. He joined RINL as MT(T) in January 1987.

VII.6. Journey Of Coke-making At Jindal Steel & Power Limited and Future Prospects

R K Sabat, B K Sahu, V M Shisani & D Mittal

1. Abstract

Jindal Steel & Power Limited introduced non-recovery Stamp Charge Coke-making of 0.8 MTPA in 2006 at the Raigarh, Chhattisgarh unit. This was the first silica brick non-recovery / heat recovery Stamp Charge Coke-oven battery, in India initially heated up with steam coal. After completing the long journey of 15 years, the plant is still meeting the requirement of rated production capacity without sacrificing the coke quality. In the late FY'16, JSPL has set up another 1.9 MTPA Recovery Stamp Charge Coke oven battery at Angul, Odisha unit. The coke oven plant is 4.3 meters tall and has four coke oven batteries of 72 ovens each. Over the years, there has been a significant change in operational philosophy for efficient operation towards excellence in the operation area.

Both the coke plants are operating well and fulfilling the requirement of metallurgical coke for the smooth operation of the blast furnaces. The major strategies had been taken so far for coke oven plants uniqueness on coke productivity and produced coke quality described briefly. The first non-recovery stamp charge battery is now fully operational through this process of excellence despite early hic-ups. Also, emphasis has been given to the steps taken to the efficient operation of both coke oven plants at JSPL, Raigarh and Angul unit.

Keywords: Non-recovery stamp charge, recovery stamp charge, operational philosophy, productivity, coke quality.

2. Introduction

It is a well-known fact that the productivity and coke quality highly depend on the properties of the charged coal blend and the operational efficiency of the coke oven plant. The quality of blast furnace coke, i.e., cold strength and hot strength, is decided based on the working volume of the blast furnace since coke is a single primary cost control driver for the production of the hot metal through the ironmaking route. The strength characteristic of coke depends to a large extent on thermal conditions prevailing during carbonization. The thermal condition of the oven is influenced by the flue temperature, wall temperature and carbonization time. The flue temperature and carbonization time are not adopted in a normal operation as they are inter-dependent factors and vary inversely.

The production of the requisite coke quality has been a problem for the iron and steel industry based on the blast furnace working volume requirement. Literature reported that low coke strength after reaction (CSR) values could lead to excessive degradation of coke, resulting in the accumulation of fines above and in front of the raceway combustion zone. This causes a change in the flow of gas produced in the combustion zone, resulting in a peripheral gas flow in the blast furnace shaft. Consequently, there is a decrease in reducing gas, increased heat load on the furnace walls and a higher frequency of burden slip. The net effect is a higher fuel rate, lower productivity, more variable hot metal chemistry and possibly reduced furnace campaign life (1). Hence optimization of the blast furnace coke rate is an essential task.

Jindal Steel and Power Limited (JSPL) is having both types of coke-making plants with non-recovery and recovery stamp charge technology for producing the metallurgical coke. Therefore, there are ample opportunities to produce optimal cost of blast furnace grade coke by utilizing the optimum percentage of non-coking coal in the coal blend. Coke makers have paid worldwide attention to achieve maximum benefits by optimizing coal blend and operating parameters. The present paper described the detailed experience of a 0.8 MTPA non-recovery stamp charge coke oven and a 1.9 MTPA recovery stamp charge coke oven built and in operation at JSPL.

3. Brief description of Cokemaking facilities at JSPL

In JSPL, coke is made from non-recovery stamp charged and by-product recovery stamp charged coke making processes to produce the desired quality of coke from a wide range of coal bases. Both coke-making technologies are located in two different geological locations, viz Raigarh and Angul, respectively. The primary difference between heat recovery coke-making and by-product recovery coke-making is the provided heat input into the coal charge. In the recovery ovens, the heat transfer is indirect through the oven walls from an independent heating system. As coal is carbonized in the absence of air inside the oven chamber, a positive internal pressure environment is created. Whereas in the non-recovery coke-making process, heat input results from the complete combustion of the volatile matter components of the coal charge within the oven crown and sole flue chambers (four-link-arch). At the oven crown, the volatile matter is combusted partially by providing a controlled quantity of pre-heated air through the oven top and upper door; remaining hydrocarbons are combusted in the sole oven flue during carbonization. As the ovens operate under negative pressure and at a temperature at which all potential pollutants break down into combustible compounds, thus eliminating air and water pollution. The sensible heat in waste gas flue passes through the heat recovery boiler (WHRB), which converts excess heat into steam for power generation. The non-recovery coke-making technology accommodates a wide variety of coals due to no impact of wall pressure or carbonization pressure; to produce similar quality of coke as compared to the recovery stamp charge coke-making; which widens the coal basket as compared to the by-products coke-making process. The schematic diagrams of the non-recovery and recovery coke oven are presented in Figure 1 and Figure 2.

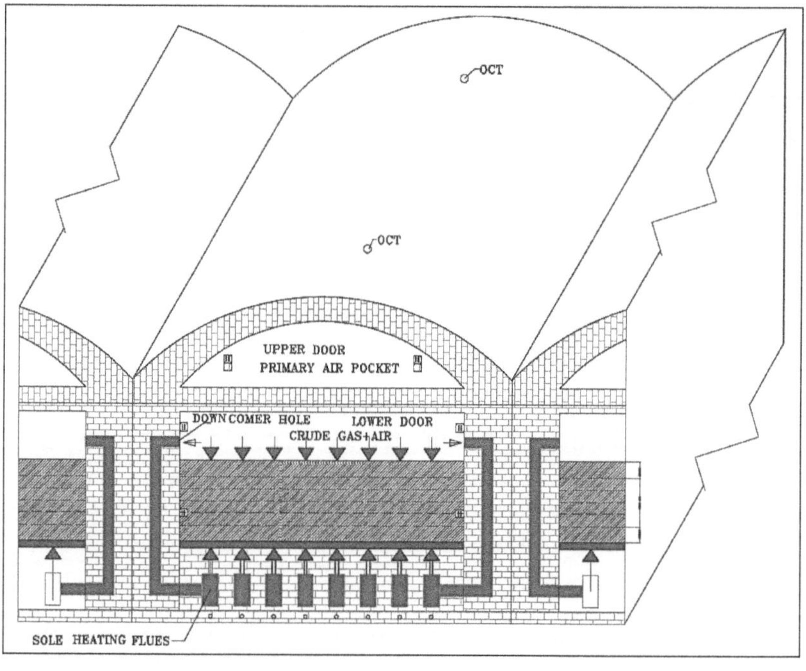

FIGURE 1: SCHEMATIC VIEW OF NON-RECOVERY COKE OVEN

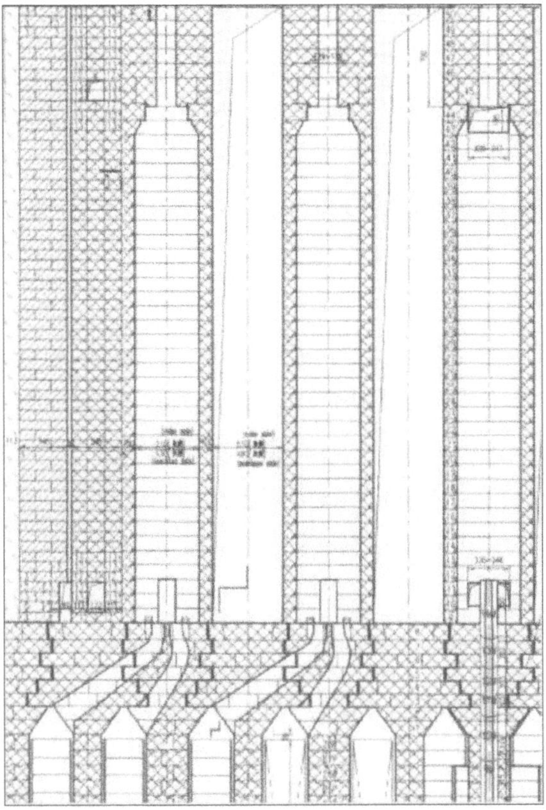

FIGURE 2: SCHEMATIC VIEW OF RECOVERY COKE OVEN

TABLE 1: COMPARISION DATA OF RAIGARH AND ANGUL COKE OVEN PLANTS

COKE OVEN	RAIGARH (C.G.), INDIA	ANGUL (ORISSA), INDIA
Capacity, MTPA	0.80	1.90
Supplier	SSIT	Sino Steel, MECC
Type	Non-recovery stamp charge	Recovery stamp charge
No. of coke oven batteries	16	4
No. of oven chambers	176	288
Avg. coal charge/oven, t	50	24
Power generation, MWH	48	-
Gross coke, t/a	850000	1950000
Gas production, Nm^3/h	-	102000
Tar production, t/a	-	81666
Sulphur production, t/a	-	3100
Crude BTX, t/a	-	26660

3.1 Non-recovery Stamp Charge Coke Oven Plant

The JSPL had commissioned the first non-recovery stamp charged coke oven plant at Raigarh, Chhattisgarh. It is a brownfield expansion project of 0.8 MTPA capacity, built with technology from SPCDI-China. The major equipment and refractories supplied by SSIT-China. It was the first non-recovery/ heat recovery coke oven plant built with silica bricks in India. It consisted of 176 ovens, two blocks of 8 coke oven batteries each, with 11 ovens in each battery. In this technology, a major part of refractory work was of silica brick of about 60,000 Ton In-addition to other refractories viz., fire clay, insulation bricks, high alumina bricks, castable, ceramic fibre blankets; used in the erection of battery proper, chimney and coke oven wharf. This mechanized coke oven battery had a set of charging-cum-pusher cars, a coke receiving-cum quenching car and a hydraulic stamping machine for each block. The hot exhaust flue gas is tapped into a waste heat recovery boiler (WHRB) for generating the steam for power generation. This non-recovery stamp charge coke oven is a clean coke oven that is environmentally friendly, without any dust and fugitive emissions.

The project groundwork was started in early 2005. The construction of the first two coke oven batteries was completed in Dec 2005 and the oven heating started in Dec 2005. The first block of the coke oven plant was put into operation in Feb 2006 and simultaneously, all coke oven batteries have been commissioned subsequently. Till today non-recovery stamp charge coke oven plant has passed a long journey of 15 years, with rated production and still striving to meet the rated capacity coke production.

In 2014 blast furnace had renovated to enhance productivity, the requirement of the metallurgical coke increased by 13%. Therefore, it was challenging for the coke oven team to produce the extra coke requirement within the existing non-recovery coke oven plant, which was already running at the rated production capacity with the optimal coke production cost. In this regard, several steps had taken in the process area to enhance the coke production from the current level. The few major steps briefly described hereafter affect the productivity and coke quality in a coke oven plant (2-3).

3.1.1 Enhancement of the productivity and coke quality

To enhance productivity, the coke oven operation team conducted several brainstorming sessions to generate new ideas for increasing productivity further and find out that the only possible way is to alter the original operational design parameters. Also, at the same time, sustaining the quality of coke with optimized cost was a significant challenge. It is well-known that productivity and coke quality highly depends on the quality of charged coal blend and operational efficiency of the coke oven plant. Hence, attention has also been paid by the coke oven team to reduce the coke cost to achieve maximum benefits in the process through coal blend optimization.

3.1.1.1 Increase The Oven Throughput

(a) Optimization Of Moisture Content of The Charge Coal

The oven throughput can be increased by maintaining optimal moisture percentage in the coal blend. The addition of moisture reduces the oven throughput and increases the coking time, which adversely affects the coke production. Since the moisture content in the coal blend directly influences the oven throughput and the heating rates, it further impacts the thermal regime and demands additional thermal energy and delayed coking time. Therefore, extra addition of moisture content is optimized in the coal blend to make the stable coal cake, except in the rainy season when moisture % in the incoming coal itself is on the higher side, which cannot be controlled beyond a certain extent (3-7).

(b) Optimization Of the Bulk Density Of The Charge Coal Cake

The bulk density of the charge coal cake is an essential factor affecting the operation of an oven, its throughput and coke quality. It has a marked influence on the size distribution and moisture of the charged coal. The bulk density in the chamber controls the local heat demand throughout the coke bed and also impacts the coke oven operation and resultant coke quality. The bulk density of the coal cake reaches an optimal level with the coal blend's optimal moisture content and crushing fines.

The study conducted elsewhere reported an increase in the stamping density, with an increase in the moisture content of coal/coal blend up to an optimal level. This is due to the closer packing of the coal particles, creating a stronger interparticle surface interaction force. This leads to an improvement in the strength of the stamped coal cake. With a further increase in the moisture level, the strength of the coal cake deteriorates. This is due to the presence of excess moisture inside the pores and between particles preventing the formation of surface forces of interaction between particles leading to deterioration in strength of coal cakes. Hence, the Moisture content in the blend was optimized and maintained at a level at which desired bulk density was achieved. Limitation to this extent of moisture in coal blend before stamping is expected to reduce the consumption of thermal energy, requisite strength and stability of coal cake and improved productivity (3, 7-8).

(c) Optimization Of the Height Of Charge Coal Cake

It is well known that an increase in the charge volume has a positive impact on productivity. Therefore, the team had decided to increase the coal cake height to enhance the productivity of the coke oven. Hence the coke oven team increased the height of charged coal cake by ~20mm after modification in the existing stamping station. Hence the input coal charge increased, which yielded more coke per pushing (2-3).

(d) Modification Of Charging Plate

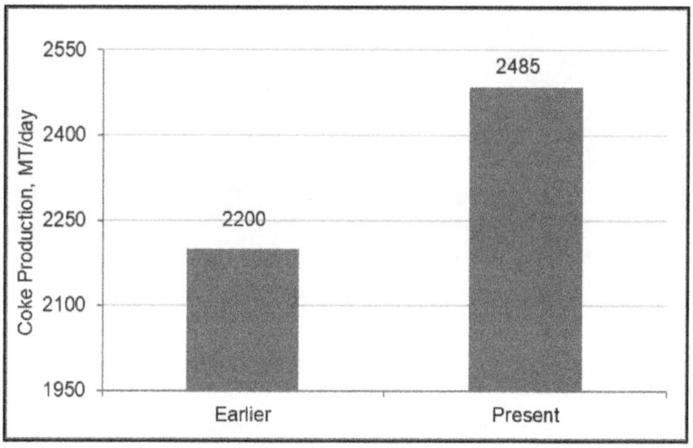

FIGURE 3: AVG. COKE PRODUCTION BEFORE AND AFTER MODIFICATION

The charging plate was made of mild steel, which gets eroded easily resulting in the breakage of the coal cake. Therefore, the coke oven team changed the existing mild steel plate with the stainless-steel plate. The surface of the stainless-steel plate is smoother than the mild steel plate and it doesn't get eroded with the moist coal blend effortlessly. The upper surface of the charging plate is softer, which helps in smooth movement and eliminate the coal cake breakage issues. With these efforts, coke production has increased remarkably. Figure 3 shows the average coke production after modification (2-3).

3.2 Recovery Stamp Charge Coke Oven Plant

JSPL, Angul is a greenfield coke oven plant with a 1.9 MTPA capacity, built with technology and major equipment supplied from MECC-Sinosteel, China. This is the first recovery stamp-charged 4.3-meter-tall coke oven battery. In April 2014, the coke oven project groundwork was started, which was commissioned in two phases. In the first phase, two coke oven batteries were commissioned in Feb' 2017 and subsequently remaining two batteries were commissioned in Nov' 2019. The coke oven battery #3 was commissioned in record time (9).

The heating-up of the coke oven batteries from SYN gas was first introduced for coke oven heating purposes in India. During heating with SYN gas, gas flow is not constrained due to some internal issues. Therefore, diluted LPG gas was also mixed with the SYN gas for the temporary heating system and the under-firing system of the first coke oven battery. The coke oven battery is equipped with the stamping machine, the charging car, the pushing car, guide car, gas transfer car and quenching car for each battery and a single gas-collecting main on the coke side. The pushing and charging emission equipment combined with the gas-transfer car and HPLA system eliminates all fugitive emissions. Also provided an auto-positioning and machine-to-machine interlocking systems in pusher-side and coke-side machines. The coke oven battery had an advanced wet coke quenching system for producing optimal coke moisture content with minimal pollution based on the environmental norms.

3.2.1 Productivity and Coke Quality

At JSPL, Angul, two key characteristics of metallurgical coke which governs the smooth functioning of blast furnaces and leads to reduction of coke rate are coke strength after reaction (i.e., CSR) and mechanical strength (M40) of LAM coke. The coke CSR index of hot strength precisely determines the coke quality because it characterizes the coke behaviour inside the blast furnace. The high value of CSR increases blast furnace productivity as it directly impacts the permeability at the lower part of the blast furnaces. The high-quality coke, in turn, helps in optimizing PCI injection in Blast Furnace. The blast furnace operation team could not produce the desired level of hot metal production due to the inadequate coke quality (especially coke CSR: 63-67; coke M40: 82-83 at coke oven end) even after the commissioning of two coke oven batteries. Also, the productivity of the two coke oven batteries was not up to the rated capacity. Therefore, the coke oven team brainstormed on low productivity and poor coke quality and started acting accordingly to achieve the rated production capacity and the desired coke quality (10-11).

3.2.2 Low Productivity and Poor Coke Quality

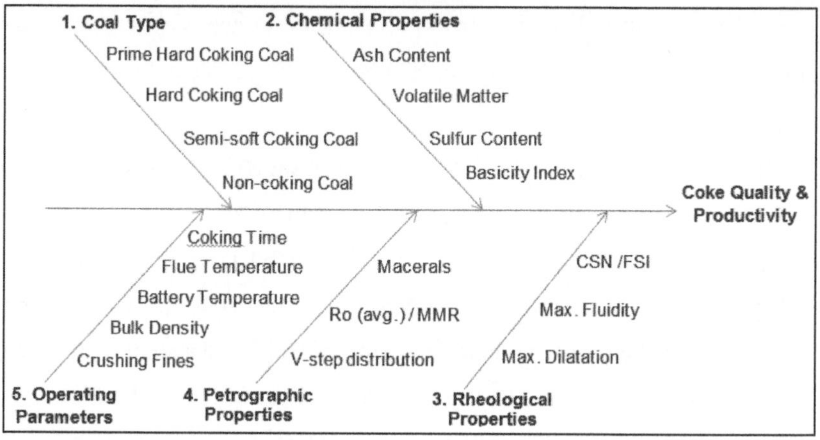

FIGURE 4: CAUSE AND EFFECT DIAGRAM OF PRODUCTIVITY AND COKE QUALITY

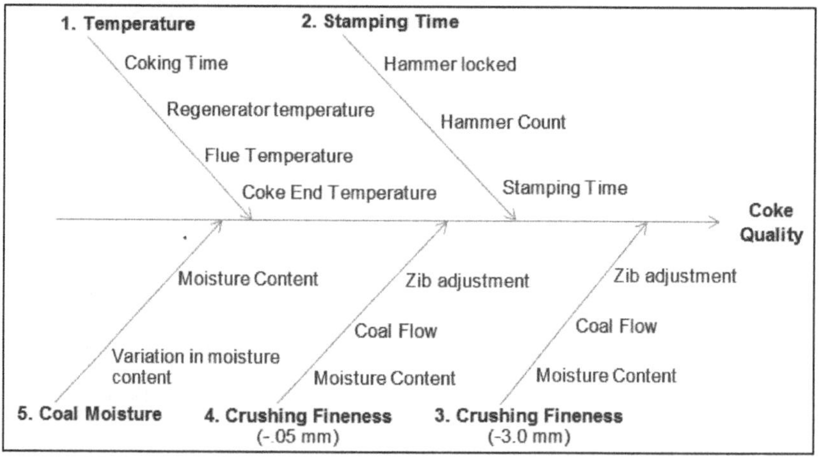

FIGURE 5: CAUSE AND EFFECT DIAGRAM OF COKE QUALITY

A detailed analysis was carried out with the help of available technical/process data to understand the reasons for low productivity and poor coke quality. The productivity and coke quality are affected by initial charging temperature, heating rate, moisture content, bulk density, coking time and carbonization temperature. The cause-effect diagram was used to describe and identify potential factors affecting productivity and coke quality. Figure 4 shows the cause-and-effect diagram of productivity and coke quality, whereas Figure 5 shows the coke quality cause and effect diagram.

(a) Effects Of Operating Parameters on Coke Quality And Productivity

Literature reported that various factors, which influence the coke quality, productivity and oven health. In fact, the characteristics of the coal blend have the highest influence (over 70%) on the quality of coke, and the operating parameters contribute the rest. Some of the critical parameters like moisture content, bulk density, the particle size of the coal/blend and battery temperature play key roles in controlling the productivity and coke quality.

(b) Battery Operations

For better control in the stamp charge coke-making process, designing a safer coal blend was the first trigger point. Quick check guides were made to ensure the desired shrinkage and the coal blend design was done with adequate safety factors.

Based on the production level, the battery temperatures were adjusted and by modulating these three process parameters battery operated within the desired push force range. Periodic auditing of battery parameters, process regime and machine time study helped improve the battery operations and maintain discipline, day in and day out. Stamping operations were optimized, and stamping time was reduced to optimum levels helping in the improvement of PRI and charging to pushing cycle time. Stringent moisture control, optimized crushing index, periodic inspection and maintenance of coal box and charge plate drive mechanism and healthy stamping machine operations helped to correct the cake breakage and return coal handling issues. Push force is maintained within norms with the help of proper blend design and regular shrinkage measurement is done to ensure the same.

For maintaining the optimum PRI level, the most important point was designing the safer coal blend for the smooth operation of the battery. A system is developed during coal selection and coal blend design to ensure desired expansion and shrinkage; blend design was done with adequate safety factors. Based on the production level, the battery temperature is maintained by modulating these process parameters and operating the battery within the desired operating parameters. The monthly average PRI of coke oven batteries # 1&2 are shown in Figure 6 (10, 12).

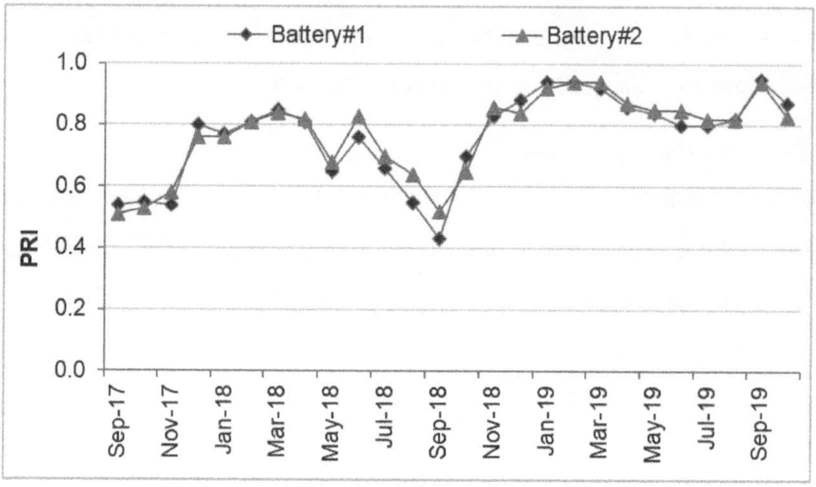

FIGURE 6: MONTHLY AVG. PRI OF COKE OVEN BATTERIES # 1 & 2

(c) Improving Performance of Machines & Equipment's

Oven machinery also plays a vital role in maintaining productivity and desired coke quality. Therefore, an ample opportunity for improvement of operational efficiency lies through the performance of equipment.

Coal cake breakage was one of the challenges, and the contribution of the charging machine and the stamping machine was very significant apart from other causes. Investigations and analysis revealed different aspects contributing to cake breakage in batteries, such as contamination in coal, coal moisture content, crushing fineness, smooth and optimum stamping operation and stamping hammer liner condition. The coal flow from feeders, the stroke of hammers, locked hammers, vibration in charge plate drive, coal box dimensions, front gate locking, and charge plate chain lubrication may contribute to the breakage of coal cake. Charging car centring position, liner conditions of the charging car, track vibrations, etc., also contribute to coal cake breakage. Improvement work was carried out in machines, maintaining optimum coal quality, moisture and crushing index to eliminate coal cake breakage. Routine checks were done to ensure proper input coal properties, machine stability and health conditions are maintained. Training of operators for perfection in stamping and charging operations helped maintain smooth charging operation and elimination of coal cake breakage (10, 12).

A lot of fluctuations were observed in performance, plant availability and product quality. As a result, the overall equipment efficiency of the coke oven was on the lower side, which affected the targeted production of the plant. Housekeeping of the battery track area and operating platforms was one of the major issues which hampered car movement and affected overall operational performance. Regular coal and coke spillage and cleaning hindered operations every time, restricting the achievement of production targets. Thus, restoring primary conditions for stringent control on operations of the battery machines was the most important task to achieve stringent daily production targets. Restriction in shutdown time made it more difficult for machine maintenance, cleaning and lifting of the spillage. With all given conditions achieving set targets was not a reality and seemed challenging to achieve the plant rated capacity.

A machine time study was carried out to enhance the machine's performance, and improvements were made to lower the pushing-to-pushing cycle time from 21 minutes to 16 minutes which helped achieve the plant's rated capacity. The incorporation of machine functions and speed interlocks helped reduce overall operation time while ensuring safe operations of the plant. This also helped increase daily shutdown time to 3 hours per day from the previous 1 hour, ultimately improving machine maintenance. As a result, higher availability of machines for the operation was achieved, which helped to sustain performance consistency. A shutdown is given in each shift for preventive maintenance and upkeep of the plant. The operational discipline to restore primary conditions, best preventative maintenance, almost zero delays, higher plant availability, achieve the best performance and produce rated capacity of the plant while producing optimum quality coke achieved by implementing the above strategies. Also, it enhanced the overall health of the battery machines and batteries, which gave an edge to take over all the issues about low performance and quality (10, 12).

Investigation work on target productivity also revealed that coal cake breakage, stamping time and pushing to push cycle time are major contributing factors. Leaving coke mass in the oven floor after pushing has a significant negative impact on coal cake breakage and oven door burning. Condition monitoring of pusher beam, pusher shoe liners and training of operators in pushing operation resolved this problem (10, 12).

(d) Improvement In Coke Quality

Coke is an important raw material for ironmaking and plays a vital role in the process and productivity of the blast furnace. The stringent demand for optimum coke quality from Blast furnaces makes it of utmost importance that day in day out, and the optimum coke quality is supplied to BF to sustain smooth operation and highest productivity. The consistency in the production of optimum coke quality is maintained on day-to-day basis. Good coal blend chemistry can significantly improve coke quality, operational discipline and optimal process regime. Machine availability and performance play an essential role in maintaining operational discipline and optimum coking cycle hours as delay in operation affects the Pushing Regulatory Index, pushing to pushing cycle time and coking cycle hours, which helped address the quality issues due to delays operation and fluctuation in coking cycle hours.

Based on the above, corrective and preventive actions were initiated modifying the operating practice, modulating in battery process, adopting new emission control system, adopting new coal selection criteria's, standard operating procedure of oven regulation, maintenance of oven machinery, technological cleaning and oven maintenance achieved excellence in recovery stamp charged coke-making process with high quality of coke in terms of coke CSR and coke M40 has improved to >69 and M40 >83 (at coke oven end) with a lesser coke cost compared to the previous figures at a higher cost. Figure 7-9 shows the effect of different proportions of prime hard coking coal (PHCC) on coke quality (10, 12).

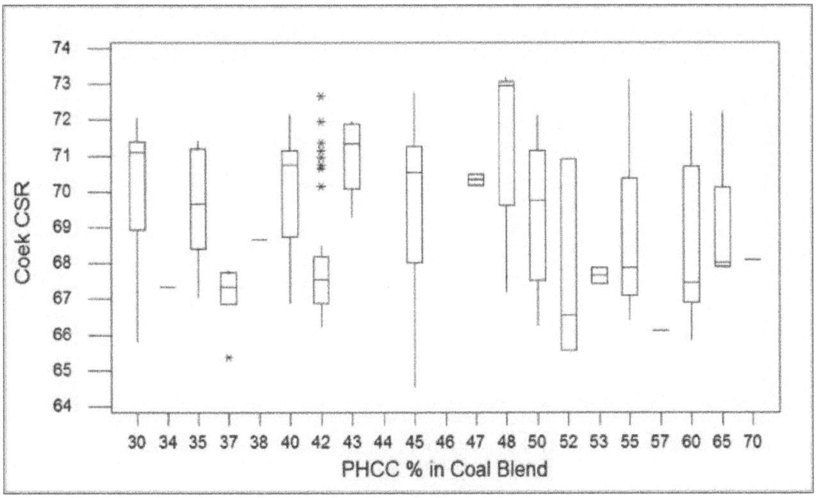

FIGURE 7: COKE CSR OF DIFFERENT PROPORTIONS OF PRIME HARD COKING COAL

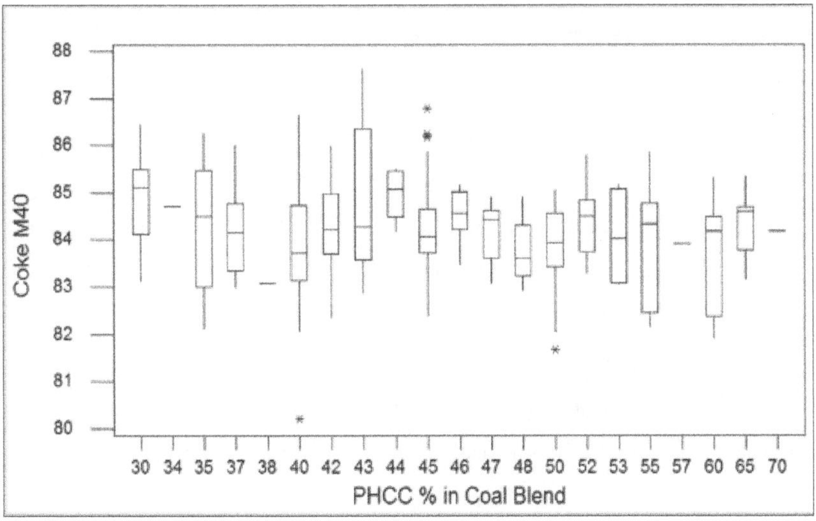

FIGURE 8: COKE M40 OF DIFFERENT PROPORTIONS OF PRIME HARD COKING COAL

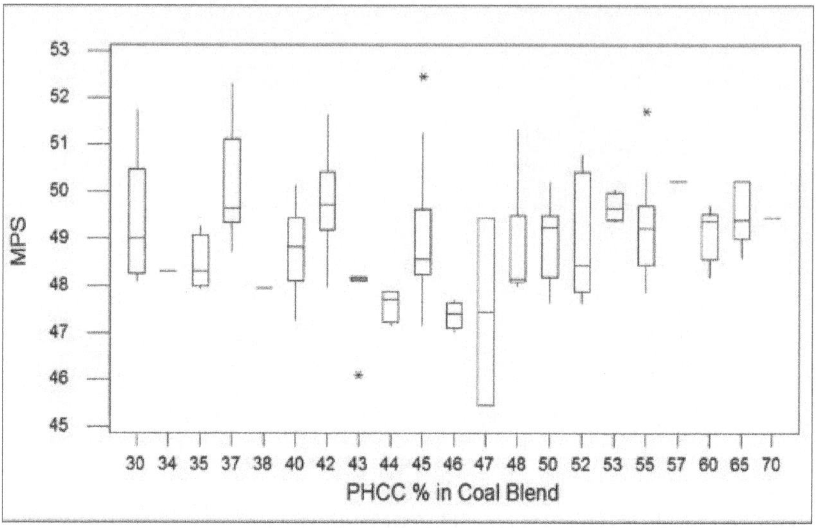

FIGURE 9: COKE MPS OF DIFFERENT PROPORTIONS OF PRIME HARD COKING COAL

4. Environmental Aspects

Since non-recovery type coke oven operated at negative pressure, no fugitive emission occurs in the non-recovery coke oven batteries. Also, in the quenching system (wet quenching), the wooden baffle and spray water system are installed to eliminates the quenching pollution.

In by-product recovery stamp charge coke oven batteries, the emissions occur at few primary sources like tall battery stacks (NOx, SO2, dust), releasing the flue gas of the battery under firing into the atmosphere, the quenching towers and some smaller stacks of dedusting equipment. The fugitive emissions occur at many sources and are caused by leaks at the closed openings of the batteries (oven doors, ascension pipes, charging hole lids, etc.) and by non-captured pushing and charging emissions. Fugitive emissions at the batteries cannot be avoided entirely, also when considering closure facilities according to state of the art in technology and being under the best state of maintenance, and contain dust, polycyclic aromatic hydrocarbon compounds (PAH) and benzene as most relevant components. Also, emissions from the coke-making by-products are of secondary importance in contrast to emissions from battery operation. This is generated from open tanks, leaks in the piping system, pressure valves, pumps, etc., and also from the Claus plant sulfur recovery unit.

Due to this complicated process, worldwide coke makers focused on controlling fugitive emissions during charging of coal into the oven. The gas evolved during charging contains a higher level of toxicity, which adversely impacts the environment and human health. Over the years, efforts have been made to reduce this emission level through installing pollution control measures and technological change, particularly in coke-making operations. In this context, JSPL Angul also paid attention to controlling possible fugitive emissions during coke making processes. Therefore, the 2nd generation of stamp charge oven battery introduced the new CGT with the facility of HPLA with the gas transfer (n-1)/(n+2). The effective utilization of these systems resulted maintained the pollution norms well below environmental norms (12).

To eliminate the gas escape in the environment, a hood is provided on the pusher side of each oven and connected with a common header to the Ground dedusting system through an actuator control valve and an induced draft (ID) fan. During charging the hood gets connected with the suction line through actuated valve operation and a small volume of gas coming out of the oven is sucked into the de-dusting system. Thus, charging emission becomes zero practically with the help of the ground dedusting system. The rigorous technological cleaning and sealing of closed openings of the batteries are ensured which helped in maintaining PLD, PLL and PLO under norms and achieving almost zero gas leakage from batteries. The particulate matter also reduced down significantly to lower levels as shown in Figure 10 and Table 2 (12).

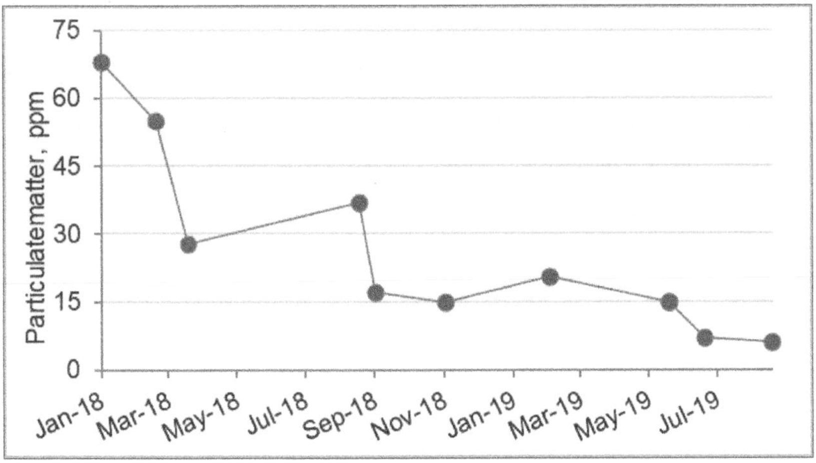

FIGURE 10: MONTHLY AVG. OF PARTICULATE MATTER

FIGURE 11: EMISSION-FREE OVEN CHARGING PROCESS

TABLE 2: COMPARISON OF STANDARDS POLLUTION NORMS WITH ACTUAL VALUE

SOURCE	ENVIRONMENTAL NORMS	ACTUAL VALUE
SOX (mg/Nm^3)	800	285
NOX (mg/Nm^3)	500	227
SPM (mg/Nm^3)	50	20
PLL	2	0
PLD	3.3	0.4
PLO	2	0

5. The Outcome of The Work

5.1 Non-recovery Stamp Charge Coke Oven Plant

The above corrective and preventive actions helped modify the operation practices by revising the work instruction to achieve the desired coke quality and productivity. The standard deviation of coke quality in terms of coke CSR and Coke M40 was substantially reduced. Adopting new emission control system/ technologies and coal selection criteria leads to the standard operating practice of oven regulation and maintenance of oven machinery, and oven maintenance achieved excellence in the recovery stamp charged coke making process.

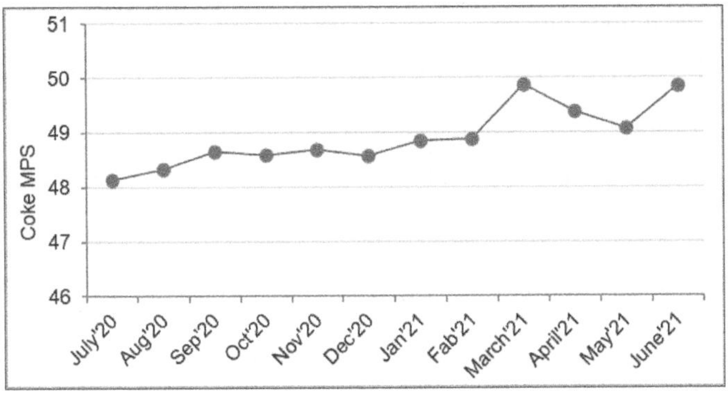

FIGURE 12: VARIATION IN COKE STRENGTH AFTER

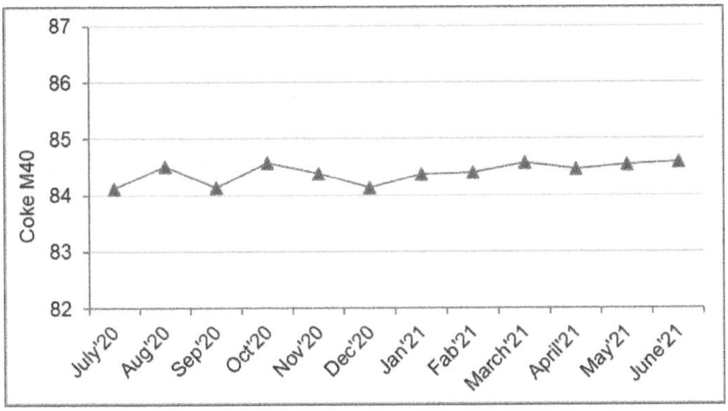

FIGURE 13: VARIATION IN COKE M40 VALUE AT

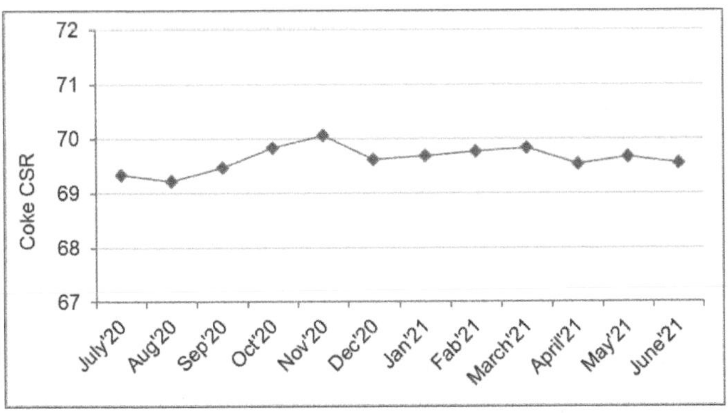

FIGURE 14: VARIATION OF COKE MPS AT COKE

With adopting all these practices, the quality of coke especially coke CSR (Figure 12), coke M40 (Figure 13), and coke MPS (Figure 14) has improved to >69, >84, and >48 respectively, with optimal coke cost (10).

5.2 Consistency In Pushing

With all the above actions taken, the standard deviation in the number of pushing in all four coke oven batteries has been reduced and maintained consistently. By using daily management tools, the performance of the coke plant, plant availability and coke quality has been improved, and a high level of customer satisfaction is achieved through the supply of desired quality of coke to the largest Indian blast furnace.

5.3 Ramp-Up of The Production Of Coke Oven Battery #3

After the successful commissioning of the coke oven batteries 1, 2 & 4, the project team start lots of brainstorming sessions for the fastest commission of the coke oven battery # 3. Subsequently, the operation team also plans for the production's quickest ramp-up to cater to the blast furnace requirement. Therefore the coke oven battery #3 pushing ramp up after first pushing was also carried out in a disciplined manner and achieved to full capacity of 72 pushing in 52 days. This is done by analysing and rectifying all barriers faced during ramp-up in previous batteries like modifications made in machines, availing sufficient coal quantity, availing heating consumables like orifice, nozzle, etc., and many more. Rigorous efforts, discipline to protocols, and safety standards by all team members result in our achievement of 72 pushing in a day in record time (Figure 15).

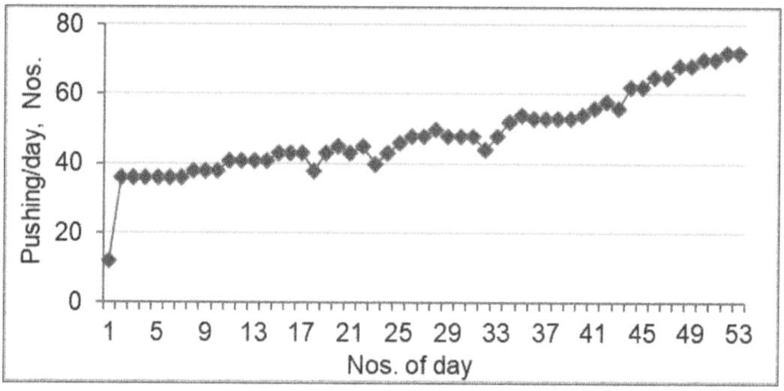

FIGURE 15: PUSHING RAMP-UP OF THE COKE OVEN

6. Future Coke Oven Battery Project

The dry coke quenching (120 tph * 2) is under the project stage and will go into operation by the end of 2021. With the argument of the 2nd blast furnace, another 2MTPA coke oven battery with retrospective CDQ is under the final stage of technology finalization, with the target of commissioning by 2023. Ongoing ahead with the journey of making

JSPL Angul as 25MTPA steel complex at a single location, further 4MTPA stamp charge recovery type coke oven battery complex will be installed as per proposed blast furnace production scheduled.

7. Summary

Despite the early learning phase of the first non-recovery stamp charge and by-product recovery stamp charged coke oven of JSPL, the plant achieved rated capacity within a short period and maintained consistency in the productivity and coke quality.

In non-recovery stamp charge coke-making, the heating of the coal cake in the oven should be uniform across the carbonization time to achieve desired coke quality and productivity. Because the heating rate of coal cake varies with varying temperatures of oven crown and sole flue, it should be maintained within the standard norm throughout the coking cycle. As the heating rate increases, the temperature corresponding to the maximum rate shifts to a higher value and coking time has been achieved according to the desired level. Controlling the oven sole flue temperature (four-link arch), oven crown temperature, and suction of battery through daily management has helped achieve homogeneous temperature distributions during the coking period are essential for productivity, coke quality and battery life. One key learning in this study is controlling operational parameters through daily management, which helps maintain productivity.

In recovery stamp charge coke-making, optimization of coal blend pertaining to the quality of coke, the health of battery and cost, along with modified standard operating practices like controlling the stamping time, crushing fineness, coke end temperature, coking time through daily management has improved the performance of coke plant with desired coke quality (Coke CSR: >69; Coke M40: >84 at coke oven end) and productivity (100%). The other important aspect of this journey of excellence in stamp charged coke-making is a systematic approach for developing and improving standard operational practices at the highest accuracy and precision adopted by JSPL; in addition to exploring new sources of coals and continuous coal blend optimization process to make cost-effective LAM coke of desired coke quality.

8. Abbreviations

Despite the early learning phase of the first non-recovery stamp charge and by-product recovery stamp charged coke oven of JSPL, the plant achieved rated capacity within a short period and maintained consistency in the productivity and coke quality.

ABBREVIATION	FULL FORM	ABBREVIATION	FULL FORM
BD	Bulk density	PRI	Pushing regularity index
BTX	Benzene, toluene, xylene	SO_2	Sulphur dioxide
CO_2	Carbon Dioxide	PAH	Polycyclic aromatic hydrocarbon compounds
CT	Coking time	PCI	Pulverized coal injection
CRI	Coke reactivity index	PHCC	Primary hard coking coal
CSR	Coke strength after reaction	PLD	Percentage leaking doors
CT	Coking time	PLL	Percentage leaking lids
GCM	Gas collecting main	PM	Particulate matters
GTC	Gas transfer car	PLO	Percentage leaking off-takes
HPLA	High-Pressure liquor ammonia system	ST	Stamping time
ID	Induced draft	GCM	Gas collecting main
JSPL	Jindal Steel & Power Limited	Kb	Coefficient of temperature
KPI	Key process indicators	OEE	Overall equipment efficiency
MPS	Mean particle size	SEC	Specific energy consumption
MWH	Megawatt per hour	SPM	Suspended particulate matter
MTPA	Million tonnes per annum	ST	Stamping time
n	Oven undercharging	t/a	Ton per annum
n+2	2nd Oven Next to n	VTP	Vertical top pressure
n-1	1 Oven before n	WHRB	Waste heat recovery boiler
NOx	Nitrogen oxides		

9. References

1. Y. Ishikawa et al. "Ironmaking Proceedings 'AIME', Vol. 42, p. 357-368 (1983).
2. R. Ranjan, M. S. Raghuwanshi, A. K. Jaiswal, H. P. Tiwari, M. G. Borkar, Productivity enhancement of non-recovery stamp charge coke oven at JSPL, Raigarh, Proc. of RACC, Dhanbad India, (2019).
3. R. Ranjan, H. P. Tiwari, P. K. Srivastava, A. K. Jaiswal, M. S. Raghuwanshi, M. G. Borkar, Enhancement of the productivity of non-recovery stamp charge coke oven plant at JSPL Raigarh, Metall. Res. Technol. 117, 617 (2020)
4. H. P Tiwari, S. K. Haldar, S. Dutta, R. K. Lingam, Heat recovery stamp charged coke making: an experience in operational excellence, Coke and Chemistry, 58(11), 432–441 (2015).
5. D. Kumar, V K Saxena, H. P. Tiwari, A. N. Rai, et al., Influence of operating parameters and raw material quality in Cokemaking process, METEC & 4th ESTAD, Germany (2019).
6. H. P. Tiwari, V. K. Saxena, 'Industrial Perspective of the Coke Making Technologies', New Trends in Coal Conversion - Combustion, Gasification, Emissions, and Coking, Edit by I Suárez-Ruiz, F Rubiera, M A Diez, Woodhead Publishing, Elsevier, 203-224 (2018).
7. H. P. Tiwari, V. K. Saxena, P. K. Banerjee, S. K. Haldar, R. Sharma, S. Paul, Study of the heating behaviour of coal during carbonization in non-recovery oven, Int. J. of Metallurgical Engineering, 1(6), 135-142 (2012).
8. H. P. Tiwari, P. K. Banerjee, V. K Saxena, R. Sharma, S. K. Haldar, s. Paul, Effect of Heating Rate on Coke Quality and Productivity in Non-recovery Coke Making, Int. J. of Coal Prep. And Utilization, 34, 306-320 (2014).
9. K. K. Mishra, D. Mittal, H. P. Tiwari, A Strategy for Fastest Commissioning of 0.475 MTPA Stamp Charge Coke Oven Battery at JSPL, Coke and Chemistry,
10. V. M. Shisani, B. K. Sahu, H. P. Tiwari, S. N. Singh, et al., Operational Excellence in Recovery Stamp Charge Cokemaking at JSPL, Angul, RACC - 2019, CIMFR, Dhanbad, India.
11. H. P. Tiwari, V. M. Shisani, B. K. Sahu, R. K. Sabat, D. Mittal, Influence of the prime hard coking coal in stamp charge coke-making: true or false, Metall. Res. Technol. 117, 412 (2020).
12. B. K. Sahu, V. M. Shisani, H. P. Tiwari, R. K. Sabat, D. Mittal, Energy-Efficient Operation of Recovery Stamp Charge Coke Oven Plant, RACC - 2019, CIMFR, Dhanbad, India.

About the Author

1. **Rabindra Kumar Sabbat,** Head of the Department, Coke Oven Plant, Jindal Steel & Power Limited, P.O. Jindalnagar, Nisa, Angul - 759111, Odisha, India, email-rksabat@angul.jspl.com

2. **Bhavendra Kumar Sahu,** AGM - Operation, Coke Oven Plant, Jindal Steel & Power Limited, Angul - 759111, Odisha, India, email- bhavendra.sahu@angul.jspl.com

3. **Vishwesh M. Shisani,** DGM - Operation, Coke Oven Plant, Jindal Steel & Power Limited, Angul - 759111, Odisha, India, email-vm.shisani@angul.jspl.com

4. **Damodar Mittal,** Executive Director and Head Iron Zone, Jindal Steel & Power Limited, Angul - 759111, Odisha, India, email - dmittal@angul.jspl.com

VII.7. Commissioning And Running of Coke Ovens In RINL/VSP

Venugopal Rao Nabhi

1. Introduction

RINL was carved out of SAIL in February 1982 to construct, commission and run India's most modern steel plant. The project was located near the shores of Vizag city at a place close to the scenic Gangavaram beach. Nearly 25000 acres of land was acquired by the state government and handed over to the Ministry of Steel for the purpose.

Though this may sound easy on paper, it was an arduous task that displaced many natives in several villages in the area. But they were given an assurance that one job would be given to a person in every household. Since this looks politically correct, it presented a huge challenge to the management of Vizag Steel, as it was in direct contravention to the objective of running the steel plant with the highest labour productivity in the country. It needs no mention that running the plant with lean and skilled manpower was a direct contributor to this objective, but as almost none of the displaced persons had any industry-specific skill worth mentioning, it was a huge challenge. To add to the woes, the number of displaced persons grew by leaps and bounds with the connivance of the state government's land acquisition officials. Therefore, the only possibility was employing them as unskilled Khalasis, but their number had to be limited to around 6000 to not impact labour productivity negatively. And the rest of the displaced persons had to be accommodated as contract labour.

Finally, when the plant was commissioned in the early 1990s, the total strength of employees was around 17000, out of which around 6000 were unskilled labour, almost all of whom were displaced persons. Training this lot was a huge challenge from the HR perspective. But adult literacy programs conducted at the shop floor with the help of the state government did yield good results. And it also led to labour productivity of 240 tons per person, which was 4 times higher than any other steel plant in the country in the 1990s.

2. Challenges in HR in VSP, Especially During the Commissioning of Coke Ovens In 1989 And The Stabilization Period Following that

The first production unit of the plant to be made operational was the Coke Ovens and Coal Chemicals Plant (CO&CCP). This was around August/ September 1989. The other units like Sinter Plant, Blast Furnace, Steel Melting Shop, Rolling Mills, etc were commissioned a few months later. Most of the senior non-executive employees in CO&CCP were redeployed from Construction departments like CERC, etc, and hardly anyone had real-time experience in a working steel plant. However, with the able guidance of experienced officers like NK Singh (HOD), TK Chatterjee (Chief Superintendent), MK Bandhyopadhyay (Superintendent) and others that had worked in SAIL plants earlier, the commencement of coke production started smoothly. They were also aided by qualified engineers who were inducted as Management Trainees and were exposed to Coke Ovens operations in other SAIL plants.

Regardless of the efficient running of the unit, there were quite a few challenges that the management faced. Some of them are listed below, including the apt measures taken by the management. For the sake of brevity, I have only listed a few of them.

1. Very strenuous work that involved daily exposure to coal dust, heat, stench and dangers of working in the Coke Oven batteries led to large disenchantment among the technicians posted there. I remember Mr Chatterjee used to confide with me by saying 'it is the dirtiest profession in the world'!

They often used to compare with other units in CO&CCP where work was less arduous. And there arose a demand from Unions to reduce their working hours. As this was a reasonable request, they were allowed 2 hours of rest after every 2 hours of work, as an informal arrangement.

2. The lack of skilled technicians in operating the charging car, pusher car, etc led to a great strain on the officers posted in the Coke Oven batteries. However, over a few months, they could successfully manage to train the operators to the required level.

3. Lack of knowledge among the technicians made them vulnerable to devious designs by several Unions. There were 13 Unions in VSP, and all of them were trying to enlist the support of the Coke Ovens staff. To cite an example, when the management decided to extend the training of a Junior Trainee by 3 months owing to poor on-the-job performance, the battery operators struck work at the behest of a couple of Unions.

A firm stand taken by the management and a few rounds of discussions with them convinced them that the action was genuine, and they resumed work.

4. Another example was a go-slow by battery operators during Diwali of 1990, demanding job rotation, better perks, more breaks from work, etc. Most of the demands were prompted by rival unions. What made matters worse was that their immediate supervisors were reluctant to identify the trouble- makers. But the slow-down was affecting production by nearly 50% and causing great concern. But when they got wind of strong actions being planned by the management, they resumed normal production.

5. During and after commissioning, hundreds of employees were being recruited. This posed a huge challenge to the local HR department, as they had to complete joining formalities quickly and issue joining orders and send them across to the Finance pay section so that they would get their first salary credited. As most recruits were illiterate, each person took a long time to have his data entered into the Personnel file. But efficient staff and a personal computer being spared by the HOD hastened the process greatly. As a result, more than 200 new recruits could be accommodated every week. In contrast to this, when I later worked in Franklin Templeton, I heard HR saying that they need 6 months to arrange for joining formalities of 300 new recruits! And those were all white-collared recruits!

6. Over time, employees in hazardous areas started demanding job rotation. The management initially baulked at the idea by stating it would affect job specialization. And this was not prevalent in any SAIL plants. The hesitation also stemmed from the apprehension that it would be difficult to rotate highly skilled technicians. And in the numerous discussions, examples of highly skilled Sardar drivers of trucks on National Highways were often quoted. In a couple of decades, even those Sardar drivers were replaced by younger drivers from across all streams of society, breaking the myth of irreplaceability. Hence it was decided to give job rotation a nod.

And the decision to go ahead with job rotation proved to be very beneficial in the long run because employees got a chance to work in all sections of CO&CCP. Job rotation was later implemented in other departments like SMS and Rolling Mills.

7. What facilitated job rotation was management's strategic planning before commissioning that comprised multi-skilling of jobs and rationalization of designations across VSP. To cite a few examples, there were only 3 standard designations across skilled employees in the Works departments, and those were Technician (Operation),

Technician (Electrical) and Technician (Mechanical). All other designations that were found in SAIL plants like Charging Car Operator, Pusher Car Operator, Guide Car Operator, etc. were rationalized as Technician (Operation), and therefore there was no hurdle in rotating employees to various positions across CO&CCP units.

8. As Time Offices in SAIL plants had proven to be epicentres of corruption, it was decided not to have them in VSP. Instead, the power was bestowed on the controlling officer of each section. While this led to better control of attendance, it also led to the officers being threatened by errant employees if they were marked absent for unauthorized absence. And as attendance details were being entered monthly, the unions had ample time to pressurize officers to condone absence without leave.

To overcome this, the HR department introduced DARS (a daily attendance recording system). This system needed attendance to be entered daily into a computer that was connected to the main IBM server. This provided a huge relief to the controlling officers because they could bluntly tell the unions that the data was already on the main server, and it was no longer in their hands to condone any unauthorized absence.

9. A plant-wide strike in October 1993 for two weeks affected all units of the plant. As CO&CCP was the most critical unit, officers of the department, with the help of officers drawn from other departments, had to operate the unit at the minimum level needed to prevent damage. This proved to be a good strategy that could prove useful in future strikes.

I could cite several other examples of HR-related challenges, but I'd like to end it here to keep it simple.

Acknowledgement: The author wishes to thank Mr TK Chatterjee for inspiring him to pen this article. The author also wishes to thank all the readers of this article for their patient reading.

About the Author

Venugopal Rao Nabhi has worked in RINL/ VSP for 21 years in Personnel Department. Moved to IT sector in 2002. He joined SAIL in 1981 while pursuing his Masters in Sociology at the Delhi School of Economics. He is a gold medallist in Anthropology from Berhampur University.

VII.8. The State-of-The-Art Stamp Charge Coke Oven Plant Of Jindal Coke Limited, Duburi. Orissa (Jindal Stainless Steel Group)

Deepak Agiwal

1. Introduction

Jindal Coke Limited is a company, engaged in the business of manufacturing, processing, finishing and dealing in all kinds and forms of coke and coke products at a facility with a capacity of **0.43 million metric tonnes p.a.** at Kalinganagar Industrial Complex, Duburi, Jajpur, Odisha. Jindal Coke Limited also aspires to elevate its capacity to **0.85 million** metric tonnes p.a by March 2023 with a second Battery equipped with a state-of-art CDQ Plant with a power generation facility. The director is **Shri Shashi Bhushan Upadhyay,** and the Head of the Department is Shri **Deepak Kesharlal Agiwal.** The existing coke ovens plant is slot type recovery Coke Ovens consisting of 64 ovens. Some of the other salient features of the Battery are stated as follows:

TABLE 1 - MAKE/DESIGN: M/S SHOUGANG LTD.' CHINA

NO.	DESIGNATION	UNIT	QUANTITY
1	Full length of coking chamber	mm	14080
2	Useful length of coking chamber	mm	13280
3	Full height of coking chamber	mm	4300
4	Useful height of coking chamber	mm	4100
5	Average width of coking chamber	mm	500
6	Conicity of coking chamber	mm	20
7	Centre distance of coking chamber	mm	1200
8	Useful volume of coking chamber	mm	26.68
9	Centre distance of vertical flue of the combustion chamber	mm	480
10	Numbers of vertical flues of combustion chambers	Set	28
11	Heating level	mm	700

2. Stamp-Charging Facility

The blended coal from the coal preparation shop, after charging into the coal tower, is charged into the coal box of a stamp-charging car by a swing feeder. A stamping machine stamps the coal in the coal box into coal cakes, which are sent by the stamp-charging car into the coking chambers of the coke oven from the coke side according to the working schedule. The coal cakes in the coking chambers are made into coke as a result of high temperature and carbonization in a coking cycle and raw gas is produced. The stamping station consists of 21 hammers.

3. By-product Plant

The By-product Plant of JCL processes and purifies the Coke Oven Gas (COG) coming from the Battery and exports the COG to internal customers at Phase#2. The plant consists of Primary Gas Coolers, Electrostatic Tar Precipitators, Gas Exhausters, Desulphurisation Plant, Ammonium Sulphate Plant, Final Gas Cooler, Secondary Purification Unit and Gas Boosters.

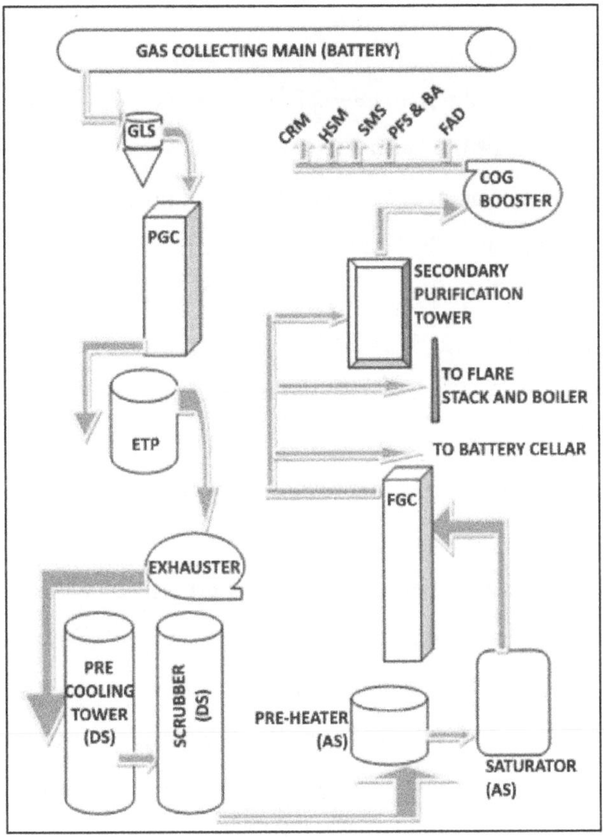

COG CIRCUIT DIAGRAM

4. Secondary Purification Unit and COG Export

In the Primary Treatment of COG, H2S present in the COG is removed by wet desulphurization process up to 0.5 ppm and tar fog up to 0.05 ppm. Next, COG is transferred to the unique Secondary Treatment unit, designed and supplied by **M/s Baosteel' China.** This unit consists of three packed bed adsorption towers with porcelain balls, hydrated Ferric Oxide and Charcoal. The porous Charcoal/Coke is meant for adsorbing **Tar particles/Tar fog up to 0.005 ppm** and Ferric Oxide reduces the **H2S up to 0.005 ppm** when COG passes through the packed towers. Porcelain balls are meant for holding the burden of the packing.

5. H2S + Fe2O3.H2O = Fe2S + H2O

This COG containing only 0.005 ppm of both Tar Fog and H2S is then compressed in Gas Boosting station and sent to internal customers like Hot Striping Mill, Steel Melt Shop, Cold Rolling Mill, Ferro Alloys, Bell Annealing Furnace and Plate Finishing System, for the purpose of furnace heating. This energy-rich COG, of premium quality, is consumed by internal customers, resulting in **reduced dependency on expensive fuels like LPF, Propane and Furnace Oil.** So, the extensive consumption of COG directly replaces costly fuels and brings a hefty organizational benefit in terms of cost and energy savings. The advantages of the Secondary unit are widespread as the clean COG decreases the frequency of choking and deposition in Gas Pipelines, Nozzles, Burners etc, leading to reduced downtime at the consumer end. Apart from that, the service life of MS pipelines also increases due to the transport of clean gas, devoid of corrosive impurities and lesser amount of deposition. The secondary Treatment unit is very common in China though yet not very popular amongst Indian recovery type Coke Ovens. This unit in JCL is one of its kind and most probably the only one installed in any Indian Coke Ovens Plant. So, we consider this project to be a "Technology Transfer" and a game-changer for smooth and uninterrupted Gas consumption at the Mills section.

6. Future Expansion Plans

As mentioned in the beginning, work is under progress to install our second Battery. The second Battery will also be similar to the existing one, with 64 ovens, identical oven specification, chimney and m-type Gas Transfer Car to cater for the charging emission.

After successful project execution and commissioning of Battery#2, we will pursue our plans for Battery#3, with subsequent expansion of By-product Units.

About the Author

Deepak Agiwal is the HOD of Jindal Coke Limited and he has been associated with this plant since its project days back in 2010. He is a bachelor's in engineering (Chemical Engineering) from Pune University having passed in the year 1998. He started his career with M/s Asian Paints and left for JK Paper Ltd. in the year 1999. He had five illustrious years in JK Paper and from there moved to Gujarat NRE Coke Ltd in the year 2004, where he worked till 2010. In 2010 August, he started his association with Jindal Stainless Limited and played the most crucial and significant leading role in successful project completion and commissioning of the coke ovens plant despite a lot of difficulties and technological hurdles. From 2014, he was assigned the role of HOD, Coke Ovens and since then he has taken the department to greater heights every year with ground-breaking records, achievements in terms of productivity, safety compliance, quality, customer satisfaction, cost optimization, QC, 5S, projects and so on.

VIII. THE REFRACTORY REPAIR AND OUTSOURCING

VIII.1. Refractory Maintenance Practice of Coke Oven Plant and Coke Drying Quenching Units

Maharaj Singh

1. Introduction

COKE OVENS are mainly made of Silica Refractories and Alumina Refractories. The quantity of refractory used in the construction of a battery depends on the size of ovens and the number of ovens in a battery. The tonnage used in batteries varies from 9,000 to 17,000 MT. In olden designs, the fire clay refractory was used up to the level where the temperature was around 7500C.ie. in the regenerator walls. The checker bricks are of HHD Fireclay quality. The oven top is of fireclay bricks with some insulation bricks.

From 1960, all Silica coke ovens came into practice. Russian designed and German batteries were in the market. Russian batteries used silica refractories right from Bus flue level, whereas others used silica from above the Bus flue level and up to oven covering tiles. As I am dealing with "maintenance practices of coke ovens and CDQ", therefore I do not want to go into details of coke oven constrictions.

It is a well-known fact that if a plant with the same drawing is installed at different locations, each plant will experience different problems. Not all plants may face the same problems, though the plants are the same. Therefore, plant engineers have to find their own solutions for the plants. They should not blame the designers every time.

Below I am narrating three problems, which we faced, and we found our own solutions and were satisfied.

2. How The Problem Was Solved

2.1 Problem 1: Modification to CDQ's Manhole For The Repair Of Pillars

Japan's NIPPON company supplied the CDQ system of coke drying unit to JSW steel ltd. As per schedule, chamber no. 3 was to be repaired as annual repair schedule, first repair. Repair to be done for Pillars. The pillars start at approx. 4.7M from the floor level. The programme was to fill the chamber up to half-meter below the start point of pillars. It means we have no approach to enter the chamber except the top mouth. And from the mouth of the chamber, the point is about M below. The main problem was that if any accident happens inside the chamber, there is no way to bring out the patient.

Also, the supply of refractory will face a problem. Therefore, after having a lot of discussion with the concerned, it was decided to fill the coke up to the 61st layer of lining. And from there the repair shall be carried out, for which pipe scaffolding shall be erected on a steel plate floor, to be made above the coke mass for repair work.

Now the problem was to find an opening from where one can go inside the chamber and supply the refractory and scaffolding material required.

On studying and having discussions with the concerned, it was decided to enlarge the existing manhole (500mm dia), provided in the wall of PDC (primary dust catcher). The landing inside the PDC was around one meter below the manhole and it was dark, difficult to know whether one is safe landing inside or not. There was no other opening anywhere from which one could safely go inside for repair work.

So, to solve this problem, I decided to make the existing manhole bigger and rectangular. We cut open the shell from outside 700 X900 mm in size with an idea to start from the floor of the PCD upward. The lining was re-laid according to these dimensions. This made the repair work very easy. One was able to enter the PCD easily and then easily walk to the chamber. It made it very comfortable to take the material for repair work inside through this manhole. (NOTE: Based on this, I noticed at Bhushan Steel's CDQ unit that Nippon has changed the round hole with rectangular opening- instead of circular and with bigger dimensions) REFER TO FIGURE 1.

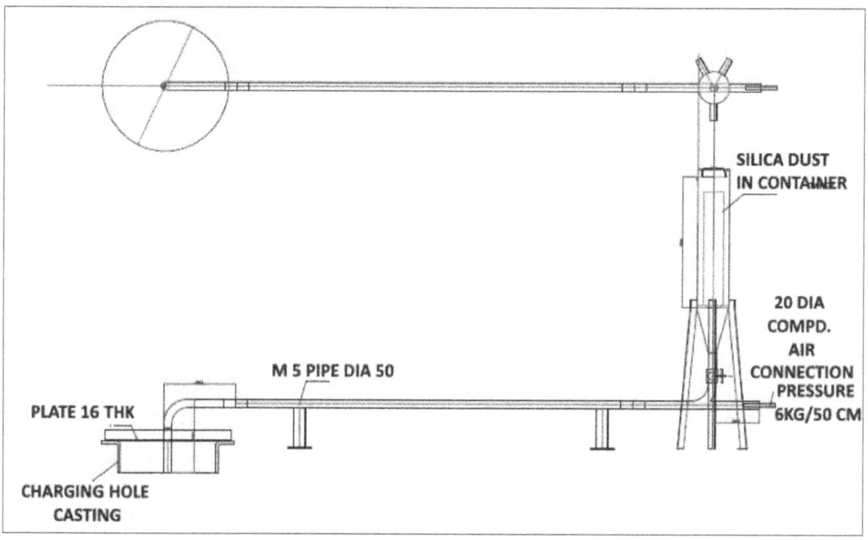

FIGURE 1 - OVEN DUSTING MACHINE

2.2 Problem 2: Introducing the Track Supports Of 7metre Tall Coke Ovens Of Soviet Design (At Vizag Steel Plant)

VSP installed three 7-meter-tall coke ovens. The charging car track was of floating design means the charging car track is resting on the top of oven top brickwork without any anchoring system. The track was supported at buttress ends only.

After few months of the operation of batteries, the floating of the track, got wavy horizontally, a lot of coal spillage was occurring, and the oven top started catching fires.

To correct this, I designed and castable blocks at every third oven of size 500X500X500mm with reinforcement in the block. The top of the block was provided with one 10mm thick plate with an anchor (reinforcement) embedded in the block and two numbers of 10dia bolts for fixing the pressure plates of tracks on both sides of the rail. REFER FIGURE 2

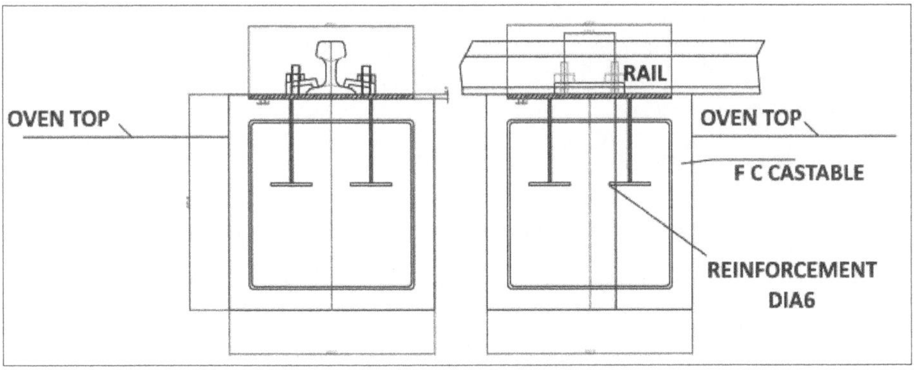

FIGURE 2 - OVEN TOP CHARGING RAIL

This modification brought the track in alignment and did not become wavy. It was a success.

2.3 Problem 3: Introduction of SS304 Plates Instead of Acid Proof Lining Of Quenching Tower At JSW Steel

This is the story at JSW steel in 1998. Coke 1, battery 4 was under heating, was in the final lap of heating, and after few days will be ready to produce hot coke, but Quenching Tower's lining was not yet started. The lining with Acid proof bricks will take a long-time. So, everyone was worried about what to do. The matter came to my notice, and I suggested using stainless steel plates, 6mm thick, instead of lining. Management agreed to it and SS304 quality plates were used. Which are in service even today. The plates were hung with a hook, 16 diameter rods bent in L shape and welded with the RCC of the concrete. The attached sketch (FIGURE 3) is self-explanatory.

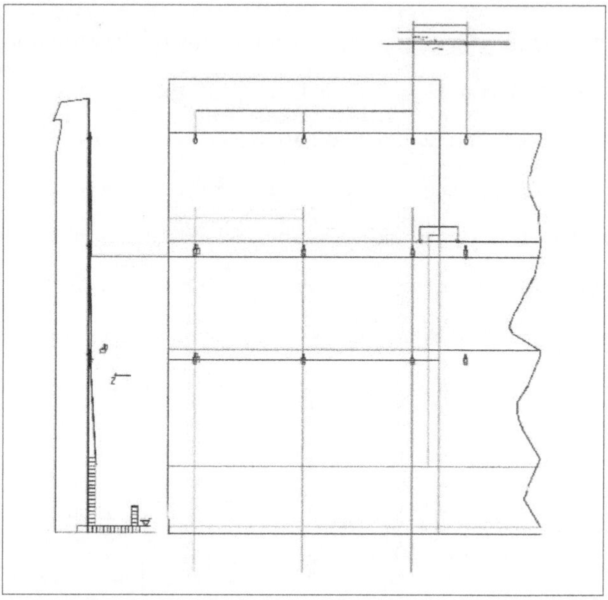

FIGURE 3 – QUENCHING TOWER SS304 LINING (VIEW Z)

2.3.1 General Refractory Maintenance Practices, Day to Day Basis

The following are the general, maintenance practices followed for maintenance of the plant in good running conditions:

- CASTABLE
- GUNNING
- HOT CRETING, and
- SILICA WELDING.

2.3.1.1 CASTABLE

Castables are unshaped refractory materials used in the construction and maintenance of furnaces. Castables have found great acceptance due to their ease of application and flexibility of applications. It also replaces special shapes and thus reduces the overall cost of construction. The installation methods include casting, gunning, pumping and injections.

For castables, generally high alumina cement is used as binding agents. Now a day's phosphate and silica bonds are also used. Most of the modern development is the low cement castables, which provide a high density of cast. Low cement castables use very fine grains of material and need less water content thus making it a more dense and high strength material and comparable to bricks.

Manufacture of castables consists of the following steps:

1. Raw material
2. Storage
3. Drying
4. Grinding
5. Screening
6. Storage
7. binder addition
8. weighing and packing
9. inspection and products.

Storage: The finished material is bagged and always kept under covered spaces. Some fine-grained components of castables may absorb moisture from the atmosphere, hence, is most important to store the bags under covered sheds. It is most desirable to use polyethylene-lined bags for packing the castables and mortars.

Installation of castables:

For installation of castables, we need:

a) Pan mixer

b) potable water

c) Shuttering-wooden or steel

d) Vibrator.

Shuttering can be made of wood, Plywood, plastic material or metallic. Shuttering should be strong enough to withstand the load of the cast material and vibration load. It is of utmost importance that the mixing instructions provided by the supplier are followed rigidly i.e., the quantity of water, mixing time, vibration time etc. the pH value of water should be between5-7. If water quality is not correct, the properties of the cast shall be hampered. Similarly, the water quantity is also a very important factor to get a good quality cast. If more water is used, the strength of castable shall deteriorate, if quantity is less then it will be difficult to cast and may result in many honeycombs in the casting. Further, the temperature of the water also plays an important role. The temperature of the water should be between 50-200C. If the temperature of the water is higher, ice or chilled water should be added to bring it down. The high temperature will accelerate the setting process and will result in deterioration of castable strength. While casting, the mass of the castable should be continuously vibrated. For vibration, electrically operated vibrators are used. The size of the vibrator depends on the thickness of the castable lining. Generally, 40 mm diameter vibrators are used.

2.3.1.2 Curing

Castables develop heat on casting due to chemical reactions with water and cement. It should be watched and as soon as heat develops, water should be sprinkled on the castable to keep it cool, or else strength shall be impaired. The supplier of the material indicates the setting time of castable, and it should be respected. Some castables need only 24 hours for setting, while others may need as much as 72 hours. Supplier indicates this in the datasheet of material.

2.3.1.3 Gunning

Sometimes castables are applied by gunning. Gunning is a very common method of applying castables. For gunning, we need a gunning machine. Many designs and capacities of gunning machines are available in the market. The most commonly used is a single chamber or double chamber machine. A double chamber machine is useful for gunning huge quantities and continuously, whereas single chamber machine is used for a batch type of application. These machines require compressed air to operate them. The air pressure should be 6kg/sq.cm.

The dry castable is fed into the machine chamber and the mouth of the chamber is closed. Below the machine, there is a neck fitted with a valve to allow the movement of the material through it by gravity flow. Below this, there is a magazine, which is connected with a compressed air supply hose and valve. From this compressed air at pressure is supplied and it carries the dry powder to the nozzle at the other end of the gunning hose. The nozzle used is generally of 40mm diameter and is tapered in construction. On the nozzle, there is an arrangement to feed water to the nozzle through a perforated steel ring. The water is mixed with the dry powder and the nozzle is directed towards the place where it is to be deposited. The nozzle must be kept at around one, meter away from the surface of the application of material and it should be rotated while gunning so that the material is uniformly and applied uniformly all over the surface. If any undulations are developed, should be scraped through with a trowel or any other means so that a uniform and smooth surface is achieved.

2.3.1.4 Shot-creting

The shot-creting is also a very popular method of application of unshaped refractories. It is similar to gunning with a difference that at the nozzle along with water some chemical is used. The purpose of this addition of chemicals is to make the grains stick to the surface immediately and rebound loss is reduced. The rebound loss in this method is only 2-3% compared to 5-6% in gunning.

2.3.1.5 Other methods

a) Spraying

It is used to spray the fine-grained refractory mass premixed with water, on the furnace walls as a maintenance practice. The best example of this practice is to spray the oven ends to keep them sealed and make them leakproof. For application, we generally use a spray machine. The machine has a small vessel made of circular section, and a removable cover bolted with top flange of the barrel. Through this top cover, the spray material is filled in the barrel, through an opening of a 75 mm dia socket. This cover is also fitted with a 15mm dia pipe, which works as the outlet of spray material. With this pipe, there is also a horizontal pipe to which the hosepipe for carrying the spray material to the nozzle, at the end of the hose through which spray material is sprayed on the wall of the oven. There is a pressure gauge also fitted to monitor the pressure. the 15mm pipe is extending up to around the bottom of the vessel, about 100mm above the bottom of the vessel. A compressed air system is attached to the horizontal portion of the pipe also to carry the compressed air inside the vessel as well as to the spray nozzle. Generally, 2kg/cm2 pressure is sufficient for spraying purposes.

b) Ramming

Ramming is another method of application of unshaped refractory. In the ramming process, the dry material is mixed with some chemical liquid in a prescribed quantity and mixed well in a pan mixer and mixed material is placed at the required place. Sometimes shuttering is also used. The material is rammed in position with the help of rammers, generally electrically operated, are preferred. Rammers can be of compressed air operated.

c) Dusting

Dusting is another special technique being used in the Coke Oven Plant. In this process, a very fine Silica powder mixed with very fine aluminium powder (premixed by supplier) is sent to the oven with compressed air through a machine. However, the machine is very simple in design, but it serves the purpose. To perform this operation, the oven is emptied, and not charged with coal. The doors are placed in position and sealed with mud if required, to stop any leakage in or out of the oven. The oven is not charged and is kept under double suction. The fine powder flows into the oven and spreads like a cloud inside the oven. It finds its way out into the flues due to excess suction, through any cracks or open joints. The fine material seals the opening and makes the oven leakproof.

d) Silica Welding

Silica welding is a special technique used to repair silica refractory in Coke Ovens in hot conditions. Silica welding is generally used as standard maintenance practice in Coke Ovens. Silica welding, as the name suggests, is silica sand properly graded and treated with very fine aluminium powder. Oxygen is used for generating high temperatures say up to 20000C. A special welding machine is used for this purpose. The machine is equipped with all required connections, hosepipe connections, welding nozzle etc.

2.4 Coke Dry Quenching and Or Coke Drying And Cooling Plant

In India, we have two designs in use. 1) Russian design and 2) Nippon Steel Corporation's design.

Russians have granted a licence to M/s MECON to install these units in India whereas Nippon Steel Corporation directly deals with the clients. Russian units are of 70T/Hr cooling capacity, Japanese plants are of 120 (JSW) and 100 mt./hr. (TISCO, JSR) and 200TPH cooling capacity at Tata Steel, Kalinga Nagar, Orissa.

The main refractory problems faced by plants are damage to PILLARS. Even the Japanese are fully aware of this problem, but so far, a satisfactory solution is not found. The pillar bricks used by the Japanese are of Mullite 50-52% fireclay bricks. They have provided 36 pillars in every chamber i.e., at 10o apart from the centre. The slope of the pillars workout to 74.6 degrees.

Almost every year when the unit is put down for boiler inspector's inspection, the repair to pillars is undertaken. As the unit is getting older, the damage to the pillars increases. And sometimes it becomes problematic to change the old, damaged bricks. The bricks are of special design with a tongue and groove system.

At JSW, we replace the damaged bricks, whereas in some other plants these are repaired with spraying, Casting and Gunning. The gunning is not able to stand the abrasion of coke. In the writer's opinion, the repair should be done with bricks only. No anchoring can be effectively used as it may damage the bricks when drilling the hole for fixing anchors.

2.5 CDQ - Food for Thought

In the CDQ unit, the pillar is the most vulnerable area and needs frequent repairs. After having some experience, the following modifications can be suggested for better performance. For existing Nippon design at JSW Steel:

1. Quality of Refractory needs improvement. The pillar bricks from layer no.65 to 70 should be changed to Mullite B quality.
2. The pillar bricks from layer no.71 to 77 should be changed to Mullite 'A' quality.
3. The two-brick design from second brick; third brick etc of the pillar (from working face) should be changed to a single brick design.
4. The slope of the pillar should be changed from 74.6 degrees to approx. 80 degrees. This will improve the stability of the pillar structure. (This will need a redesign of shapes)
5. The pillar lining should be brought forward to match the working face i.e., 150mm offset should be dispensed with.

6. The cantilever effect of 450mm at layer no.65 should be reduced as much as possible. This will need design modifications.

Further future design changes suggested are as follows:

1. The number of pillars should be spaced at 12 degrees instead of 10 degrees as at present. This will increase the pillar width, which in turn increase the stability of the pillar.
2. The refractory quality should be changed to 80% Alumina bricks, for the 900 series of pillar bricks. This will increase the abrasion resistance of bricks in many folds.
3. The lining thickness at the cold zone should be increased from 620mm to say approx. 800mm. As at present, the erosion of refractory in zones below pillar and above pillar is not balanced. OR change the Quality of bricks to 70% Mullite bricks, which will have higher abrasion resistance.

Note: - A thorough discussion must be held before implementing suggested modifications for future designs in full or part (Any takers?)

About the Author

Maharaj Singh, Ex G.M (MECON) & Ex Consultant JSW. Maharaj Singh, Dip Refractory Engineering, Dip Mechanical Engineering, M. Sc (Engineering), AMIE (Engineering), Qualified in 'Furnace Brick Work' from the UK, joined Tata Steel in 1954 as a mason in the Refractory Department. He gradually got promoted to the post of General Foreman. After serving for 22.2 years with Tata Steel, he joined M/S. MECON as construction i/c in 1976 and rose to the position of GM. After a successful tenure of 20 years in MECON, he joined JSW Steel, Bellary, Karnataka, as the GM, in 1996. After retirement from JSW Steel in 2015, he has kept himself engaged in private consultancy.

Maharaj Singh has a total work experience of 62 years. He has vast experience in the construction and maintenance of furnaces of Steel Plants, coke ovens, as well as 'relining ' of Blast Furnaces. His hands-on skill has been further honed by continuous up-gradation of educational qualifications and periodical exposures to the steel manufacturing units in the UK, USA, Soviet Union, Germany, Netherland and Italy.

VIII.2 Repair & Maintenance Practices of Coke Oven Batteries

Tilak Kanta Chaudhuri

1. Introduction

In an integrated steel plant, Coke Oven Batteries play one of the vital roles in the production chain. It provides coke, the major energy input for the blast furnaces but also supplies the coke oven gas as fuel to all the different Mills & shops apart from its own consumption. In view of the above, the health and life of coke oven batteries are very important for attaining high production levels in an integrated steel plant.

During the operation of a coke oven battery for the production of coke, the whole installation is more or less exposed to wear & tear. For example, at least once within 24 hours, the coke is pushed out of the coke oven chambers due to this process, the heating walls are subject to a significant thermal shock & mechanical load, which cannot be fully avoided, even if the battery is operated under best possible conditions.

Measures to prevent premature failure of coke ovens and increase their life can contribute a lot towards this endeavour. The prolongation of coke oven service life to 40 years or more is becoming now realistic in Coke Ovens abroad because the operational practices have improved, control devices are developed, new repair technologies are available and total quality management is promoted. Unfortunately, the Indian Coke Oven lags behind in these areas and has a lower service life.

With the recent introduction of stringent norms for Coke Oven Batteries by CPCB, Coke oven health plays an important role from the point of view of pollution control as oven health and environment pollution are directly related.

2. Factors Influencing Service Life of Coke Ovens

If we, the Coke Oven operator gives special attention to some of the fundamental principles, it can significantly influence the Service life / Expected lifetime of a coke oven battery.

2.1 Design Of the Battery

2.2 Work Culture

i. A well trained and motivated team for the production.

ii. Operation of the battery as constant as possible.

iii. Adherence to technological & operational parameters.

iv. Maintenance of proper anchorage, Hydraulic & Heating regime.

v. Operational, Mechanical & refractory maintenance practices.

2.3 Heating & Hydraulic Regime

i. Technical control during initial heating up & commissioning.

ii. Heating of the battery according to the recommendations of the designer.

iii. Optimal coking time.

2.4 Quality Of Coal Blend

i. Good choice and preparation of the charging coal include crushing, blending, moisture etc.

3. Symptoms For Deterioration in Battery Health

The system of monitoring and diagnosing battery conditions involves the identification of the relevant parameters through observations and measurements. Using these parameters, the ageing of ovens can be characterized. Ageing of coke ovens is manifested by the following:

i. Increase in pollution

ii. Reduction in the heating quality

iii. Deformation of the Structural Member.

iv. Degradation of the refractory

3.1 Visual Symptoms

3.1.1 Door Gassing/Blazing

Door gassing starts when door knife-edge adjustment and sealing surface leave a gap, which is large enough to not allow condensed tarry vapours to form a gasket. Door blazing is the worst type of menace in battery operation as it leads to deformation in the Structural elements resulting in loosening of brickwork and the accelerated mode of deterioration starts.

Cleaning of doors, doorframe and door knife-edge adjustment is absolutely necessary. Even slight imperfections will not allow the formation of a gasket for sealing.

Hence, a lot of emphasis has to be laid on the proper cleaning of these components. Mechanical door/frame cleaners and hydro-jet cleaners should be in working order.

3.1.2 Chimney Smoking

This indicates the compactness of brickwork. Unless this problem is liquidated in time, battery heating and coke quality will be adversely affected. Normally there should not be any visible smoke from the chimney.

3.1.3 Carbon Depositions in Oven Roof & Oven Top Equipment

There are two simultaneous processes of carbon growth, cracking and carry over. The finer size dust present in the coal blend is carried over along with coke oven gas during charging. The higher the finer size fraction in coal blend, the higher is the risk of getting ascension pipes blocked. For carrying over of fines, HPLA to be maintained at optimum level. For cracking of gas, the flue temperature is an important factor. The carbon deposit on the oven roof, charging holes and IV, Gooseneck, Ascension Pipe is not the only problem concerning the gas extraction from the oven chamber, but also concerning the charging and pollution norm. When deposits reach such a problematic level, it becomes necessary to de-carbonize the oven chamber. This de-carbonizing is also detrimental to the service life of a coke oven. Therefore, the operating conditions must be found to limit as much as possible the carry-over and the gas temperature, besides the choice of the coal.

3.1.4 Green Pushing and Pushing Emissions

These are mainly related to heating inefficiencies, Work culture, Equipment design and are generally associated with poor coke quality.

3.2 Technical Symptoms

3.2.1 Tar Yield and Quality

A direct reflection of battery performance is noticed in tar yield and quality. It is well known that tarry vapours undergo cracking leading to more free carbon, identified as primary quinoline insoluble (QI) having a particle size of fewer than 20 microns if the under-roof space of the oven is over-heated i.e. undercharging or improper heating. Hence, the QI content of tar is a clear indicator of the working of coke ovens. Generally, for a 4–5-meter battery with both side-collecting mains, QI should be within 4-5% and for a 7-metre tall battery, it should be within 2-3%.

3.2.2 Coke Oven Gas Quality

Any deterioration in battery condition and the hydraulic regime is directly reflected in coke ovens gas quality. Nitrogen content in coke oven gas tends to increase in case of air infiltration. The composition of coke oven gas also changes. Moreover, the Calorific Value of gas is also reduced.

TABLE 1: CAUSE – MANIFESTATION – EFFECT SYNDROME

CAUSES	LEADS TO	MANIFESTATION	EFFECT
Door leakage or, Improper door boxing.	Damage to Door, D/Frame, Reinforcement elements i.e., Buck-stay, Tie Rod, Springs.	Loosening of brickwork	Damage to battery
Prolonged door opening. Gap in-between door knife-edge-d/frame and jamb brick – D/frame	Air infiltration, Slag / Clinker formation on oven wall, Burning out of graphite layer.	Hard pushes	Sticker oven
Inadequate collecting main pressure	Air infiltration, Burning of the graphite layer	Hard pushes, Chimney smoking	Sticker oven
Excess micro fines in the blend	Choking of GN / IV Box, Carbon deposits, Oven pressurization	Pushing & charging problem, High amperage	Brick lining damage
Coal blend	Excess wall pressure, Less shrinkage	Hard pushes, High amperage	Damage to brickwork, Sticker oven
Variation in oven/battery temperature	Poor coke, Pushing emissions	Hard pushes, High amperage	Damage to brickwork,
Low EVT, Air infiltration in bass flue/regenerator	Poor coke	Pushing emissions	End brick damage
Over/ improper heating, Improper blend	Fissured coke	Hard pushes, High amperage	Damage to brickwork
Under Charging	Localized high Free space temperature, cracking & burning of tarry vapour	Poor tar quality	Damage to brickwork

4. Typical Damage on Coke Oven Wall

During operation, the lining of heating walls is subjected to thermal, mechanical and physio-chemical influences, which lead to gradual wear, chipping, pitting, slag formation on the wall etc. Cracks develop without the displacement of individual bricks. Operational, heating and hydraulic regimes are to be maintained to minimize the effects of this wear. In case of delay in liquidating the faults developed in the initial stages, rapid deterioration starts with widening of cracks, shifting of bricks, erosion, chipping etc. under the influence of various stresses acting and the ovens ultimately become unserviceable.

Initially, just after commissioning, the physical and performance characteristics of any battery deteriorate very slowly. It is often difficult for operation personnel to recognize the deterioration until the problems become serious. Once the critical point is reached, further deterioration becomes accelerated. This critical point can be delayed by regular battery evaluation and inspections followed by corrective repairs.

5. Refractory Maintenance

However, despite the best attention, damage to the refractory brickwork cannot be avoided. Variations of temperatures, mechanical stress and other negative influences are causing damage to the heating walls.

Already after the commissioning of a battery, leakage and hairline cracks can appear, which can be treated by airborne sealing.

In most cases, the first small damage will occur in the area of the frame joints, such as vertical cracks right beyond the chamber frame and the first heating flue.

At this early stage, the damage can be repaired by gunning/ Ceramic welding and with selective maintenance, batteries with this kind of damage can reach a high age.

Unfortunately, however, many operators do not realize the coming up of this kind of damage and continue the production of coke without doing any maintenance. Thus, in time, the initial small cracks are followed by edge spalling, surface damage, small holes and disjointing.

These kinds of damage require other repair methods to continue the operation of the battery. Technologies like gunning and ceramic welding are suitable to repair this damage and to avoid their extension.

Further decay may result in bigger holes or even in the collapse of complete heating flues, making the production of coke impossible. Even such a big damage can be repaired successfully with suitable technology, such as the reconstruction of end-heating flues or repair of complete heating walls to obtain a good condition of the ovens.

6. Various Loads on Coke Oven Brickwork

In the course of coke oven battery operation, the masonry brickwork undergoes loads of various natures, which are as follows:

Transverse load - as a result of pressure on the heating walls during the process of carbonization.

Static load - from brickwork and structural elements of the ovens/ battery.

Dynamic load - induced by the movement of coal charging cars.

Mechanical load - appearing during coal charging into the ovens and coke discharging.

Thermal stress - occurring in the refractory brickwork during the coking cycle in a battery.

These ageing symptoms lead to operating difficulties and problems. It can be seen that each operating problem leads to another operating problem and finally leads to impairment of coke production due to frequent failures of pushing etc.

7. Common Oven Lining Defects

The most common oven lining defects are:

1. Vertical cracks in first, second and third vertical flues from both pusher side & coke side.
2. Cracks in the central part of the ovens, principally under the middle charging holes.
3. Damage to the lining in the zone of middle charging holes.
4. Deformation – Movement of brickwork inside the oven under the middle charging holes.
5. Spalling of bricks especially in the end vertical flues.
6. "Undercuts" in the oven walls in the first two courses from the sole bricks.
7. Wall deformations (Convexity, concavity) opposite various vertical flues.
8. Chipping of jamb bricks.

9. Falling out of bricks from between two parallel cracks in the end verticals.
10. Cracks and deterioration of end roof bricks.
11. Damage to sole bricks up to various depths.
12. Damage in the face walls of lintels, inclined zone bricks, regenerators etc.
13. Hole in the sole in gun flue type oven.

The types of preventive maintenance most often used are spraying, pointing, dusting, gunning or ceramic welding of oven walls, repair and replacement of sole bricks, repair of end heating flues, relining of face walls of lintels, packing of leaking joints, repair of regenerator face walls etc.

8. Repair Methods

1. the airborne sealing (Dusting) of hairline cracks and small leakage
2. the gunning of cracks, edge spalling or surface damage
3. the ceramic welding of major cracks, holes or damage towards the centre of the oven
4. the reconstruction of end-heating-flues and
5. the reconstruction of complete heating walls

will be described in more detail.

Generally, we can say that in all cases we are talking about a "hot repair" affecting only a single oven chamber or, if necessary for safety reasons, neighbouring oven chambers, while the remaining ovens continue the production in their normal rhythm.

9. Typical Damage on Coke Oven Wall

During operation, the lining of heating walls is subjected to thermal, mechanical and physio-chemical influences, which lead to gradual wear, chipping, pitting, slag formation on the wall etc. Cracks develop without the displacement of individual bricks. Operational, heating and hydraulic regimes are to be maintained to minimize

10. Airborne Sealing (Dusting)

This is used to seal small leakage and hairline cracks, which may appear already in young ovens. But it is also used after repair by gunning or ceramic welding hot/cold repair to seal remaining leakage.

The material is very fine grain refractory powder, which expands under high temperatures after penetrating small holes and hairline cracks.

The following preparations have to be made for the airborne sealing:

1. the oven doors, the leveller doors, the ascension pipe lid and the charging holes have to be sealing to be checked.
2. the inspection holes of the two adjacent heating walls have to be opened/ or wall to be made under suction.
3. the airborne sealing device is introduced in one of the central charging holes and is connected to the compressed airline.

After these preparations, the airborne sealing material is filled into the airborne sealing device and is blown into the oven chamber by compressed air until a pressure of approximately 120 mm water column is reached inside the oven chamber.

Due to the overpressure, the material penetrates the cracks where it expands and seals the leakage.

Existing leakage is proved by dust coming out of the inspection holes. The quantity of dust will be reduced during the airborne sealing and will finally disappear.

The blowing of material into the chamber is made 3-4 times in intervals of about 10~15 minutes. The intervals are required to allow the expansion of the material.

4. The whole operation will take 1½ to 2 hours, and in the end, no more dust should come out of the inspection holes.

11. Gunning

The technology of gunning is well known for many years and is used by almost all the coking plants in the world.

The advantages are –

1. the suitability for the end 4 6 flue area of the heating wall from both coke side to pusher side and from the sole level up to the roof
2. the short period of intervention. A skilled team will be able to repair one chamber within one or maximum 2 shifts
3. the lifetime of a repair may be up to 6 months

For better result & good life:

The oven chamber must be taken off production 4-5 hours before starting the repair.

After pushing, the charging holes and the ascension pipes are opened, allowing the graphite to burn off.

Then the oven doors are dismounted and, if available, a dummy door with sectional windows is installed.

Alternatively, ceramic fibre blankets can be glued onto the chamber frame to avoid a drop in temperature.

11.1The Procedure of Gunning

When the preparation work is finished, the oven can be cleaned by blowing air or by using cleaning tools.

Cleaning is very important for a successful repair by gunning.

The heating walls are cleaned by using a pneumatic hammer having either one or three "teeth".

It is used to remove spalling, slag or old repair material. After cleaning, the brickwork is blown with compressed air to remove also the dust.

While a part of the team is busy with cleaning, the rest of the team is preparing the gunning machine and the equipment.

Gunning is made with a special gunning machine with an Electro-mechanically operated rotor system to convey the material. Normally used M/c are VELCO ROTAMAT or, ALIVA type gunning machines with very good results

The dry gunning material is filled into the funnel of the machine, the adjustable rotor conveys it to the material hose, and compressed air is pressing the gunning concrete through the hose towards the lance. Through the mixing unit at the beginning of the lance, water is added through a controllable valve and is mixed up with the material on its way to the gunning nozzle.

The humid gunning concrete is now gunned onto the heating wall, starting to repair deep damage first, until the whole damaged area is treated.

While gunning and immediately after, one of the operators has to smooth the surface using a scraper. This is important to obtain an even surface allowing to push the coke without problems.

After gunning, the rebound is removed from the oven sole and the original door is reinstalled, and the same schedule is followed up on the opposite side of the oven if required.

12. Ceramic Welding

Ceramic welding is a supplement to the conventional repair techniques and is suitable for the repair of major cracks, holes or other important damage on the refractory brickwork.

Of course, it can also be used for the repair of normal wear of the surface. however, the price is relatively high compared to other repair methods, but a very effective method and good ceramic welding give better life without affecting the side ovens.

12.1The Ceramic Welding Procedure

Preparation and cleaning are made in the same way as for gunning. The ceramic welding is based on the principle that a dry mixture of refractory material and a metallic powder is projected onto the heating wall. Air is used as a transport medium and Oxygen is inserted near the tip it acts as a catalyser for the ignition.

The burning metallic powder is creating temperatures of approximately 18000C + and is melting the refractory material as well as the surface of the brickwork.

After cooling down to operating temperature, the brickwork and the repair material will be from a homogeneous block of a crystalline structure. Holes, cracks and joints between old and new brickwork can be sealed with this technology.

The ceramic welding requires skilled workers because the surface cannot be treated later. For this reason, a high-quality standard of work is necessary to assure an invariable quality of the repair.

Before Cleaning After Cleaning After Welding

STAGES IN CERAMIC WELDING

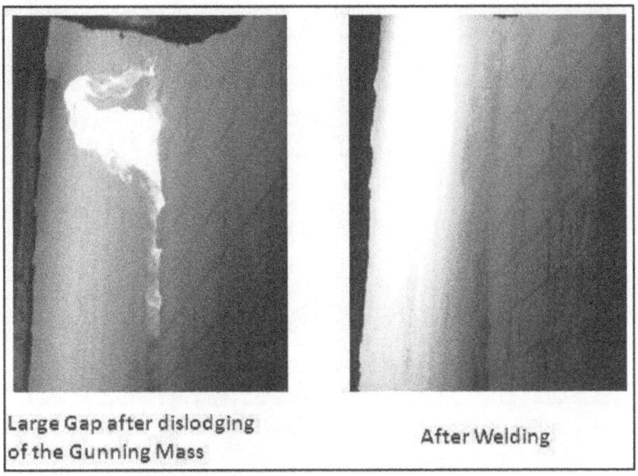

OLD TO NEW JOINT WELDING IN BATTERY 3

THE DAMAGE IN OVEN NO 16 AT DPL (PUSHER SIDE)

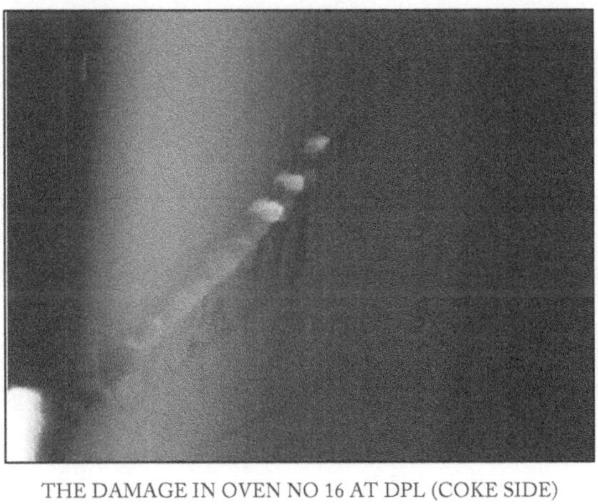

THE DAMAGE IN OVEN NO 16 AT DPL (COKE SIDE)

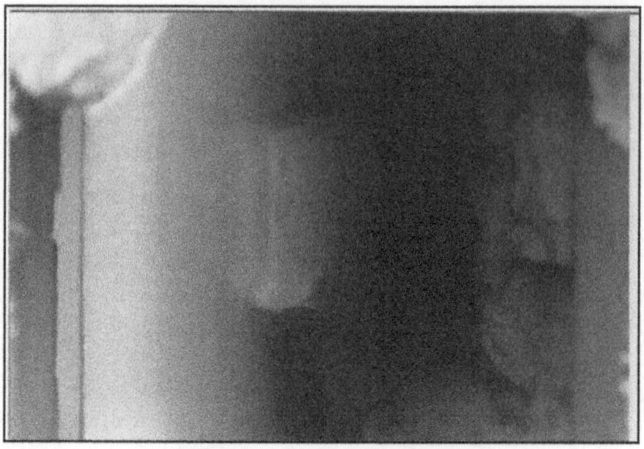

WELDING IN PROGRESS

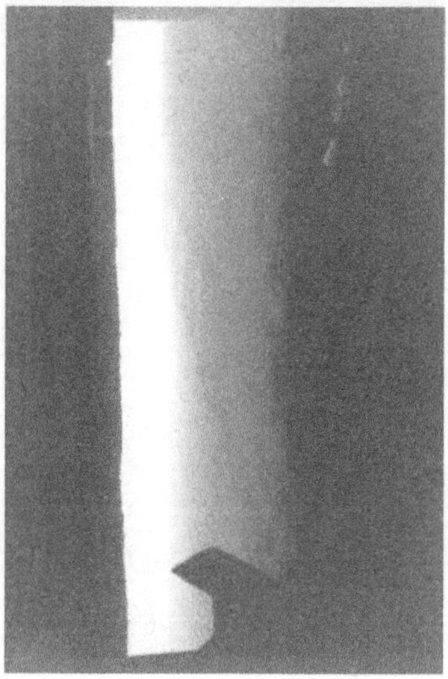

REPAIR COMPLETED

UNDER CUT

13. Partial or Complete Reconstruction of Heating Walls

During the operation of a coke oven battery, in some cases already after 15-20 years, important damage of the brickwork can occur. In this case, normal maintenance or small repairs as described above, will not be sufficient to secure a smooth operation with low emissions and constant quality.

At this stage, it is in many cases necessary to reconstruct single heating flues or even a complete wall.

The partial or complete reconstruction of a wall or a group of neighbouring walls is made according to the following schedule:

1. taking measures to protect the remaining brickwork and to keep their temperature as high as possible
2. cooling down and stabilization of the ends of the remaining brickwork
3. demolition of the brickwork to be renewed
4. Jacking up / loading of un-dismantled heating wall & subsequent reconstruction of the wall
5. heating up and commissioning

To give an example, we suppose that a group of end-heating walls will be reconstructed up to a depth of four flues or, more.

The following preparations are required:

1. reduction of the temperatures in the heating wall to be repaired as well as in the neighbouring walls
2. pushing of coke out of the oven chambers affected by the repair and disconnection from the collection main
3. the neighbouring walls are isolated by a specially prepared construction covered with ceramic fibre blankets to protect these walls as well as the remaining parts of the repair walls from cooling down. For the same reason, the sole channels will be closed, and the regenerator faces will be isolated with ceramic fibre.

If it is necessary, the regenerators can be heated externally.

14. Isolation of the Area To Be Repaired

A dummy wall will be installed into the oven chamber on the level of the first binder brick beyond the repair area to protect the chamber from cooling down and to stabilize the heating wall.

The isolation for the neighbouring walls will be installed from the oven sole up to the vault and from the chamber frame up to and including the dummy wall.

Isolation is also required on the last flues of the remaining brickwork.

It is made of ceramic fibre (13500C, 50 mm thick, 128 Kg/cbm), installed directly onto the heating wall and covered by a glass fibre fabric with an Aluminium coating to protect the ceramic fibre from abrasion.

After having installed the isolation, supports will be mounted on the level of the first binder brick beyond the repair area.

15. Demolition

To avoid a blockage of the air and gas channels, the flues are opened and cleaned, and the top of the flues port will be covered with ceramic fibre.

The demolition is started on the roof down to the oven sole.

During the demolition, the brickwork in the alternate header is cleaned and not damaged brick of the last remaining flue to obtain a jig jag joint.

To avoid the cooling down of the brickwork, it is immediately covered with ceramic fibre, and supports are installed to stabilize the heating walls and to prevent disjointments normally done by giving support with brickwork.

If it appears necessary, the oven sole is also demolished to repair the damage in this area as well.

The remaining brickwork of the bottom and the wall will then be cleaned to secure smooth expansion of the new brickwork.

16. Bricklaying

The first layer of the new bricks needs a horizontal sliding joint for the expansion of the silica bricks.

For this reason, it is necessary to clean very thoroughly the tongues and groves before laying the bricks.

The reconstruction of the new brickwork with silica bricks of the original size up to the level of the hairpin bricks, in consideration of exact measurements and of the thermal expansion.

On each flue, on one side, the Panel bricks of the first two layers will only be laid after having cleaned the flues.

To obtain proper pushings, the new brickwork will be approximately 5 10 mm thinner on the coke side respectively 5 10 mm wider on the pusher side than the remaining brickwork.

For better, the connection to the remaining brickwork should be made with longer Panel bricks made of fused silica. These bricks have a better resistance against changing temperatures and cover the complete flue without vertical joints.

For better, the connection to the remaining brickwork should be made with longer runner bricks of fused silica.

These bricks have a better resistance against changing temperatures, and unnecessary joints are avoided.

Between the flash plate and the first heating flue, A ceramic fibre blanket are pasted which will be pressed during heating up, leaving a joint of a maximum of 10 mm.

During the bricklaying, supports are installed at approximately 1 m distance to stabilise the new bricks and to prevent deformation.

They also serve as a scaffold.

17. Heating Up and Finishing Work

Before heating the new brickwork, the following work still has to be made:

1. adjustment of the support installed on the level of the header bricks
2. installation of thermocouples into the heating flues and the oven chamber
3. heating up of end flues is made by heat transmission from the neighbouring walls by conduction & radiation. Small parts of the isolation are removed to transmit the heat. While the temperature is rising, the openings in the isolation are enlarged to obtain a constant rise of temperature according to the heating curve for silica bricks.

When a temperature of approximately 350°C is reached, most of the expansion took place and the oven top bricks can be installed.

The door openings are covered with ceramic fibre blankets or are closed by dummy doors until reaching a temperature of 850 to 900^{0}C under firing system are connected, allowing to correct the temperature whenever necessary.

At a temperature of approximately 1100^{0}C, all supports, isolations and other auxiliary materials are removed from the oven chamber.

If necessary, the connection between old and new brickwork will be sealed, either by ceramic welding or by gunning/spraying

The oven chamber will be cleaned of debris bypassing Pusher.

Finally, the doors are mounted, the temperature is brought to operating temperature and the oven is given to the production.

TABLE 2 - COMPARISON OF VARIOUS TYPES OF REPAIRS OF COKE OVEN

	HOT REPAIR	COLD REPAIR	REBUILDING
Cost	Least	More	Most
Time	Least	More	Most
Durability	Least	More	Most
Expected life before next repairs	3 to 5 years	Above 8 to 10 yrs.	15 to 30 yrs.
Working condition during repair	Difficult	Normal	Normal
Possibility of retrofitting of new technology items during repair	Usually not possible	Possible	Possible
Possibility of attending repairs of Chimney, tunnel etc.	Not possible	Possible	Possible
Possibility of thorough cleaning of jammed ports.	Usually not possible. Only end ports can be revived.	Possible	Possible
Waste heat box repairs	Usually not possible.	Possible	Possible

18. Maintenance Practices of C.O. Batteries

18.1Refractory Maintenance

1. Spray pointing of end brickworks of oven.
2. Hot patching and pointing of end brickwork.
3. Semi-dry gunning.
4. Silica Dusting of the oven.

5. Ceramic Welding.
6. Charging hole brickwork pointing.
7. A.P. Standpipe brickwork patching & pointing or changing.
8. Charging hole frame changing.
9. Oven top brickwork grouting.
10. Door repair: Partial repair / Full repair as per requirement.
11. AP base & spigot joint sealing with ceramic fibre rope: as per requirement.
12. Regenerator face wall pointing: Breeches pipe joint sealing, WH Box Base sealing.
13. Inspection flue ring grouting/changing.
14. Top face Casting/ pointing.
15. Door frame rope caulking in the ceramic rope at the back and face with Ceramic / Aluminium rope and pointing.
16. Jamb/end Flue cleaning.
17. A.P. Shell bricking
18. I.V. gland leakage sealing.
19. Riser Grouting.

18.2 Mechanical Maintenance

1. Door regulation, diaphragm knife edge & spring, latch screw/ spring etc changing.
2. AP cap/ Water seal cleaning & changing.
3. AP shell repair & changing.
4. G. Neck/I.V. changing.
5. Door frame changing.
6. Door Repair/Servicing.
7. Tie Rod repair/changing.
8. Anchorage springs measurement & adjustment of load: once every six months.
9. Flushing liquor and HPLA distribution pipe servicing/changing.
10. Buckstay repair/changing.
11. Tie rod spring changing as per requirement.

12. Checking & changing of mechanical loading of the door frame by Pr. Bracket or, bolt Changing of flushing & HPLA liquor nozzle, valve & water valve
13. G.C. Main butterfly valve changing

18.3 Heating Maintenance

1. Manifold and riser cleaning/servicing/changing along with checking of gas leakage
2. WH Box Valve, Bus flue Air flap etc. airing, cleaning, servicing.
3. Chimney draught, WH Tunnel draught, WH Box draught etc. regulation.
4. Control Vertical flue temperature measurement, cross wall temperature.
5. Coke mass temperature measurement & Control.
6. Free space temperature measurement & Control.
7. Flow Governor (Askania) Butterfly valve of foul gas and Foul Gas Main rotation checking.
8. Gas cock (I.V. & R.V.)) greasing, servicing, cleaning.
9. Regenerator airing and adjustment.
10. Oven observation (Charge height & Sole pressure at the end of coking).
11. Nozzle Plate cleaning/changing.
12. Regulation of Top Pressure, Draught.
13. Burning checking.
14. Main Header cleaning.
15. B.F gas Riser cleaning.
16. Cellar, floor & drain cleaning.
17. End flue temp. measurement/month
18. Finger plate adjustment fortnightly
19. Waste heat box spindle & disc etc checking every day
20. Waste heat gas analysis 4 times/week or, as per requirement
21. Pre-heater to be checked & maintained at 85°C+ for better CO gas at the header
22. Cross leakage monitoring / checking all heating walls once a month

23. Reversing mechanism/system checking for proper opening/closing of reversing gas cocks & waste heat boxes monthly once

18.4 **Operational Maintenance**

1. Door frame & Door cleaning/changing and knife-edge adjustment after fixing in an oven.
2. Gooseneck Isolation Valve Box cleaning & Liquor flow checking.
3. Charging hole cover/hatch & Frame cleaning & looting.
4. Setting free of I.V.
5. Progressive notching of IV in the Shift.
6. Measurement of Ampere reading of pusher: all ovens in the Shift.
7. Oven top, service platform - pushing spillage back-fire, equipment & cars cleaning: daily.
8. Ram car and quenching car track cleaning: daily.
9. A.P. Water Seal Cap cleaning.
10. G.C. Main chasing daily & top cleaning by blowing & painting as per the requirement to prevent corrosion.
11. Butterfly Valve checking daily.

19. Work Culture

Compared to other countries, in India, the life expectancy of Batteries, in general, has been less due to the following factors:

1. Labour unrest and law & order problems affecting the steady working of the batteries at the desired level of production.
2. Inadequate upkeep of batteries and equipment.
3. Irregular working schedules warranted by fluctuations in demand of coke and the vagaries in the supply of desired coking coal.

As per soviet standards, from the 5th year of operation of a battery, around 7% of the total investment cost is ploughed back for maintenance and repairs to sustain the battery during its normal life span.

They have also evolved some wear and tear indices as a guideline suitable for determining when a Coke Oven Battery ought to be shut down for rebuilding. They are defined in the following terms.

1. The average coking period should not exceed the planned value by more than 15%.

2. Not more than 15% of the ovens should be operated outside the pushing cycle (series).

3. Not more than 10% of the ovens should be operated with or, require raking up (Stickers).

4. Not more than 20% of the ovens should be operated with a reduced amount of coal charge.

5. The expansion of the brickwork should not exceed twice the theoretical value of Silica brick.

6. The average resistance of the regenerator checker work should not exceed an 8mm water gauge.

The reinforcements/ anchorage should be in good condition. It is further insisted that, if either of points 2 & 3 is contravened, the battery should be shut down without any question.

Whereas the life of a Coke Oven Battery is normally expressed by the number of years of its working, a more pertinent parameter for judging its life is the Dry Coal Carbonized /Oven from the Battery before it is taken down for Re-building.

Taking into consideration the conditions prevailing in India, an average throughput of 150,000MT of Dry Coal Carbonized /Oven can be taken as a measure of satisfactory performance of a Battery during its lifetime.

About the Author

Tilak Kanta Chaudhuri (Ex GM-COCC-DSP/SAIL) is a Ceramic Technologist who graduated from CCT of Calcutta University in 1978. Joined Mysore Porcelain, a high-tension insulator manufacturer under BHEL. In 1979 switch over to SAIL DURGAPUR STEEL PLANT Refractory Department. Initially worked in an open-hearth furnace, Blast furnace, Rotary Kiln and all types of Reheating Furnaces for its rebuilding and maintenance. Since then, for the last three decades worked in

the COCC complex in various capacities and took a leading part to introduce ceramic welding technology at different steel plants in India. Presently working as an independent consultant and associated with manufacturers to develop refractory products for SMS, COCC and helping to develop or, introduce the innovative process in India to reduce pollution with some leading global coke oven repair group.

VIII.3 Outsourcing Service of Operation & Maintenance of Coke Oven Plant

J K Choudhary

1. About SRG Earth Resources Private Limited

SRG Earth Resources Private Limited was established in the year 2005. The basic purpose of the formation of this company was to provide Mining Consultancy. Later, the Company entered into the field of Operation and Maintenance of Coke Oven Plant both non-recovery type with heat recovery system and Recovery Type Coke Oven Plant, By-product Plant, Material Handling System, Civil Construction, Mechanical Fabrication & Erection and Coal Mining.

SRG Earth Resources Private Limited is rendering services in the field of Operation & Maintenance of Coke Oven Plant and Material Handling Systems for the last 16 years. The Company started Operation & Maintenance Services from the year 2005. SRG is also undertaking heating and commissioning of Coke Oven Plants.

SRG Earth Resources Private Limited is one of the leading service provider companies in the field of Coke Oven Plant and Material Handling Plant in India. SRG Earth Resources Private Limited Started Operation and maintenance services of non-recovery coke oven plant from Jindal Steel & Power Limited, Raigarh.

SRG Earth Resources has rendered Operation & Maintenance Services to the following Plants in India:

1. Jindal Steel & Power Limited, Raigarh – 0.8 million ton non-recovery coke oven with material handling from 2006 to 2013.
2. Bhushan Power & Steel Limited, Jharsuguda – 0.45 million ton non-recovery coke oven with material handling from 2007 to 2021 and 1.09 million ton Stamp Charged Recovery coke oven plant with By-product plant and allied material handling system till date.
3. Bengal Energy Limited, Kharagpur – 0.6 million ton non-recovery coke oven plant with material handling including wagon tippler, stackers and reclaimers since 2009 to date.

4. Electrosteels steel Limited, Bokaro – 1.0-million-ton Vertical Non-recovery coke oven plant with material handling from 2013 to 2019.
5. Bhushan Steel Limited, Angul – Heating and commissioning of 2 nos Recovery Type coke oven plant. Chinese design.
6. Bhushan Steel Limited, Angul – Heating, Commissioning and Operation & Maintenance of 7.6-meter-tall Asia's tallest coke oven with by-product plant, CDQ and allied material handling system from 2013 to 2020.
7. Tata BSL Limited, Angul - Mechanical Maintenance of Coke Oven, CDQ, By-product Plant, Material Handling System and Utilities of both the Units i.e., Coke Oven 1 and Coke Oven 2 from September'2020 / December'2020 to till date.
8. Jindal Steel & Power Limited, Angul – Operation & Maintenance of 1.9 million ton Stamp Charged Battery with By-product Plant and allied material handling system from November'2016 to date.
9. SAIL – IISCO, Burnpur – Heating, Oven Anchorage, Oven Top maintenance, Refractory Maintenance and Door Repair and Cellar Maintenance of Battery No 11 from December'2016 to December'2018.
10. SAIL – Bokaro Steel Plant, Bokaro – Operation and Maintenance of Battery No 7 from January'2021 to date.

2. Strength Of SRG

1. SRG has a highly experienced team of Coke Oven and By-product having experience of 2 years to 30 years.
2. SRG has 100-150 nos Degree engineers having experience from 2 years to 10 years.
3. SRG has 100 plus nos Diploma Holders from 0 – 5 years of experience.
4. SRG has 300 plus Highly skilled manpower.
5. SRG has 1800 nos of Skilled manpower.
6. SRG has 1600 nos of Semiskilled and Unskilled manpower
7. SRG has all kinds of Tools & Tackles required for operation & maintenance.

8. SRG has earth equipment like Hydra, Tractor with Trolley, Hyva, Tipper, Payloader, Pick Van etc; which are mainly used for handling of Coal & Coke and maintenance.

3. Achievements Of SRG

1. SRG has achieved 100% targeted production in all plants.
2. SRG has got best Service Provider Award
3. SRG has got the best Safety Award.
4. SRG has got best House Keeping Award
5. SRG has achieved ZERO Break down during a month so many times during Service Period in different plants.
6. SRG always undertake developmental Projects in all plants.
7. SRG has always maintained Industrial Harmony.
8. SRG has always followed the SOP provided by the Client.
9. SRG has undertaken TPM, 5S, HIRA, TQM etc in all the plants.

4. ISO

SRG is ISO 9001: 2008 and OHSAS 18001: 2007 registered Company. The company is maintaining the ISO policy.

5. Training & Development

1. SRG is providing training to its employees for their development on a continuous basis.
2. SRG is providing training in-house and through institutes for their development.
3. Toolbox training is given by SRG.
4. Process training is given by SRG.
5. Safety talks and training is given by SRG.
6. TPM, TQM, 5S, HIRA training is given by SRG.

6. Why Outsourcing

Outsourcing is not just a way to reduce costs; it is an essential element for achieving business excellence.

Third-party management providing the best people, processes and technology for the execution of non-core business functions.

7. About SRG Earth Resources Private Limited

SRG Earth Resources Private Limited was established in the year 2005. The basic purpose of the formation of this company was to provide Mining Consultancy. Later, the Company entered into the field of Operation and Maintenance.

About the Author

JK Choudhary is the Director of SRG Earth Resources Private Limited.

VIII.4 Outsourcing in Coke Plants In India

A C Patnaik and P K Mohanty

1. Outsourcing

Materials and services not available IN-HOUSE and hence procured from other agencies are called Outsourcing in common parlance. If an entrepreneur has vision and financial capability all activities of setting up an Industry like Feasibility-Report, availing infrastructural facilities, design backup, project management, construction, commissioning and finally operation & maintenance of a plant in production can be outsourced. But in the present article, outsourcing "Operation & Maintenance" of coke plants during the production stage is discussed. If proper attention is paid by Industries to the points discussed, "Outsourcing" will have better prospects.

2. Recent Trends in Outsourcing

Previously operation & maintenance of the coke oven complex was mostly being taken up departmentally. However, during the last few years, integrated steel plants having coke making facilities like all units of the Steel Authority of India, Rashtriya Ispat Nigam Limited (Visakhapatnam) and private sector units like Tata Steel (Jamshedpur & Kalinganagar), JSPL (Angul & Raigarh), Bhusan Steel (Jharsugada), JSW (Bellary), JSL (Jajpur) etc are preferring to outsource different units of coke oven complex in totality or piecemeal basis.

The advantage being accrued are:

1. Operation & Maintenance is highly Labour intensive and calls for statutory clearance of PF, ESI, Factory, Electricity Board, Insurance, Depts of Government. Recruitment of this workforce attending to all statutory bodies, taking measures for strike & labour unrest, accommodation & transport, providing medical facilities, Disaster management are an additional burden on plant authorities and can be dispensed with outsourcing.
2. A lot of Investment & procurement activities are required in procuring material handling equipment, machinery, tools & tackles and consumables, their maintenance and periodic testing.
3. In India remuneration of contractual labour is less compared to workmen doing similar jobs in organized sectors of Industrial establishment.

All the above results in less burden huge cost saving and hence minimising production cost. The plant can run with a few skeleton engineers supervising the work of the outsourced agency.

3. Operation & Maintenance on coke oven complex O/M

Activities in any unit mainly comprise of

1. Inspection, condition monitoring, health diagnostic, report preparation and suggestion for remedial measure
2. Operation of plant, machines by skilled operators to achieve production schedule
3. Preventive maintenance of equipment as per SOP prepared by the equipment manufacturer
4. Breakdown or planned shutdown maintenance
5. Repair/replacement/sub-assembly of equipment, structural, mechanical parts
6. Technological upkeep, housekeeping etc
7. Electrical/Instrumentation/Automation

Outsourcing can be done on item wise rate contract basis or fixed time-period basis. In a coke oven complex, the main units are:

1. Coke oven battery proper with oven machines
2. Coal & Coke handling
3. Conventional water quenching system or modern coke dry cooling plant
4. By Product plant
5. Pipelines & auxiliaries

Over a period of time, plants & outsourcing agencies have developed micro-planning of sub-activities with their SOP & frequency and manpower requirement.

4. Eligibility Requirement

The job of operation and maintenance of a coke plant is enormous and hence the contracting agency has to be resourceful.

1. Technology in coke oven plants is advancing day by day especially in areas of process control, condition monitoring, hydraulics and automation, By product. The agency has to absorb and implement the same by recruitment of specially trained engineers.
2. Safety, environmental and health standard are becoming stringent globally. Priority has to be given to this aspect. All standard PPE and job-specific items are to be provided.

3. The machinery and tools requirement are huge. To name a few, tractor trolley, hydra crane, dumper, payloader, trailer, bobcat, welding and gas cutting sets with accessories TIG, MIG welding sets, chain pulley blocks, hydraulic and screw jacks, hook chucks, grease pumps, drilling and grinding machines, pyrometer, hydraulic torque wrench, measuring instruments, hand tools, electrical instruments etc are required. The agency should possess sufficient stocks.
4. The job involves fabrication, erection & subassemblies for which special skilled persons are required.

5. Manpower Requirement

Manpower requirement is dependent on the scope and criticality of work. Deployment of the required number of skilled manpower for round the clock work considering leave vacancies and absenteeism is a must.

We have estimated different categories of manpower required for comprehensive operation and maintenance of coke oven complex at NMDC's nagarnar, Jagdalpur 3.0 MTA steel plant against a tender submitted which will give an indication of manpower requirement for coke oven complex of all public and private sector plants of the country is enormous.

Salient features of NMDC' s coke oven complex.

1. 2 nos. recovery type top charged coke oven batteries, 67 ovens 7-meter height of Gipprocks/MECON design. Capacity gross coke= 0.88 x 2 MTA. Coking period= 16 hours.
2. 2 nos. CDCP each with 4 chambers of 52 TPH each.
3. 2 x 4 nos. waste heat recovery boilers of 25 TPH each Steam generation= 2272 t/day at 66 at 500 C.

The scope of comprehensive operation and maintenance covers battery proper with oven machines, CDCP, Coal & handling system, By product, Pipelines, Utilities and Auxiliaries.

The number of skilled manpower in coke oven areas is decreasing year by year because of retirement of old hands and reluctance of new generation to work in the dusty and hazardous area so efforts to be made to attract young hands through good pay packages and other facilities.

SL NO	CATEGORIES	REQUIREMENT (OPERATION + MAINTENANCE) NOS
1	Senior Engg with at least 5-15 yrs. experience in coke ovens	37
2	Shift Engg degree holder with 2-5 yrs. experience	69
3	Admn/Safety/HR officers	10
4	Supervisors Diploma Engg	37
5	Operators (Machines etc) SOT, Diploma or ITI	78
6	JOT (ITI)	50
7	HSW Highly skilled (ITI)	207
8	SW (ITI)	271
9	SSW (Experienced)	126
10	USW (Experienced)	147
11	Admn & other office staff	15
	Total	1047

About the Author

A C Patnaik (Mobile 9861087262): BSC (Engg) from U.C.E., Odisha in Mechanical Engineering (now VSS University of Technology). Worked in Coke Oven construction sites of Otto India at SAIL units of Rourkela, Burnpur, Durgapur, Bhilai, Bokaro and also TISCO, Jamshedpur in areas of Construction, Commissioning, Operation & Maintenance. Joined Neelachal Ispat Nigam Limited since inception and retired as GM (Coke Ovens). Now associated with Utkal Projects (P) Ltd.

P K Mohanty (Mobile 7682846012): B. Tech (Hons), IIT (Kharagpur) in Metallurgical Engineering. Worked in Otto India at their corporate office, Kolkata and manufacturing unit, Rourkela in areas of manufacturing, Quality assurance, Sales and Contract management. Also worked in Neelachal Ispat Nigam Limited. Now associated with Utkal Projects (P) Ltd

About the Company

Utkal Projects Private Limited is a private limited company incorporated under the Companies Act 1956 in the year 2009. Also, it is registered with Govt Bodies like NSIC and Director of Industries, ELBO, Govt of Odisha for Electrical work.

Area of Operation: Comprehensive operation, repair maintenance (mechanical and electrical), technological upkeep, refractory work of coke oven complex both recovery and non-recovery type of batteries. We have also taken up maintenance jobs in Coal Handling units of Power Plant and Plate rolling mills.

Assets: The company has a group of professionals having vast experience in turnkey execution of coke oven complex for SAIL units at Rourkela, Bhilai, Durgapur, Burnpur and Bokaro, RINL (Vizag), NINL (Odisha), Tata Steel (Jamshedpur). They have a skilled workforce with years of experience. Over the years the stock of machinery, tools & tackles, instruments have increased.

Experience: During the last ten years of the company's incorporation, it has executed substantial jobs in its specific field at NINL (Odisha), SAIL (Rourkela), Jindal Stainless steel (Odisha), OPGC (Odisha). It has made its presence felt at other units of SAIL, RINL, JSPL, Bhusan Steel by regularly participating in tender enquiries and being qualified techno commercially.

LAST WORD: The Company's policy and motto have always been providing hassle-free services to the clients for their maximum satisfaction by adhering to quality standards, safety, environmental, health and statutory norms and timely completion of the job at the most competitive rate. Also, the welfare and congenial working environment of our workmen always gets priority.

The experience of providing service to their clients is by and largely satisfying. But what worries them now, is cut-throat competition among outsourcing agencies through a system called Reverse Auction for arriving at the minimum and at times unworkable price. Award of such order has adversely affected longevity and productivity. Besides cost consideration what is most required is quality, dedication and transparency in transactions. The old adage "If you throw peanuts, you get monkeys only" is also true for Outsourcing.

Address: Utkal Projects Private Limited, Flat no. 102/B, Supratik Exotica Apartment, Paikraypur Road, Gothapatna, Bhubaneswar, Odisha-751003. E-Mail: utkalprojects2009@rediffmail.com Key persons: Mr Manmohan Samal, Managing Director, Mob: 7682846010.

IX. The Merchant Coke Making Plants

IX.1. Factors Affecting the Cokeries Of Gujarat

Dipen N Chanchani

1. Introduction

GUJARAT has been a major supplier of Met Coke to Indian and Overseas industries from mid-1990 and had mushroomed from 2004 onwards.

These cokeries had supplemented the import of coke by the coke users. The major raw material i.e., coking coal was imported from Australia, Russia, the USA etc., whose pricing is normally governed by the JSM index. As, these cokeries, were not a part of integrated steel mills, the problems faced by these cokeries were consistent and predominantly factored due to the following reasons:

The steel mills and related industries were using less domestic coke particularly from the Gujarat region as:

1. Ispat Industries Ltd., located at DOLVI, PENN, in Maharashtra region, who'd was one of the main consumers, using approx. 3300 MT per day, was taken over by JSW in 2010, establish their own cokery and near 2013 was completely out of the open market.
2. ESSAR Steel Limited, at Hazra, Surat, has been facing financial crises and the domestic cokeries were unable to cater for their demands.
3. Numerous Coke plants were established in Southern India and started catering for the local market and even supplying lower sizes to the Central and Northern Part of the Country.
4. Northern India, which was also a major market for Gujarat Coke manufacturers, started being serviced by the coke ovens in Jharkhand / Odisha belt, with the production of low ash coke, with Ash ranging in the region of 12/24/28% and was delivered at a much cheaper price.
5. The Chinese Coke was relatively being sold at a lower price due to the availability of indigenous coking Coal at subsidised rates and dumping coke in the Indian market. At times the prices of Coke being dumped by China were only marginally higher than the prices of imported coking coal.

6. Indian cokeries have been protected by raising Anti-dumping duty on Chinese, Russian and other origin Cokes, which has made them possible to survive over the years with thin margins.
7. Over a decade most of the cokeries were operating with only 20%-30% capacity.
8. Gradually the coke ovens in Gujarat are now switching over from regular TATA type ovens to Pusher type ovens, based on various designs. Few of them have also been modified to stamping charging of Coal Cake.
9. Even after all a decade of struggle the coke industries in Gujarat have managed to survive all ups and downs, until recently, when the sudden demand of Met Coke, has renewed the hope of the existence of cokeries and this sector is now being revived.

M/S SU MANGALA COKE

We, at M/s Su Mangala Coke Pvt Ltd., have made it possible to blend coal due to palletisation, using a higher percentage of soft/semi-soft coking coal and obtain coke of required parameters, even in the oldest oven TATA type technology. This has resultantly reduced the input cost of raw material and also reduces emission, by front charging and reduces spillage of coking coal etc.

2. Plant Overview

COAL PALLETS

CHARGED INSIDE THE BEE-HIVE TATA TYPE OVENS

FINISHED COKE

FEW OF THE MERCHANT COKE PLANTS IN GUJARAT IN 2021

COKE OVEN NAME	LOCATION	COMMENCEMENT	OPERATIONAL STATUS	TYPE	COKE CAPACITY-MTPY	NUMBER OF OVENS	COKE WEIGHT PER OVEN
United Coke Pvt Ltd	Vedi, Anjar, Kutch.	2005	Operational	Stamp charging, non-Recovery	60000	26	20
Shree Electromelts Limited	Bhavnagar	2004	Operational	Stamp charging, non-Recovery	60000	30	16
Su Mangala Coke Pvt Ltd.	Lunva, Bhachau, Kutch.	2004	Operational	Stamp charging, non-Recovery	30000	36	6
Vimla Coke Pvt Ltd	Bhachau, Kutch	2004	Operational	Top charging, non-Recovery	30000	36	6
Balaji Coke Pvt Ltd.,	Nani Chirai, Kutch.	2000	Operational	Stamp charging, Recovery of Tar	120000	30	165

About the Author

Dipen N Chanchani is engaged in the coke industries, since the last twenty-eight years and has also developed a technology, which enables the use of lower grades of coking coals to make quality coke, even in the basic coke oven design, which helps in coke making more viable. Presently, based in Ahmedabad, Gujarat and operating a unit in Kutch under the brand name of M/s Su Mangala Coke Private Limited. He can be contacted through email: dnc@sumangalacoke.in and Mobile: +91 9879607222.

IX.2. Development Of Non-Recovery Coke Oven for Coke Making in India

Subhendu Bagchi

Coke is manufactured commonly by two methods that are:

1. Indirect Heating in The Absence of Air, Known as By-Product Recovery Oven

In this system, heat is supplied to the coal mass in an indirect way that is through the heating chamber (flue) in adjoining walls. As there is no air (Oxygen) inside the oven, the volatiles come out during the coking process collected, and after cooling and cleaning the same gas is injected into the heating chamber of the oven wall along with air. As this is a combustible gas, it generates heat after combustion and supplies it to the coal mass through conduction.

During cooling of coke oven gas during subsequent stages different martial collected.

The by-product systems have value addition due to the recovery of coal chemicals.

These are mostly tall ovens with Silica bricks, or a few are of medium height but with Silica bricks.

One design of by product type coke oven indigenously developed as “SEMET-SOLVAY & SEN” Technology for 2.5 m tall recovery oven and installed at M/s RakeshMaruti Ltd., Gujarat, which is in operation, but oven installed at M/s Adhunik Metaliks in Rourkela is not in operation.

Presently it is the only 2.5 m tall recovery indigenously developed non-silica technology of coke making design available in India.

During the process, some toxic materials are generated.

The stringent pollution laws and the high cost involved in the installation of batteries as well as of high-cost pollution control equipment and complex operation procedure restrict the small and medium coke producers to adopt this technology.

2. Direct Heating with Restricted Air Known As Non-Recovery Or Heat Recovery Coke Oven

In this system, the heat is generated by the combustion of volatiles inside the oven above the coal bed. The heat generated due to combustion penetrated the coal bed after radiating from the oven arch.

The system has vertical passages in-wall known as the downcomer and at the base structure known as the sole flue.

There are provisions for air injection in different locations of the flue system to meet the oxygen requirements for combustion. These are known as secondary air, tertiary air, etc. Since there is restricted air injected into the coke chamber known as primary air, partial combustion of the volatile occurs in space above the coal bed. The residual unburnt volatile matter along with burnt gas passes through the downcomer and sole flue. During the travelling process of the above-mentioned mixture, the air is added for further combustion of residual volatile gases and heat is generated, this heat is transmitted to coal mass through conduction.

After completion of this travel the hot gas enters one common tunnel and is sent to the chimney. This gas is known as waste gas because there is no residual unburnt hydrocarbon, and this gas has only sensible heat.

Sensible heat is used for power generation using Waste Heat Boiler. This is the reason this system is called Non- Recovery Heat recovery (NRHR) type oven.

Non-recovery types of coke ovens are energy efficient and environmentally friendly.

These types of ovens have a minimum investment on pollution control equipment and the installation cost is also low. Operation procedures of these ovens are simple and easy.

3. History Of Indian Non-Recovery Coke Oven

In India, the concept starts from Dhanbad (Jharkhand) due to the availability of coking coal in the adjoining area at that time.

Initially, coke was made in a stack burning covered with mud. The coke quality with this system was not uniform, loss of yield and pollution levels were very high besides social problems.

In the second stage coke was made in ovens but without a flue system, it was known as Sabjee Bhatta. There was a quality complaint, and the coking time was high.

In the third stage, the beehive with the sole flue concept with multiple chimneys came into effect. This Design was also known as Tata Type Oven.

In this design, coal was charged from the top, and coke was pulled out from the side of the oven with drag (a T-type mild steel structure manufactured with M.S. Round and Rail). The drag was placed in the oven before charging and after complete carbonization, the drag was pulled out and the coke mass came out with the drag.

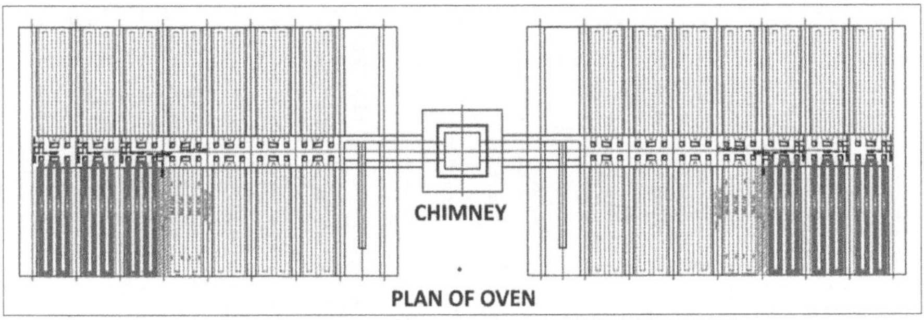

BACK-TO-BACK/ TATA TYPE OVEN WITH DOWNDRAFT

In the fourth stage, some modifications like a single chimney in place of multiple chimneys, downdraft tunnel, double stage tunnel, Puller in place of drag were introduced.

This design is still in operation for small coke producers in many parts of India. The major functioning of ovens was performed manually with large safety challenges.

4. Mechanization Of Beehive Ovens

In the fifth stage, as per available information, the first mechanized pusher-type non-recovery coke oven was constructed by Late Ashoke Ranjan Dasgupta in and around seventy when he was in Bird and company.

After the nationalization of coal mines, he resigned and installed a small coke oven battery at Bokaro industrial area, Dhanbad, Jharkhand, in the name of Basudha Udyog and also designed small coke oven batteries for some other small coke producers. At the time, the late A. P. Banerjee was his associate. After four to five years, he left his association and joined one plant at Gujrat

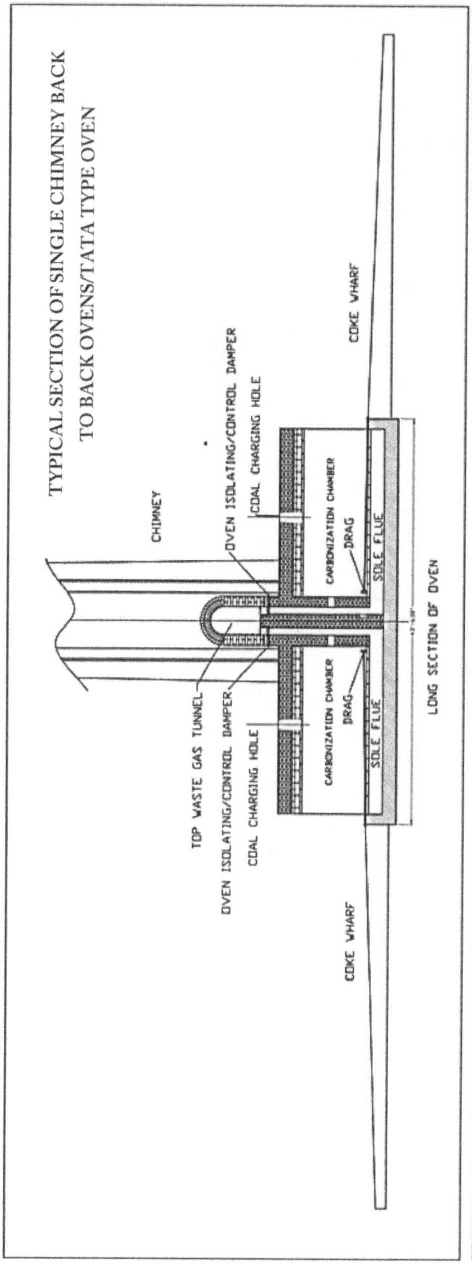
TYPICAL SECTION OF SINGLE CHIMNEY BACK
TO BACK OVENS/TATA TYPE OVEN
CHIMNEY
OVEN ISOLATING/CONTROL DAMPER
COAL CHARGING HOLE
CARBONIZATION CHAMBER
DRAG
SOLE FLUE
COKE WHARF
TOP WASTE GAS TUNNEL
OVEN ISOLATING/CONTROL DAMPER
COAL CHARGING HOLE
CARBONIZATION CHAMBER
DRAG
SOLE FLUE
COKE WHARF
LONG SECTION OF OVEN

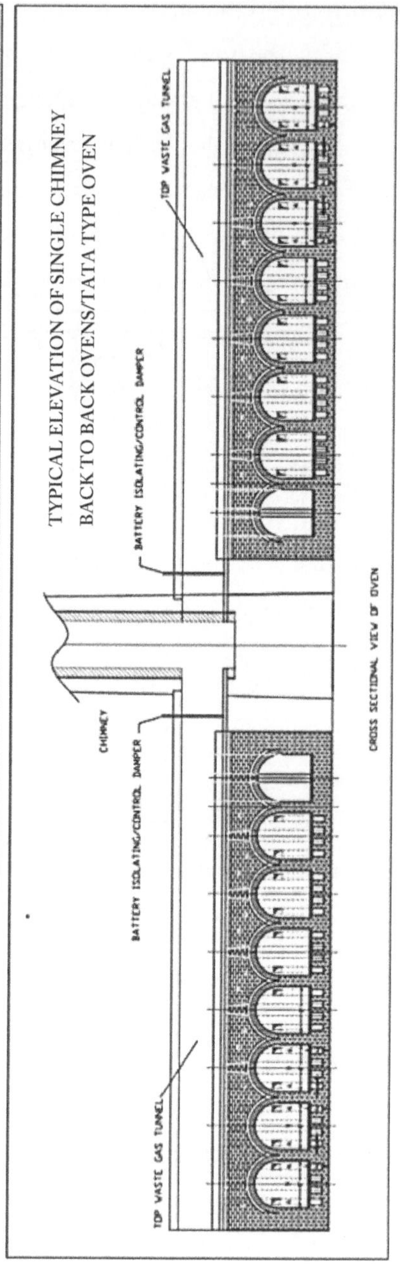
TYPICAL ELEVATION OF SINGLE CHIMNEY
BACK TO BACK OVENS/TATA TYPE OVEN
TOP WASTE GAS TUNNEL
BATTERY ISOLATING/CONTROL DAMPER
CHIMNEY
BATTERY ISOLATING/CONTROL DAMPER
TOP WASTE GAS TUNNEL
CROSS SECTIONAL VIEW OF OVEN

After that CFRI (presently known as CMFRI) developed one design in the name of Kumbhraj, team head was Mr Kunda Singh and that was installed at Katras, Dhanbad, Jharkhand, and after a long gap two plants at Gujrat. This design has not been widely marketed.

Initially, this design had top charging with a coke puller. Later pusher cars were included in this design.

In the late eighties, Late A. R. Dasgupta and his associates developed a mechanised non-recovery coke oven design with pusher and charging car that had both horizontal flue and vertical downcomer on the wall with sole heating and that was installed in Sitaram Sing Hard Coke, Katras, Dhanbad.

An old waste heat boiler was installed for power generation. The power was generated but the efficiency of it was too low so that could not be continued.

This was the first attempt at waste heat power generation through waste gas in a non-recovery coke oven in India.

Coke-oven of this design was installed at Bihar Hard Coke and Foundry Fuels Products Ltd. Basudha Udyog, Dhanbad. Usha Ispat, Sindhudurg district. Southern Fuels, Tuticorin, Walman, Nirgundi, Kharagpur and many other places.

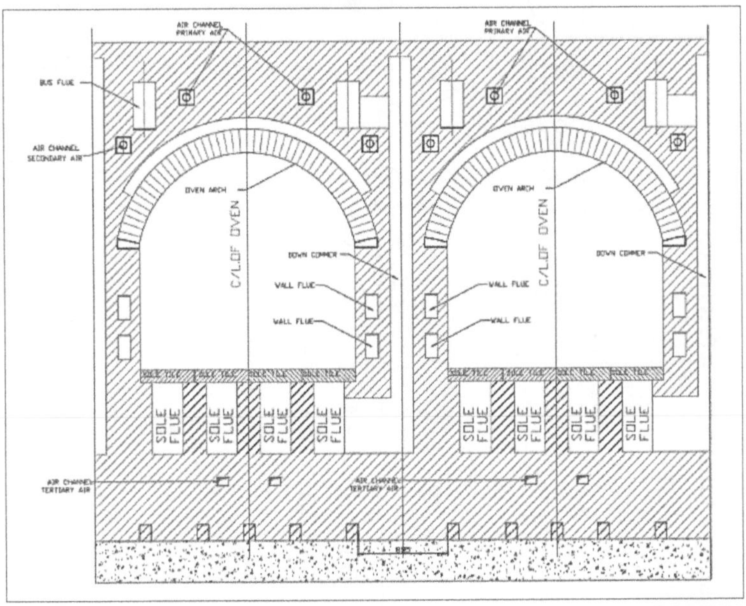

TYPICAL SECTIONAL VIEW OF NON-RECOVERY MECHANISED POWER RECOVERY TYPE COKE OVEN WITH HORIZONTAL WALL FLUE AND VERTICAL DOWNCOMER

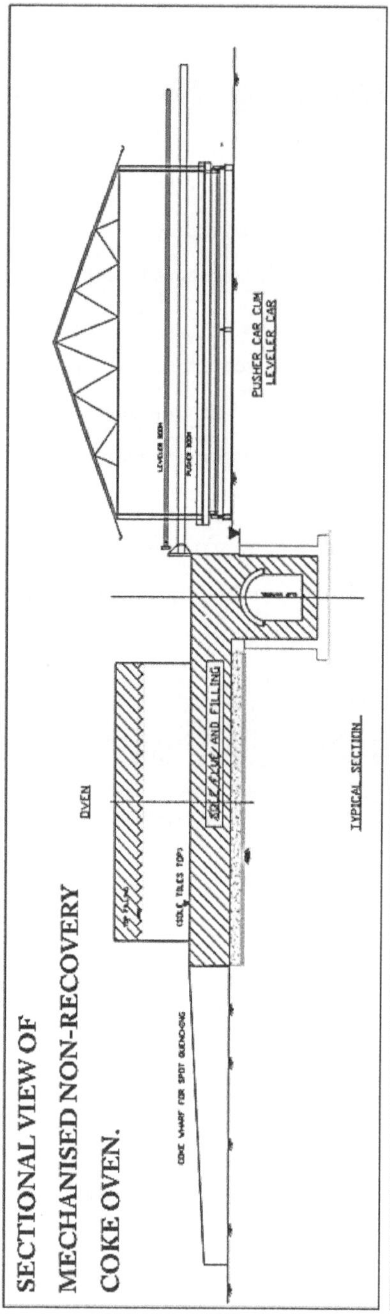
SECTIONAL VIEW OF
MECHANISED NON-RECOVERY
COKE OVEN.
PUSHER CAR CUM
LEVELER CAR
OVEN
SOLE FLUE AND FILLING
TYPICAL SECTION

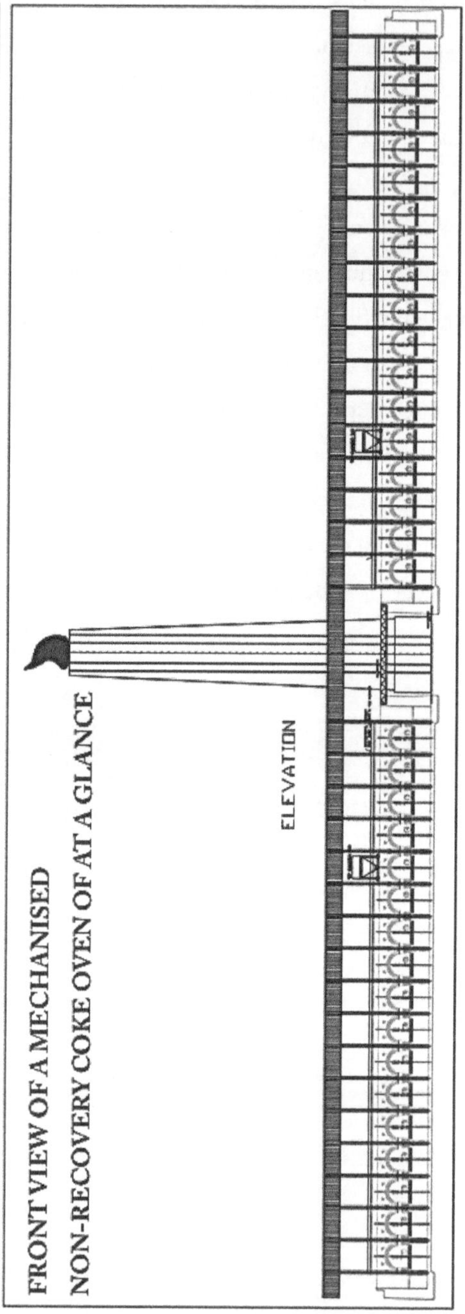
FRONT VIEW OF A MECHANISED
NON-RECOVERY COKE OVEN OF AT A GLANCE
ELEVATION

This design is still widely accepted for installation and operation with modifications by his associates and some other pupils today also, in the name Dasgupta Design or Modified Dasgupta Design with the inclusion of quenching car with advanced quenching system and with other advanced equipment

Oven with modified design installed at Bhatia Power and energy and Basudha Udyog Gumpdi Pundi, Chennai. Metalik Fuels, Kharagpur, and many other places.

Another design was developed in that period for non-recovery pusher-type coke oven with horizontal flue by Late A.P. Banerjee, two plants on that design were built in Gujarat and one was in Odisha. After the demise of Mr A.P. Banerjee, the availability of that design was discontinued.

There was another design of a pusher-type oven supplied by CFRI. The Batteries of that design were installed in some of the places like S.B.Q. Steel, Gudur, Usha fuels Raipur. and some other places. This design or its type is also in operation and implementation in some places till today.

Some other designs are also developed by some other pupils and implemented but not widely accepted.

5. Associated Development in Coke Oven

In India, in the year 2005 first waste heat power generation for industrial purposes was started at Electrosteel, Haldia, West Bengal.

Since then, power generation has been included in many Coke oven plants in India

In the year 2006 stamped coal cake charging system was included in the system

It was a single pressure plate with longitudinal travel in house design of Electrosteel, Haldia.

This type of stamping system was adopted by some of the coke oven plants of India.

The Coke-oven plant of Electrosteel was of Dasgupta Design. The plant is still in operation.

In the year 2007 – 2008 one multi-press three-stroke, variable pressure Stamped Coal Cake Charging design was developed by M/s Souam in association with M/s Indpac and M/s. Verma Engineering all were Dhanbad based and included at Bhatia

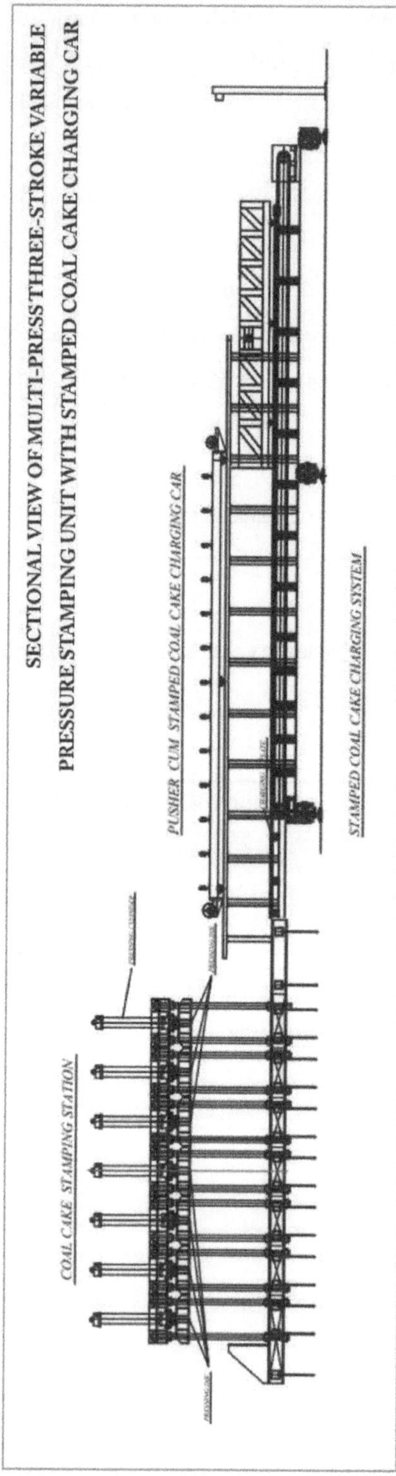

SECTIONAL VIEW OF MULTI-PRESS THREE-STROKE VARIABLE PRESSURE STAMPING UNIT WITH STAMPED COAL CAKE CHARGING CAR

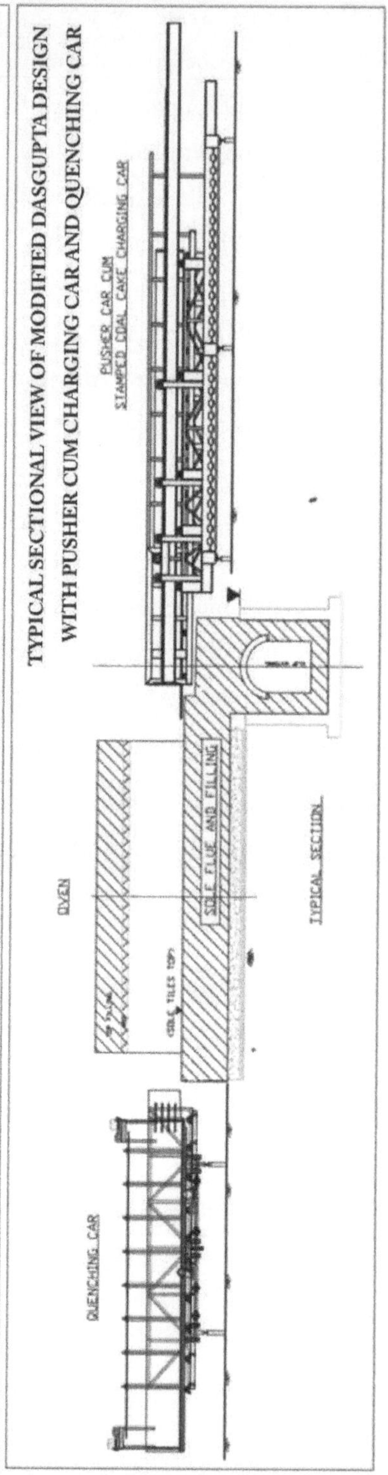

TYPICAL SECTIONAL VIEW OF MODIFIED DASGUPTA DESIGN WITH PUSHER CUM CHARGING CAR AND QUENCHING CAR

Coke, Gumdipundi, Chennai. in place of their existing top charging system. This was one unique design of this type. This plant is modified Dasgupta Design.

This type of stamping system has been included in many plants of Dasgupta Design or Modified Dasgupta Design and many other Indian designed coke-oven plants in India.

Presently many developments are working on non-recovery power recovery coke oven is going on in the different area and we are in the process of continuous improvement of non-recovery coke plants to match with modern days requirement.

About the Author

Subhendu Bagchi an Associate of Late AR Dasgupta and Late AP Banerjee worked with Late Bonomali Sen (Advisor Coke, BCCL, CIL) in Kusunda Pellet Plant, Dhanbad, for trouble shooting. Presently associated with Metalik Fuels Pvt. Ltd. (Kharagpur), Harsh Fuels Pvt Ltd. Gujrat, Coke-oven, Balajee Hitec Rolling Pvt. Ltd, Giridih. Maruti Udyog Pvt. Ltd. Dhanbad

X. THE WORLD OF COAL CHEMICALS AND GROUND REALITIES IN INDIA

Jayanta Sen & Tarun Chatterjee

1. Introduction

Extraction of industrially significant chemicals from coal may be characterised by the processes of coal carbonization, coal gasification and coal liquefaction. The roots of the coal chemicals industry can be retraced to the beginnings of coal carbonization undertaken to produce coal tar. The coking of coal and distillation of coal to produce coal gas was possibly the most significant developments that established the coal chemicals industry.

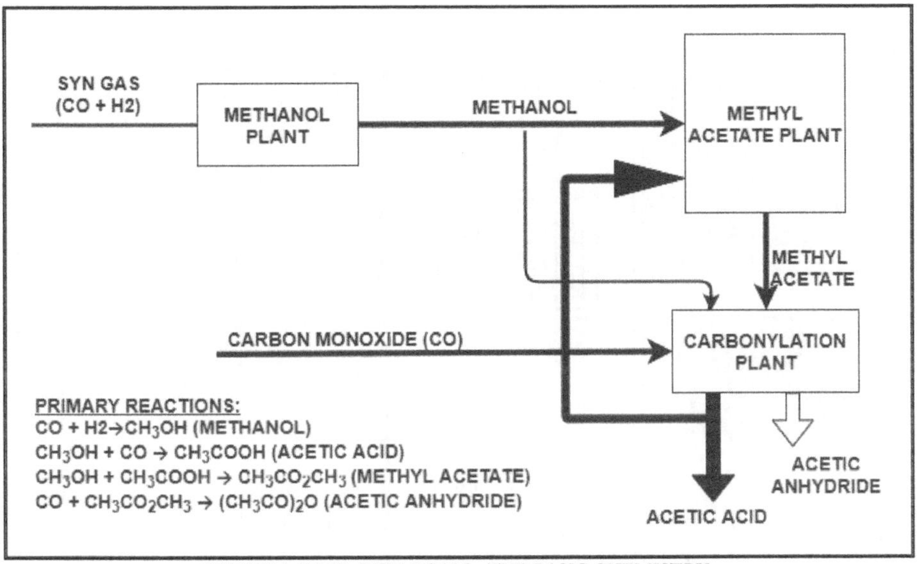

FIGURE 1: COAL CHEMICALS - THE BASIC CHEMISTRY

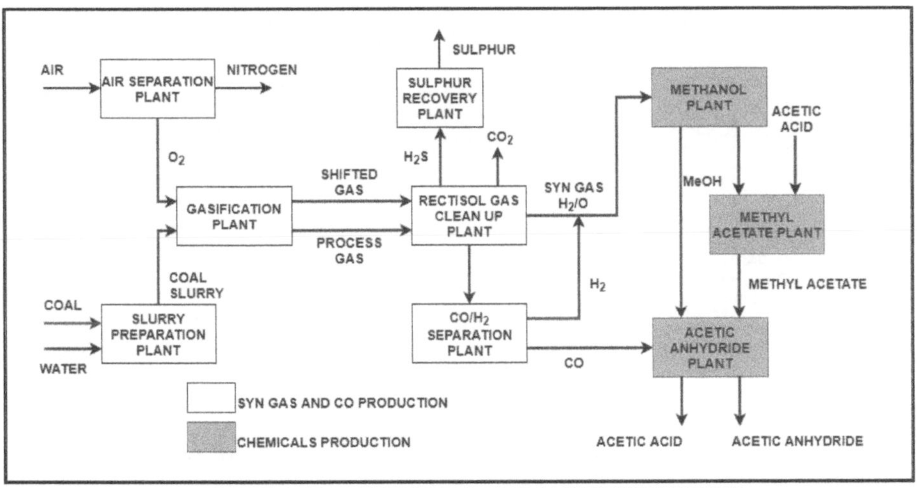

FIGURE-2 COAL CHEMICALS - THE GASIFICATION ROUTE

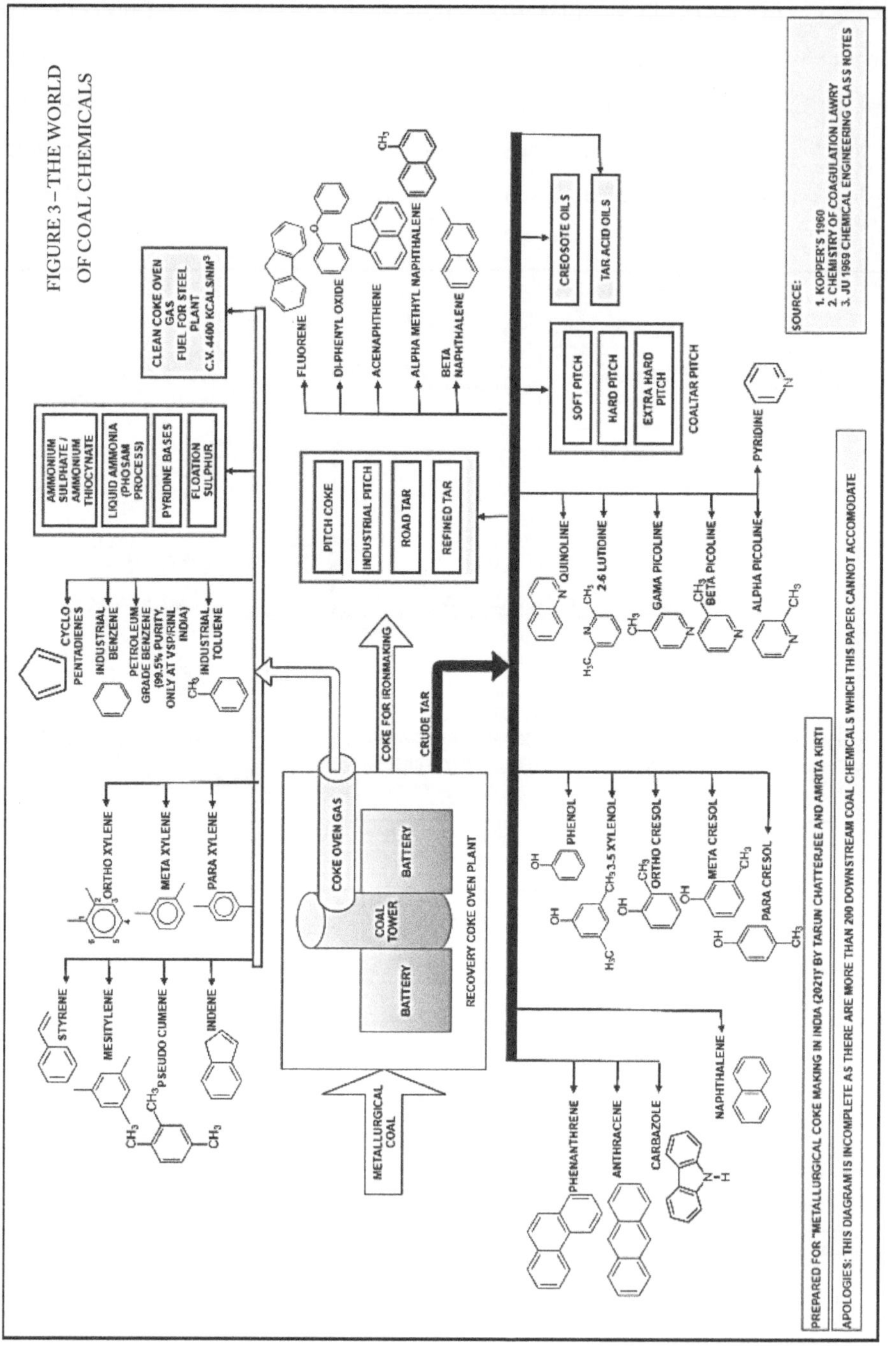

FIGURE 3 – THE WORLD OF COAL CHEMICALS

Coal gas began to be widely used for illumination during the Industrial Revolution in England. Both these production processes yield coal tar, from which a range of compounds essential for establishing an organic chemical industry can be isolated. For hundreds of years, coal remained the only source of aromatic chemicals that were essential for producing a range of dyes and solvents.

Coke is traditionally produced by heating coal to temperatures above 1000°C in an anoxic environment. This creates opportunities for collecting volatile components and leachates of the production process to use as feedstock to extract a range of value-added products. Typical by-products of the coal coking process are ammonium sulphate, coal tar, coke gas, crude benzole and sulphur. Crude benzole and coal tar are among the most significant raw materials for coal chemical plants.

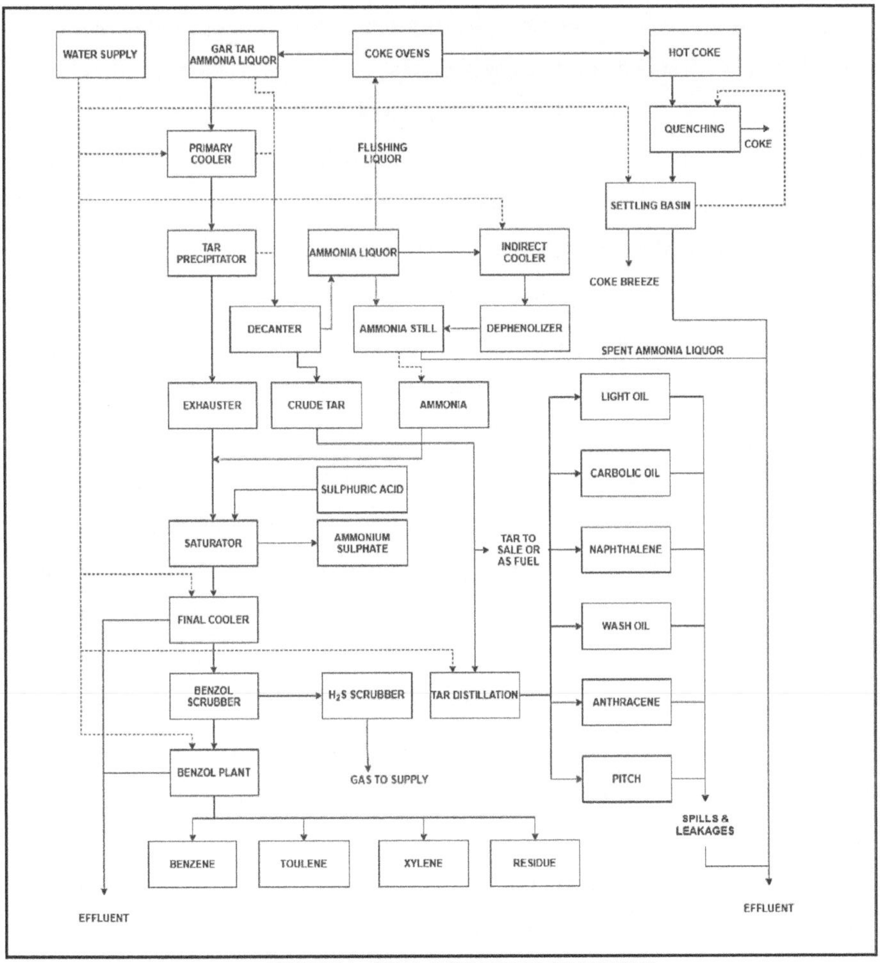

FIGURE 4 - COAL CHEMICALS RECOVERY

Owing to their aromatic composition, compounds derived from the coking of coal have molecular structures that are not easily found in other sources. More than 200 aromatic hydrocarbons can be commercially recovered as value-added products during the destructive distillation of coal. The typical production process flow and the wide range of industrially valuable chemicals synthesized from the coking of coal are presented in Figure-2, Figure -3 and Figure -4. Typical usages of some of these by-products are presented in Table-1.

In colonial India, the coal chemicals industry was essentially driven by British industrial demands, including those for the war efforts of World War II. Jamshedji Tata identified the demand for toluene during the War and Tata Steel's toluene recovery plant was the first in India to extract benzole from coke oven gas.

However, in post-independence India, the significance and economic value of coal chemicals largely remained neglected by planners, technologists and plant operators. The government put together plans for massive industrialization in the country. Three new steel plants were commissioned - at Bhilai using Soviet technology, at Rourkela using German technology and at Durgapur with British technology. The new steel plants had metallurgical coke-making and chemical recovery units along with sulphuric acid plants making ammonium sulphate fertilizer, as the country needed locally produced fertilizers to drive agricultural growth. The by-products plant designed and supplied by the Germans was technologically superior and included a Pitch Aeration Reactor to make the extra-hard pitch. This pitch fetched a premium price due to its demand in the aluminium and graphite electrode manufacturing industries.

The fertilizer complex set up at Rourkela began producing newer fertilizers such as calcium ammonium nitrate and ammonium phosphate. The plants at Bhilai and Durgapur struggled economically in manufacturing government-subsidized ammonium sulphate, incurring excessive repair and maintenance expenses at the corrosive saturator units and sulphuric acid plants. The Bokaro steel plant established in the early seventies with Soviet collaboration continued the practice of producing ammonium sulphate without tar distillation or benzole distillation. The RINL Vishakhapatnam steel plant set up with Russian collaboration in the late eighties replicated the production principles of Bhilai and Bokaro. However, for the first time in India, this plant implemented HRED (Hydrogen Refining and Extractive Distillation) technology under license from M/s Krupp-Koppers of Germany to manufacture caprolactam-grade benzene of 99.5%

purity. At that time, there were only two other plants in the world capable of producing benzene of such purity. In the late nineties, a joint venture between MMTC and the Government of Orissa created the NINL plant at Duburi, with MECON as the technology consultant. But even here the plant was designed to take the age-old ammonium sulphate route, without tar and benzole distillation capabilities.

TABLE 1: ILLUSTRATIVE DOWNSTREAM APPLICATIONS OF METALLURGICAL COKE, COKE OVEN GASES & COAL CHEMICALS *(Source: 100 Years of Metallurgical Coke, 2018, Tarun Chatterjee)*

PRODUCT	APPLICATION
Coke Of Various Fractions	Foundries, iron making, ferroalloys, lead/zinc smelters, soda ash etc.
Coke Oven Gas / Town Gas	Common industrial fuel, secondary fuel in blast furnaces, power plants
Ammonia (As Ammonium Sulphate)	Fertilizer
Anthracene Oil	Industrial fuel, carbon black
Benzene	Caprolactam, benzene hexachloride, DDT, chloro-benzene, malic anhydride etc.
Creosote Oil	Wood preservative, industrial fuel, carbon black.
Crude Coal Tar/Pitch	The aluminium industry, graphite electrode, carbon paste, secondary fuel in blast furnaces, power plants, furnace fuel
Naphthalene	Beta-naphthol/insecticide, dye intermediates, alpha naphthalene, domestic use.
Phenol	Resins, antiseptics, disinfectants
Toluene	Paints, thinners, explosives, rubber chemicals, dye stuff
Xylene	Printing ink, paints, thinners etc.

2. The Post Liberalisation Scenario in India - Widening Technological Options

2.1 The Private Sector

TATA STEEL: Both the Jamshedpur and Kalinganagar plants have wisely avoided extraction of ammonia from coke oven gas using the obsolete ammonium sulphate route and have instead chosen to incinerate ammonia, releasing nitrogen and hydrogen. None of these plants has invested in tar distillation, benzole recovery or any further distillation. The bulk of their coal tar is consumed internally as secondary fuel in blast furnaces, aimed at reducing the consumption of expensive metallurgical coke. The coal tar is also utilized as a binder-cum-fuel in the pelletizing plant. Any surplus crude tar is sold to external customers. Crude benzole remains in the coke oven gas, enriching its energy value and is widely used in furnaces as a fuel of choice. At Jamshedpur, a recovery plant has been commissioned to extract sulphur from hydrogen sulphide available in the coke oven gas.

The acquired Tata Steel BSL plant at Angul came with an ammonia incineration unit without tar and benzole distillation. Subsequently, an ammonia cracker and a Claus Plant for sulphur recovery were commissioned at the 7.6m tall, stamped charge battery complex.

JSW/JSPL/JSSL: Their plants located in Karnataka, Maharashtra and Orissa have adopted simple configurations and layouts with minimum investments in coal chemical complexes. None of the units has an ammonium sulphate plant or tar distillation plant. The JSPL unit at Angul has a hydrogen sulphide scrubbing unit that recovers sulphur from coke oven gas.

The Claus Process: It enables the cracking of NH_3, H_2S and HCN vapours scrubbed from coke oven gas. NH_3 and HCN vapours are cracked to produce N_2 and H_2 which are then fed back to the coke oven gas header before PGC. H_2S is cracked in a Claus Reactor to produce elemental sulphur. This process has replaced the traditional ammonium sulphate plant, ammonia incinerator and sulphur recovery by the wet process. The process produces HP and LP steam as by-products.

Under the top pressure of the de-acidifier and ammonia stripping system, an H_2S and NH_3 vapour mixture is fed to the burner system in the catalytic convertor. Air for combustion is supplied to the burners by air blowers. The hot gases enter the catalytic convertor bed resulting in the cracking of NH_3 vapour. Downstream, in the catalytic reactor, the vapour mix enters the Claus Reactor where a series of reactions take place involving H_2S and SO_2 to produce elemental sulphur and water. The hot liquid sulphur is processed to produce solid sulphur.

2.2 The Public Sector

The SAIL / RINL / NINL plants commissioned before 2000 did not have access to the know-how of extracting ammonia from coke oven gas as the incineration or cracking technology was not available in India and MECON did not take much interest in promoting the new technology. Obsolete technology continued to be implemented even during expansion projects RINL and other SAIL plants.

A naphthalene crystallization plant to produce naphthalene flakes was later installed at the Durgapur Steel Plant. An ammonium sulphate plant to recover ammonia and a Claus Plant for sulphur recovery was also commissioned at the new 7.0-metre-tall coke oven battery at the IISCO Steel Plant, Burnpur. MECON introduced an ammonia incinerator at the NMDC integrated steel plant in Chhattisgarh, but the chemical recovery plant installed here was too elaborate. This led to investments that may have been avoided by adopting a rationalised technical approach.

3. Case Study

The Coal Chemical Plant at Vishakhapatnam (RINL)

A case study outline is presented as an Appendix with this article, referring to technological, production and market data from the coal chemical complex at Vishakhapatnam Steel (RINL). As may be observed, a host of downstream chemical industries across India are engaged in value-added production using primary coal chemicals sourced from RINL.

4. Environmental Challenges of Coal Chemicals

In a paradox of nature, the elements hydrogen (H), carbon (C), nitrogen (N) and oxygen (O) are individually friendly to humans but may become lethal when they combine to form compounds such as hydrogen cyanide or thiocyanates. Raw coal contains all these elements and also sulphur, in the form of numerous complex compounds. Coke and coal gas obtained during the coking of coal, therefore, contains many highly toxic and carcinogenic compounds that pose serious health hazards. Consequently, the design and deployment of coal chemical production units have to address and overcome substantial challenges concerning environmental contamination and pollution.

TABLE 2 - QUALITY OF EFFLUENT AT INLET TO EFFLUENT TREATMENT PLANT

PARAMETER	TYPICAL CONCENTRATION
pH	7.5 - 10.0
Temperature	60 - 80 °C
Oil & Grease	100 - 300 ppm
Total Ammonia	500 - 1000 ppm
Free Ammonia	300 - 600 ppm
Fixed Ammonia	200 - 400 ppm
Cyanide	10 - 100 ppm
Phenol	500 - 1200 ppm

Liquid effluents and gaseous emissions from coke oven plants are severely contaminated with lethal pollutants. The contamination of effluents generated by coking plants may be represented in Table 2.

These effluents and emissions have to be effectively treated prior to discharge to reduce contamination to desirable limits, in order to prevent acute pollution of water bodies, the air and lands adjacent to the plants. The current effluent and emission standards defined by regulatory authorities for coke oven plants are as follows in Table3.

TABLE 3 - REGULATORY STANDARDS OF EFFLUENT QUALITY

PARAMETER	MAXIMUM CONCENTRATION
pH	6.0 - 8.5
Total Suspended Solids	100 mg/l
Biochemical Oxygen Demand (BOD) (3 days at 27°C)	30 mg/l
Chemical Oxygen Demand (COD)	250 mg/l
Oil & Grease	10 mg/l
Ammoniacal Nitrogen (as N)	50 mg/l
Cyanide (as CN)	0.2 mg/l
Phenolic Compounds (as C6H5OH)	1 mg/l
Source: GoI, MoEF Notification Dated 31 Mar 2012	

TABLE 4 - REGULATORY STANDARDS OF EMISSION QUALITY (*Source: GoI, MoEF Notification Dated 31 Mar 2012)*

PARAMETER	MAXIMUM CONCENTRATION		
	NEW BATTERIES	REBUILT BATTERIES	EXISTING BATTERIES
Fugitive Visible Emissions			
Leakage From Doors (PLD)	5	10	10
Leakage From Charging Lids (PLL)	1	1	1
Leakage From AP Covers (PLO)	4	4	4
Charging Emission (HPLA)	16	50	75
Stack Emissions			
SO_2 (mg/Nm^3)	800	800	800
NOX (mg/Nm^3)	500	500	500
Particulate Matter (mg/Nm^3)	50	50	50
Particulate Matter During Stamp Charging (mg/Nm^3)	25	25	25
S In Coke Oven Gas Used for Heating (mg/Nm^3)	800	--	--
Fugitive Emissions (Benzo(a)pyrene			
At Battery Area (µg/m^3)	5	5	5
At Other Units of Coke Oven Plant (µg/m^3)	2	2	2
PLD - Percent Leaking Doors PLL - Percent Leaking Lids			
PLO - Percent Leaking Off-Takes HPLA - High-Pressure Liquor Aspiration			

5. Coal Chemicals - Plant Safety and Risk Assessment

In the chemical process industry, as with many other hazardous production facilities, plant safety is paramount. Adequate knowledge of safety procedures, appropriate training of all staff involved in production processes and implementation of safe practices at all times are essential to prevent industrial accidents that may injure personnel and damage plant & machinery. Periodic risk assessments and regular safety audits prepare and aid production plant operators and managers in developing safety protocols in order to minimize damages to staff, process equipment and the environment. Assessment, planning and review of safety protocols have to clearly focus on:

1. Accident Analysis.
2. Classification of plant hazards and their control.
3. Deployment of personal protective equipment.
4. Human error analysis.
5. Onsite and off-site emergency management.
6. Plant safety and safety regulations.
7. Principles and procedures of safety audits.
8. Risk assessment techniques (HAZOP, HAZON, Fault Analysis, Consequence Analysis).
9. Risk management principles.
10. Safe storage and transportation of chemicals, including hazardous ones.

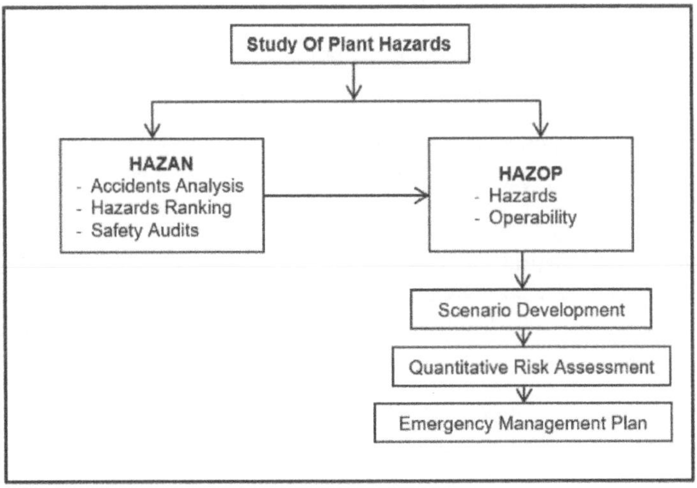

FIGURE-5

A hazard and operability study (or HAZOP) is a structured, systematic and critical evaluation of a complex existing process or a planned new process in order to assess the risk potential of malfunction or operational breakdown of individual equipment and the resultant effects it may have on the entire facility, including workers at the facility. A HAZOP is undertaken primarily to identify dangers and may be carried out several times during the lifetime of a plant or process. The timing of a HAZOP is usually identified by the objective of the study, which in turn determines the benefits that may be derived.

Hazard Analysis (HAZAN) is a technique that focuses on jobs and tasks to identify hazards before they can occur. The analysis typically takes into account the relationships between the staff, the tasks to be carried out, the tools available and the surrounding environment. Once uncontrolled hazards are identified, action may be initiated to either eliminate the risks or reduce them to an acceptable level.

Adequate and appropriate training of staff is an important approach in reducing risks. Such training programs have to focus on:

1. Basic safety and use of personal protective equipment.
2. Chemical process safety.
3. Electrical safety.
4. Fire and explosion hazards at the workplace.
5. Handling toxic materials.
6. Hazard evaluation techniques.
7. Hazard identification and risk assessment.
8. Review of industrial accidents.
9. Safety protocols and shutdown processes.
10. Working safely in confined spaces.

6. Concerns & Conclusions

The discovery of large underground petroleum reserves in the Middle East and Arabia after World War II led to the commissioning of numerous petroleum production facilities in the region. The widespread availability of fuel oils drove the vicious cycle of production and consumption in a world hungry for energy. Massive petroleum production generated funds for the development of petrochemical industries, making crude oil the raw material of choice to produce industrially significant aliphatic and aromatic compounds that were traditionally extracted from coal. This new manufacturing trend relegated the previously dominant coal-chemical industry to a secondary producer.

Owing to many positive attributes, coal may be considered the favoured source for new engineering materials including environment-friendly uses of graphite, speciality chemicals and aromatics. A recent synthesis of complex advanced polymers - especially those incorporating poly-aromatic units in their main chains - has created new opportunities for developing value-added products either from coal or from tars obtained by the carbonization of coal during coke production. Coal-sourced chemicals, therefore, become a viable option for countries with abundant and cheap sources of coal.

Conflict and political instability in the Middle East in tandem with the tussle for production dominance within and beyond OPEC member-states, have fuelled volatility in crude prices and steadily escalated the costs of petroleum products. It has been calculated that coal-sourced chemicals are economically profitable when crude oil prices soar above US$50 per barrel but become uneconomic when crude slides below US$35 per barrel.

By currently estimated rates of consumption, global coal reserves are expected to last significantly longer than those of crude and natural gas. While the coal chemical industry is dependent on the economics of scale and deployment is capital intensive, technological breakthroughs may drive fundamental change. An estimated 45 large coal-to-chemicals production units are currently operational worldwide, with more than half of them in China. A similar number of new production units are either planned or under construction, with a significant number of proof-of-concept plants of different sizes.

- Owing to the convenience of use and economic considerations, coal tar is often burnt as a secondary fuel in blast furnaces in many of our steel plants. It is also used as a primary fuel in lime calcining plants. Coal tar burning releases many hazardous chemical compounds. We must deploy technologies to control and prevent the release of poly-aromatic hydrocarbons (PAHs) to the air and water.
- Scrubbing ammonia from coke oven gases using the ammonium sulphate route should be phased out. There is no longer much economic value in producing ammonium sulphate and the process is maintenance intensive and carries safety and health risks.
- Towards the end of the 1990s, RINL made a half-hearted attempt at forming

strategic business units to look after the marketing of by-products. Economics and business models are ever-changing. The focus for steel plants being steel production, there have to be specific business models for coal chemicals, preferably involving private participation.

- Coal chemical departments in our integrated steel plants do not attract adequate attention from top management, both at the plant and corporate levels. For example, organizations like POSCO (South Korea) & China Steel (Taiwan) have recognized the commercial potential and have formed separate entities like POSCO Chemical and China Chemical & Pharmaceutical Co. to exclusively produce coal chemicals.
- Unlike tar obtained from petroleum, coal tar is very complex in chemical composition and contains high molecular weight complex resins, particularly 1/3rd iso-cyclic aromatics and 2/3rd hetero-cyclic poly-nuclear resins (PAH) with high C/H ratios. The use of coal tar as fuel by customers beyond the steel plants themselves has to be strictly restricted. A detailed study by state Pollution Control Boards and subject matter experts is necessary given the serious health impacts of effluents on humans, animals and plants.

The full potential and market value of chemicals that may be synthesized from effluents and by-products of coking coal batteries at our steel plants has not yet been adequately capitalised. As noted above, in addition to the loss of economic potential, the burning of crude tar produces effluents containing poly-aromatic hydrocarbons that are toxic air and water pollutants. There is scope for our country to emulate technology leaders such as China, Germany, S Korea and the US that have established aggregated secondary chemical processing plants for producing coal chemicals on a commercially viable scale.

While large deposits of coal are available in India, the high ash content of our coal (nearly 30% on average) makes it commercially unsuitable for industrial production processes. The NITI Aayog has set up a technical committee to evaluate and advise on a technical roadmap to improve the usability of indigenous coal reserves, including the development of surface coal gasification processes suited to Indian conditions. New disruptive technologies may also address current environmental concerns of hazardous residues and gases, including greenhouse gases that are generated during coal-chemical production processes.

With this perspective, it will be prudent for the government and the private sector to commit investments and to support R&D for production processes and technologies to make the synthesis of coal chemicals economically attractive in India in the years ahead.

7. Appendix

Case Study: The Coal Chemical Plant at Vishakhapatnam Steel (RINL)

7.1 RINL - Coal Chemicals Produced

Public Data Source (Some Specs Vary): https://www.vizagsteel.com/code/products/byprod.pdf

AMMONIUM SULPHATE	NITROGEN (BY WEIGHT)	MOISTURE (BY WEIGHT)	FREE ACIDITY (BY WEIGHT)	PACKING
	20.60% (min)	1.00% (max)	0.04% (max)	50kg. Bag
Earlier used as a fertilizer due to its Nitrogen content. Presently used as the Nitrogen-bearing component in mixed fertilizers.				

HOT PRESSED NAPHTHALENE (HP NAPHTHALENE)	COLOUR	CRYSTALLIZATION POINT	ASH (BY WEIGHT)	MOISTURE (BY WEIGHT)	TOLUENE INSOLUBLE
	White or Light Brown	78.50° C	0.20% (max)	0.50% (max)	0.20% (max)
Used as dye intermediate, insecticide, dispersing agent, tanning agent and as a raw material for Beta-Naphthol, refined Naphthalene etc.					

DRAINED NAPHTHALENE OIL (DNO)	SPECIFIC GRAVITY (AT 20°C)	MOISTURE	NAPHTHALENE
	1.02 (max)	4.0% (max)	35% (min)

PHENOL FRACTION	SPECIFIC GRAVITY (AT 20°C)	PHENOL	NAPHTHALENE	MOISTURE
	0.99 – 1.02	20 – 30%	30% (max)	4.00% (max)
Used as Naphthalene-bearing feedstock. Naphthalene and Coal Tar Oils are separated for their specific end-uses.				

COAL TAR WASH OIL	SPECIFIC GRAVITY (AT 38°C)	MOISTURE	DISTILLATION UP TO 230°C	DISTILLATION 230°C–300°C	DISTILLATION ABOVE 300°C	NAPHTHALENE
	1.04 – 1.08	4.00% (max)	10.00% (max)	85.00% (max)	5.00% (max)	15.00% (max)
Used as a preservative for wooden sleepers, as feedstock for Carbon Black and as a disinfectant.						

ANTHRACENE OIL	SPECIFIC GRAVITY (AT 20°C)	MOISTURE	DISTILLATION UP TO 300°C
	1.12 (max)	**3.00% (max)**	**20**
Used as a feedstock for Carbon Black, as it gives better yields and quality of the finished product.			

COAL TAR FUEL (PITCH CREOSOTE MIXTURE)	SPECIFIC GRAVITY	MOISTURE	VISCOSITY (AT 93°C)
	1.15 – 1.27	4.00% (max)	2 – 3.5 (E°)
Used as fuel in various furnaces.			

HARD PITCH	SOFTENING POINT (R&B)	COKING VALUE (MIN CONRADSON)	TOLUENE INSOLUBLE (MIN)	QUINOLINE INSOLUBLE (MAX)	ASH (MAX)
	90° – 115°C	48 – 51	25 – 28%	12%	0.3 – 0.5%
Used for making electrodes in the Aluminium industry, ladle repair in blast furnaces and for manufacturing graphite.					

CAPROLACTUM-GRADE BENZENE (CG BENZENE)	SPECIFIC GRAVITY (AT 15.6°C)	TOTAL SULPHUR	CRYSTALLIZATION POINT	NON-AROMATICS	PURITY (BY GLC METHOD)	THIOPHENE CONTENT
	0.879 – 0.886	5 ppm (max)	5.3°C (min)	500 ppm (max)	99.92% (min)	10 ppm max.
Used as the raw material for various drugs, dyes, synthetic rubber and Styrene. Also used for manufacturing pesticides, Monochlorobenzene, Caprolactam, Phenol, DDT, LAB, Nitro Benzene, Aniline, Maletic Anhydride, Cumene, Cyclohexanone, Nylon-6, resin intermediates and nylon intermediates.						

TOLUENE (NITRATION GRADE)	DISTILLATION RANGE (1-96%)	SP. GRAVITY (AT 15°C)	RESIDUE ON EVAPORATION	H_2S & MERCAPTANES	PURITY (GLC METHOD)
	0.6°C (Including 110.6°C)	0.870 – 0.874	5 mg/100ml (max)	Negative	99.2% (min)

TOLUENE (INDUSTRIAL GRADE)	SPECIFIC GRAVITY (AT 15°C)	DISTILLATION UP TO 105°C	DISTILLATION UP TO 120°C	RESIDUE ON EVAPORATION	H_2S & MERCAPTANES	PURITY (GLC METHOD)
	0.860 – 0.875	5ml of 100ml	90ml of 100ml	10mg/100ml (max)	Negative	92.0% (min)
Used as the raw material for some solvents and artificial sweeteners. Also used in the paint industry, rubber industry, in rubber chemicals, printing ink, Benzoic Acid, Sodium Benzoates, synthetic fibres, Benzyl Chloride, thinners and varnishes.						

LIGHT SOLVENT OIL (LSO)	SPECIFIC GRAVITY (AT 15°C)	DISTILLATION UP TO 125°C	DISTILLATION UP TO 170°C	RESIDUE ON EVAPORATION	H_2S & MERCAPTANS
	0.850 (min)	5ml of 100ml (max)	90ml of 100ml (max)	10mg/100ml	No positive rex.

SOL–110 (NON-AROMATICS)	SPECIFIC GRAVITY (AT 15°C)	DISTILLATION RANGE (1-95%)
	0.84 (min)	55 – 110°C
LSO & Sol-110 are used as solvents, in manufacturing printing inks, as a thinner, as a starting material for different dyes and manufacturing resins for rubber compounding.		

7.2 RINL - Saleable Coal Chemical Products

PROJECTED QUANTITIES OF SALEABLE COAL CHEMICAL PRODUCTS *(based on rated capacity):*

SL#	PRODUCT	PRODUCTION (M TONS)	IDENTIFICATION
1	Ammonium Sulphate	54,000	
2	HP Naphthalene	2,250	Tar Products
3	Drained Naphthalene Oil	1,050	
4	Phenol Fraction	1,200	
5	Medium Hard Pitch	28,000	
6	Benzene	10,000	Benzole Products
7	Toluene	2,000	
8	Non-Aromatics	500	
9	Light solvent Oil	1,200	
10	Still Bottom Oil	50	

However, actual production quantities depend on the quality of the raw coke oven gas (i.e., the content of ammonia, crude tar and crude benzole in the gas mixture).

7.3 RINL - Production Data (2017-18):

SL#	PRODUCT	PRODUCTION (M TONS)		IDENTIFICATION
		Projected	Target	
1	Ammonium Sulphate	47,300	51,000	
2	HP Naphthalene	2,200	2,250	Tar Products
3	Drained Naphthalene Oil	1,500	1,050	
4	Phenol Fraction	1,000	1,200	
5	Medium Hard Pitch	26,000	26,425	
6	Benzene	11,500	10,660	Benzole Products
7	Toluene	2,850	2,040	
8	Non-Aromatics	1,050	670	
9	Light solvent Oil	1,100	1,250	
10	Still Bottom Oil	20	30	

7.4 RINL - Illustrative Downstream Customer Base for Coal Chemicals

As may be observed from the data presented below, the raw feedstock is procured by organizations located near the RINL plant and supplied to processors located all over the country that produce secondary and tertiary value-added coal-chemical products.

PRODUCT	MAJOR CUSTOMER LOCATIONS
AMMONIUM SULPHATE	ANAKAPALLI
	CHODAVARAM
	ELURU
	HYDERABAD
	KOYYALAGUDEM
	NIDADAVOLE
	PEDDAPURAM
	RAJANAGARAM
	RAMANAKKAPETA
	THADEPALLIGUDEM
	THIRUVALLUR
	VEERISETTIGUDEM
CRUDE COAL TAR	VISHAKHAPATNAM
MEDIUM HARD & LIQUID PITCH	NELLORE
	VISHAKHAPATNAM

PRODUCT	PRODUCTION LOCATION
BENZENE	AHMEDABAD
	CHENNAI
	HYDERABAD
	MUMBAI
	VADODARA
DRAINED NAPHTHALENE OIL	BHILAI
	VIZIANAGARAM
	VISAKHAPATNAM
HARD-PRESSED NAPHTHALENE	BHILAI
	KAKKALUR
	MUMBAI
	PUNE
	VAPI
	VISHAKHAPATNAM
PHENOL FRACTION	DURG
	MUMBAI
	VISAKHAPATNAM

About the Author

Jayanta Sen has spent over 25-years in the Information Technology industry with experience involving Business & Operations Management, Sales & Marketing, Customer Support Services and Systems Development. Educated in Chemistry, Jayanta takes a keen interest in renewable technologies and environmental conservation.

XI. BUSINESS & LOGISTICS

XI.1. Supply Chain Logistics: Coking Coal Trade

Wing Cdr Amit Goswami

1. Introduction

After firewood, coal is the oldest energy source from ancient times, and it has still retained its importance worldwide. India is one of the largest producers of steel in the world. Fortunately, India has the fourth-largest coal reserve in the world. As of March 31, 2019, India had 326.49 billion metric tons (359.89 billion short tons) of coal. However, India's reserve of high-quality coal, which is suitable for making coking coal is limited. India has a reserve of 19082 million tons of coal. Out of this limited reserve, only 4649 million tons of Prime Coking coal is available. The greater ash content of Indian coal is a point of concern for the steel industry. At best the steelmakers can use Indian coal mixed in a small proportion with imported coal. Therefore, Indian Steel Manufacturers depend a lot on importing coal from foreign countries, such as Australia, Canada, the USA, and some other countries for making coking coal. Australia is the top supplier of coal to India. India imported a record high of 5.36mn tons of coking coal import from Australia in Dec 2020. The trend continues. India imported 5.62 million tons and 5.47 million tons of coking coal from Australia in Jan and March 21, respectively.

Coal is imported from other countries by sea, loaded in the holds of large cargo vessels. Therefore, this discussion will be restricted to Supply Chain Logistics in Coal Trade from Australia by sea.

2. Supply Chain Management

In commerce, supply chain management (SCM) denotes the management of the flow of goods and services between businesses and locations. It includes the movement and storage of raw materials, work-in-process inventory, and finished goods, as well as the end-to-end order fulfilment from the point of origin to the point of consumption.

3. Background

Coal is mined mostly in the states of the East Coast of **Australia**. The largest black coal resources occur mainly in Queensland and New South Wales. [About 70% of coal mined in Australia is exported, mostly to eastern Asia. Australian coal is of very high quality. Australia produces Hard coal, soft coal and PCI (Pulverized Coal). Australian coal is found to be more suitable than Indian coal for making coking coal.

4. Dispatch Activities at the Supplier's End

Indian buyers negotiate and place orders on Australian coal suppliers through normal channels. They open an LC in a suitable Bank. The buyer also negotiates with a shipping company for the 'voyage charter' (Chartering a vessel for a single point to point voyage) of a cargo ship to carry coal to India. Panamax (> 70,000 tons capacity) and Superpanamax (> 90,000 tons capacity) categories of cargo vessels are usually hired for carrying coal from Australia to India. Ships carry Goods by sea under a contract of carriage between the ship-owner and the buyer. The ship-owner may employ a loading broker to control the allocation of space, advertise the service, make the loading arrangements, and prepare documents on the ship owner's behalf. Documents that are required before loading cargo include:

1. NOR.
2. An Export Permit.
3. A certificate of origin issued by the Chamber of Commerce (A certificate of quality).

5. Issuance of NOR

When the vessel arrives at the load-port, the master of a vessel issues a NOR (Notice of Readiness) to the Shipper and the Port Authority. NOR clearly states that the vessel has arrived at the load port and is ready in all respects for loading the specified cargo. In this context, there is an interesting clause called '**Laycan**'. The word is a hybrid between 'Lay time' and 'Cancellation Date' in a charter party agreement. This clause defines the time window in which the seller is obliged to accept the vessel in the loading port.

6. The Action Begins at The Seller's End

On receipt of NOR, the dispatch activities begin at the Shipper's end. The flow chart placed in Appendix A to this paper shows the step-by-step activities.

A few things here need special mention:

1. A sample is collected while loading on the ship just before loading a ship. It is

sent to a laboratory for testing and then retained at the Shipper's end with the test report for the next six months. It is kept as a reference point in the event of any future dispute. It is called 'Umpire Sample'. This sample is retained along with its laboratory test report for six months at the shipper's end, as a reference point in case of a dispute.

2. The entire cargo handling operation is mechanised at Australia's end. The sample collection is also done mechanically.
3. Loading of coal directly from wagons to ship is automated.

7. Post Loading Activities

1. Cargo stowage plan, cargo plan, or hold distribution plan are produced by shore loading staff or the Ship's staff to the Port Authority.
2. The specialist hatch-sealing operatives seal the Ship's hatches after the whole consignment gets loaded on board. They issue a hatch-sealing certificate.
3. Compartments are sealed to prevent unauthorised entry after loading and fumigation.
4. A certificate of fitness to proceed to the sea is obtained from Marine Administration at the loading port state. It certifies that the stowage of the cargo conforms to the appropriate regulations or approved practice.

At this stage, the Ship Owner representatives, the seller's and the buyer's representatives jointly conduct a Draught survey. The draft of a ship is the vertical distance between the surface of the water and the lowest point of a vessel. Draught survey is done twice – before and after loading the vessel. The result of this survey establishes the weight of bulk cargo. Therefore, the net weight of the cargo is the corrected loaded displacement, less the corrected light displacement.

8. Documentation By the Seller

After loading the vessel, the Shipper prepares the following documents for export:

1. Invoice.
2. Shipper's Letter of Instruction for the captain to take care of the cargo during the voyage.
3. Bill of Lading.

9. Details Of Bill of Lading

Out of the above documents, the Bill of Lading is crucial. It contains the following information:

1. Details of the cargo i.e., Name of the shipper and the quantity/weight of the cargo.
2. Details of the ports of loading and discharge, place and date of issue of the bill of lading and number of bills of lading.
3. The Purchase orders or unique reference/invoice numbers helps the Shipper and the consignee to release the goods for pickup or accepted at delivery

A copy of the Bill of Lading is handed over to the Captain of the Ship. Once he puts his signature on it as having received the cargo, he becomes totally responsible for the cargo on board during the entire voyage.

The sailing time from Australia to India is 18 to 20 days.

10. Activities At Buyer's End

At the Importer's / buyer's end, activities for cargo clearance starts as soon as he receives dispatch documents from the Shipper. An importer is allowed two to three free days to clear his goods from the port area on arrival - free of demurrage charges. Therefore, the buyer's agent keeps track of the possible date of arrival of the Ship and keeps several documents ready before the ship's arrival. He tracks the Ship's movement with the help of a satellite-based 'Vessel Monitoring System' (VMS). The 'Automatic Identification System' (AIS) helps track bigger vessels crossing waters worldwide by offering real-time information about the vessels. Currently, several websites provide ship tracking using the automatic identification system to display the real-time location of the ships. The documents which the importer keeps ready are:

As the Ship approaches the discharge port, the Port Authority/buyer gets a NOR. It is a document sent by the ocean freight forwarder, freight carrier, or agent to the consignee or Notifies Party indicating the shipment's arrival date at a specific location (usually the destination). On receipt of the arrival notice, the importer prepares a Bill of Entry. To facilitate, streamline and expedite customs clearance for imports, electronic processing of the Bill of Entry has been introduced. If the goods are cleared through the EDI system, no formal Bill of Entry is filed. It is generated in the computer system. The importer does not submit documents as such for assessment. However, if he does not opt for EDI, he keeps the following documents ready:

1. Bill of Lading or Delivery Order.
2. GATT declaration form duly filled in.
3. Importers/CHA's declaration.
4. Import License wherever necessary.

Simultaneously he submits a requisition for a rake of wagons for dispatching the cargo to the final destination. Placement of wagons precisely at the time of discharge from the Ship is necessary to avoid demurrage.

At this juncture, customs clearance is one of the prime activities. A flow chart in this connection is attached as Appendix B. Currently, for coking coal, the basic customs duty in India is 2.5%. Once the Taxes and Customs duties have been paid, samples of coking coal are drawn and sent to the laboratory. Then the cargo is discharged, loaded on wagons and dispatched to the ultimate consignee.

However, if any qualitative or quantitative discrepancy is revealed the clearing agent makes a report and the buyer takes up with the shipper or raises an insurance claim, as deemed fit.

In conclusion, it may be seen that the process of importing coking coal from abroad is systemised. Therefore, there may not be any room for discrepancies if the steps are followed meticulously.

About the Author

Wing Commander Amit Kumar Goswami is an Air Force Veteran. After serving 26 years in the Logistics Branch of the IAF, he took voluntary retirement in 1996. He made a second successful inning of his life as an author, with three published books to his credit. In this article, Wg Cdr Goswami has contributed an article on the export/import formalities of importing coking coal from Australia by Sea.

11. Appendix A

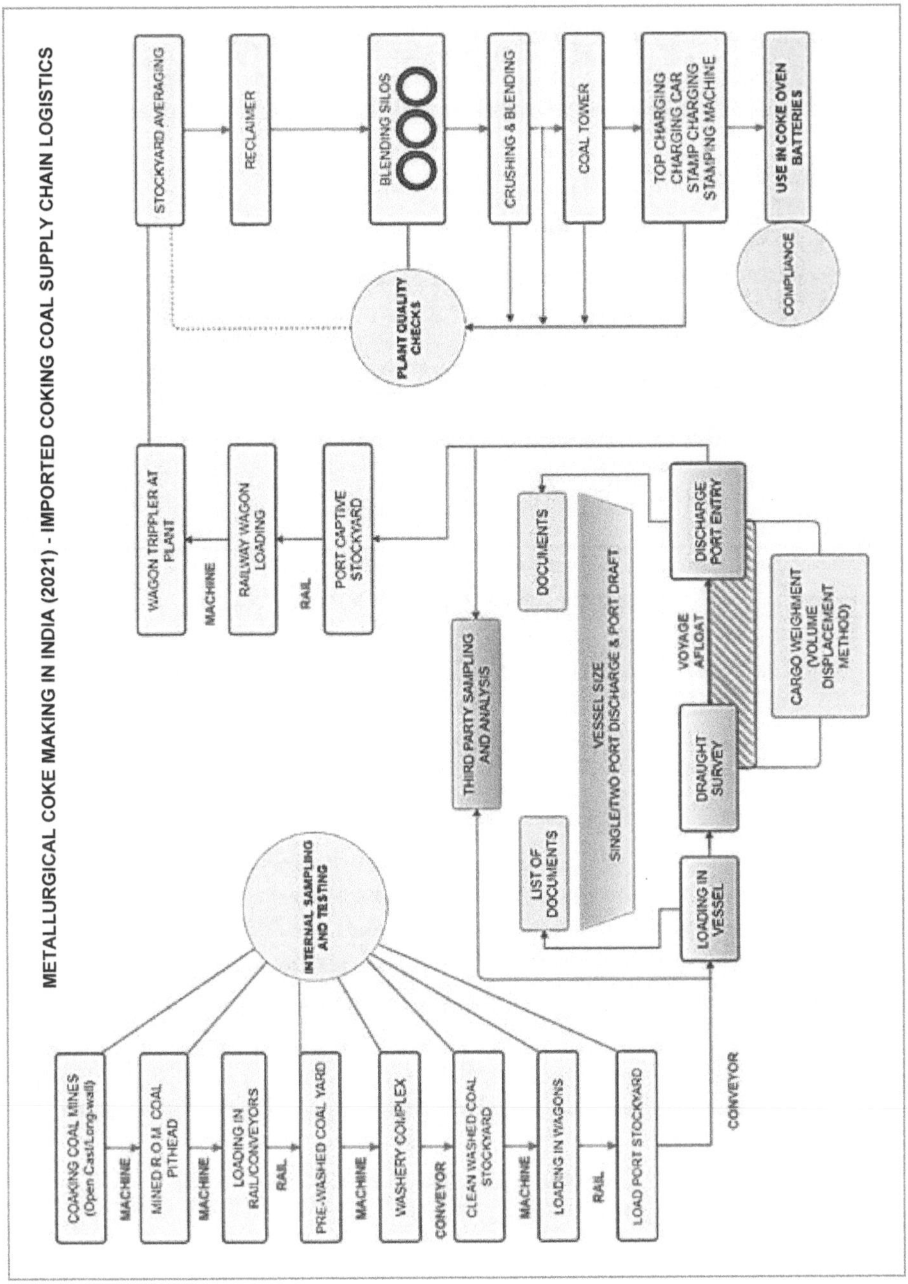

12. Appendix B

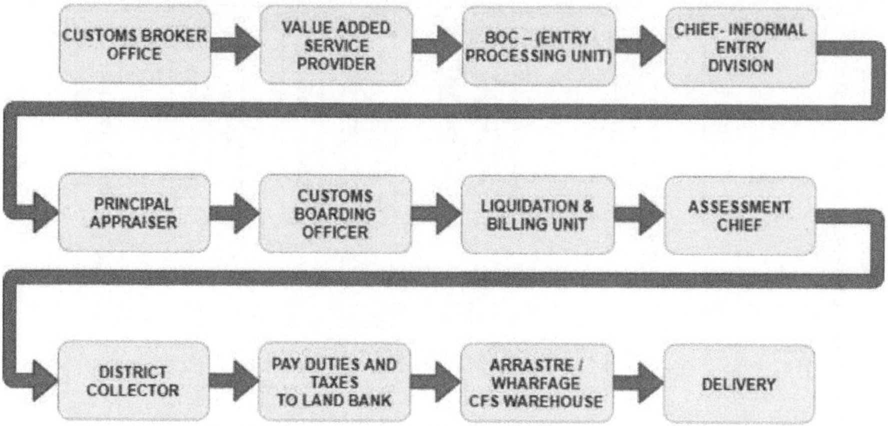

FIGURE 8: CUSTOMS CLEARANCE OF IMPORTED BULK CARGO

13. Bibliography

After loading the vessel, the Shipper prepares the following documents for export:

I am grateful to Mr Sanat Gupta for his valuable guidance while writing this article.

1. http://bulkcarrierguide.com/cargo-documentation.html.
2. https://www.twill.net/knowledge-hub/logistics-know-how/important-shipping-documents/.
3. https://www.twill.net/knowledge-hub/logistics-know-how/important-shipping-documents/.
4. https://www.shippingandfreightresource.com/documentation-involved-in-a-sea-freight-shipment/.
5. https://www.scribd.com/document/209707994/Dry-Bulk-Cargo-Import-and-Export.
6. https://www.shippingsolutions.com/blog/documents-required-for-international-shipping.
7. https://www.prsindia.org/content/cargo-handling-major-ports-3.
8. https://www.globalsecurity.org/military/library/policy/army/fm/55-17/ch8.htm.
9. https://carrerstudies.blogspot.com/2020/07/import-customs-clearance-procedures.html.
10. http://timesofindia.indiatimes.com/t10-may-27-2019/why-does-india-import-coal-despite-enough-reserves/articleshow/69507484.cms.
11. https://www.iea.org/reports/coal-2020/trade.
12. https://www.telegraphindia.com/jharkhand/dhanbads-coking-coal-industry-upbeat-after-australia-import/cid/1812844.
13. http://www.bcmcoal.com/blog/coal-supply-chain-management/.
14. https://en.wikipedia.org/wiki/Supply_chain_management.
15. http://coal.nic.in/major-statistics/coal-reserves.

** A glossary of Shipping Terms has been attached as an Annexure to this book.

XI.2. Risk Management in The International Trade of Coal, Coke and Steel

P P Mohharikar

1. Introduction

There are several risks associated with the international trade of any commodity. These risks include the risks associated with the variations in the rate of exchange of foreign currency, the risk of not getting paid for the items exported, risks associated with the variations in the shipping rates and demurrage and dispatch settlement. There are country-specific risks also. In this article I have tried to put forth how such risks were managed in the organisations I worked for.

2. Foreign Exchange Risk

Most of the Integrated Steel Plants in India import Coking Coal, Limestone and some Ferroalloys. The effort is made to export the steel to the extent to which foreign exchange is required for items that are needed to be imported. The foreign exchange risks are also managed by hedging using appropriate tools.

3. Credit Risks

It is always better to export against advance payment. If it is not possible the buyer is requested to establish a letter of credit which may be confirmed by the local bank.

In one case during 1995 Chinese Ambassador in India spoke to the Chairman of the organisation I was working for and requested that we should not insist on confirming Bank of China's LC by SBI as the Bank of China was 60th in the world while SBI was 360th in the ranking. Our organisation approved as a trial case to go ahead with the shipment of 20,000 mts of rebars without getting the Bank of China's LC confirmed. We had sold the material on an FOB basis. The Bank of China did not pay for 45 days and raised a discrepancy in the documents that the date of Certificate of Origin was later than the BL date. This discrepancy was not tenable as per UCPDC500 the governing law for LC transactions worldwide which does not prohibit the date of the certificate of Origin to be later than the BL date. In fact, in India, we have to submit a copy of BL to the Chamber of commerce to get a Certificate of Origin.

To resolve the issue a two-member team of which I was a member was sent to China. Bank of China suggested that they will organise the meeting with the end customer. When we insisted that the discrepancy raised by them is not tenable and we exported the material on the strength of Bank of China and not on the strength of Chinese customers, the representative of Bank of China started talking to us in the Chinese language which we did not understand. Finally with the help of another influential Chinese customer Mr X who flew from Singapore to Shantou we met the Bank of China official the next day. Mr X had advised us to agree to the meeting with the end customer. During the meeting, the Chinese customer showed us a certificate from SGS which was indicating that the material was not as per the specifications of the order. We told the Chinese buyer that if he does not want this material, we will sell it to Mr X and the customer said that in such case we should reimburse him freight @USD 20 pmt which was paid by him. We agreed to this but mentioned that we will have to take RBI clearance for making the remittance. However, as the payment for the consignment was not made the title of the goods was with us and Mr X will receive the material and sell it. Chinese customer knew that he may not get freight amount from us. He then told us the real story that the prices had come down by 15 USD/mt and he will make a big loss. He requested some face-saving discount and we agreed to 0.38 USD/mt of discount and the case was resolved.

4. Shipping Risks

The shipping rates are very sensitive to the cargo availability and availability of the ships to move them. As such annual budgeting for the shipping bills becomes very difficult for an organisation unless they enter into annual COA (Contracts of Affreightment). PSUs in India have to go through the Ministry of Transport and Shipping, Transchart wing to enter into COA's. During the year 2007 RINL entered into COAs between 44 to 48 USD/mt for shipment of coking coal from Australia to Gangavaram Port Visakhapatnam. However, during the year there was a sharp dip in the freight rates and RINL could get a 150000 DWT Cape vessel at USD 5 pmt for transporting coking coal from Australia to Gangavaram port. It was difficult to pay shipping rates as per COA and survive while shipping companies insisted that we honour the contracts. To resolve the issue, we could negotiate with the shipping companies and they agreed to extend the period of COAs to 2 years instead of one year.

In export contracts, we have seen that some international customers are chartering vessels on back-to-back basis by offering shipping companies the RINL's shipping terms and conditions. This safeguards them from demurrage and dispatch.

5. Country-Specific Risks

Certain countries are politically not stable or in war zone etc. It is always advisable to take suitable insurance cover while trading to these countries. Some countries like Sri Lanka presently are facing the problem of availability of foreign exchange and their LCs are not being discounted in other countries. In some cases, in small lots material can be shipped to these countries by containers.

Success in international trade requires meticulous planning. The specifications of the products required by the customers should be properly understood by the supplier. The timely supply and the quality as per specifications agreed is essential for building a long-lasting relationship.

About the Author

P P Moharikar a Chemical Engineer by qualification, was associated with the greenfield project of RINL-Vizag Steel from the Project approval stage reporting directly to the Managing Directors and co-ordination with the Ministry of Steel. Served as General Manager (Marketing) before leaving the Company in 2015, Presently runs his own company dealing with International Trading operating from Vizag.

XII. FUTURE OF COKE MAKING IN INDIA

XII.1. Future Of Coke Making

Dr Mukesh Kumar

1. Introduction

India is the 2nd largest producer of steel having a production capacity of around 142 million tonnes. The steel demand is increasing continuously and is expected to grow @CAGR 7-8% in the next decade i.e., up to 2030. It is expected that during this period steel production capacity in the country will be increased to around 300 million tonnes with an increase in per capita consumption from the present level of 74 Kg to around 158 Kg. The demand is mainly arising due to the government continuous focus on the infrastructure and construction sector. Thus, India will see large brownfield expansions as well as Greenfield steel plants in time to come. India will be the single largest driver of global steel demand growth over the next 20 years.

The Indian steel industry is structured in between three broad categories based on route wise production viz. Blast Furnace-Basic Oxygen Furnace (BF-BOF,) Electric Arc Furnace (EAF) and Direct Reduced Iron-Induction furnace (DRI-IF). The recent trend is to go large size blast furnaces and hence most of the expansions are being planned considering the BF-BOF route. Although presently around 45% production is through BF-BOF route and 55% is contributed by other sectors, but in 2030 it is expected that BF-BOF route shall contribute to around 65% and only 35% may be contributed by other sectors i.e. EAF/IF route. This is mainly due to the non-availability of adequate scrap in the country.

Coke is the main constituent of the BF-BOF route, and its production is an important part of the integrated iron and steel plants based on blast furnaces. Coke is a reducing agent and a source of thermal energy and provides physical support for the burden in the blast furnace (permeable grid, supporting a column of input materials). Although the BF-BOF route is the most preferred route for large expansions like proposed in India, it has the major challenge of reducing greenhouse gas emission (GHG). The iron and Steel industry is regarded as one of the largest CO2 emission industries globally and contributes around 7-8% of the total emission. The coke ovens and blast furnaces are considered as the main sources of CO2 and thus new designs shall have to be adopted to curb the emissions. Hence, technological interventions are going to play a major role in the future expansions of the steel plants.

Alternatives have been developed and new concepts are being examined for making steel without coke. Current initiatives like sustainable steel making, low carbon steel making etc. all envisage steelmaking without coke. But blast furnaces are likely to continue as main iron-making technology for at least 20 years more and are likely to exist for at least 30 years more, and so will be coke making.

2. Future Trends in Blast Furnace Technology and Coke quality

The majority of the Blast Furnaces in the existing plants are of small size and have volumes varying between 1000 M3 to 2000 M3. However, new plants, as well as expansions, are being executed with modern large size furnaces having capacity ranging between 4000 M3 to 5000 M3 mainly on account of higher productivity, lower operating cost and better environmental performances. In India also, a similar trend is continuing, and all new expansions are being planned with large blast furnaces. A survey undertaken by CRU[4] about future expansion shows that increases in steel demand will drive Indian steel production to higher efficiencies and this will lead to a change in the BF size profile and productivities, both of which will impact the demand for higher quality coke and thus more hard coking coal. Small size furnaces shall be gradually phased out as they reach the end of their functional lives to be replaced by larger, more efficient BFs in the 3,000-4,000+ m3 working volume range as can be seen from Figure-1.

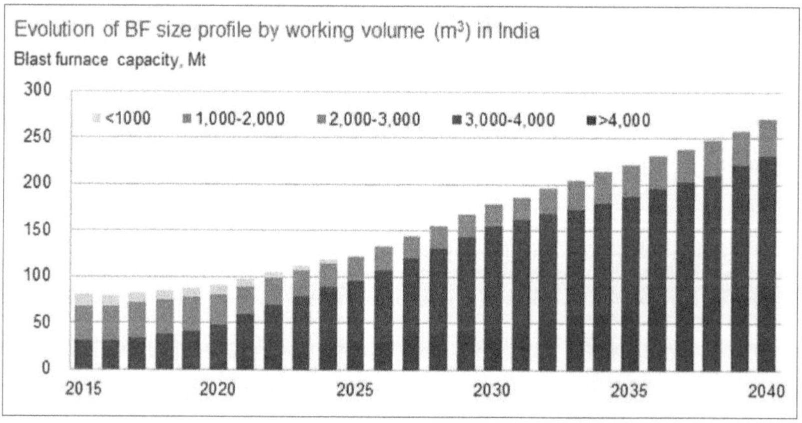

FIGURE-1: FUTURE TRENDS IN BLAST FURNACE PROFILES

[4] *Hard coking coal to fuel steel growth in India by CRU*

Increasing the size of the furnace will increase the weight of the burden and higher loads will be exerted on the bottom of the burden. Under these conditions, the key to productivity is the structure of the burden, which needs to be sufficiently strong to encourage increased and more evenly distributed gas flow, allowing more efficient reduction. In addition, better operating performances can allow lower slag volumes meaning less energy and, therefore, coke is required for melting. Poor quality, low CSR coke breaks down in the BF if subjected to excessive loads and disintegrates, reducing the permeability of the burden, and so coke quality becomes more important as BF size increases. Therefore, the requirement of higher CSR coke is expected to increase in Indian BFs in the medium- and long-term and this may change the technology of coke making from Top Charge to Stamp Charge battery.

3. Metallurgical Coal / Coking Coal

India is having large reserves of thermal & coking coal but due to high ash content, utilization of domestic coking coal is very limited. Although total coking coal resources in India is around 34 billion tons, as can be seen from Table-1, still more than 85% requirement is met by Import. Since the cost of coking coal is the major constituent in the overall cost of steel production and contribute to the tune of 40-45%, efforts are being made to minimize coke consumption by adopting various initiatives like PCI injection, use of coal briquettes, use of biomass, coking technology to produce high coke strength after reaction (CSR), blending with low-cost coal, and recovery of heat in quenching and use of the same for power generation etc.

TABLE 1 - COAL RESOURCES OF INDIA

TYPE OF COAL	PROVED	INDICATED	INFERRED	TOTAL
Total	125909	142506	33149	301564
(A) Coking: -				
-Prime Coking	4614	699	0	5313
-Medium Coking	13303	11867	1879	27049
-Semi-Coking	482	1004	222	1708
Sub-Total Coking	18400	13569	2101	34070
(B) Non-Coking:-	106916	128838	30249	266002
(C) Tertiary	594	99	799	1493

Some of the plants are blending using indigenous coal to the tune of 30-40% but the majority of others are blending to the tune of 10-15% only. Efforts are being made to maximize the use of indigenous coking coal but looking into the continuous increase in steel demand, it is expected that the scenario may not change much, and India has to depend upon the import of hard coking coal as large BFs are going to be dominated in the industry in future. Hence, Coke making technologies are going to play a major role as most of the expansions are being planned with the BF-BOF route only due to the non-availability of any other proven steel making technology in such large sizes.

4. Coke Oven Technology

Coking coals are the coals which when heated in the absence of air(called Pyrolysis Process), first melt, go in the plastic state, swell and solidify to produce a solid coherent mass called coke. When coking coal is heated in absence of air, a series of physical and chemical changes take place with the evolution of gases and vapours, and the solid residue left behind is called coke. The equipment where such reaction takes place is called Coke ovens. Two types of systems are used in the industry for converting coking coal into coke:

1. Top Charging Coke Oven
2. Stamp Charging Coke Oven

5. Top Charging

Top Charging i.e., Conventional Coke making is done in a coke oven battery of ovens sandwiched between heating walls. The carbonisation takes place at a temperature around 1000°-1100° C up to a certain degree of de-volatilization to produce metallurgical coke of desired mechanical and thermochemical properties. During carbonisation, coking coals transform the plastic state at around 350°-400° C swell and then re-solidify at around 500°-550° C to give semi-coke and then coke. In coke ovens, after coal is charged inside the oven from the top, plastic layers are formed adjacent to the heating walls, and with the progress of time, the plastic layers move towards the centre of the oven from either side and ultimately meet each other at the centre. During coke making, two opposite reactions take place, viz. condensation and pyrolysis. The quality and quantity of the plastic layer are of extreme importance, and it determines the inherent strength of the coke matrix, also known as Coke Strength after Reaction (CSR).

6. Stamp-Charging

Stamp charging is a variation of conventional coke-making wherein, at the pre-carbonisation stage, mechanical forces are used to reduce the air component between the coal particles in a fixed form – the stamping box. Thus, a compacted mass with a higher density is pushed into the oven as a block instead of in-situ charging in the top oven and thus having less compaction/density. The process of carbonization is the same in both cases as explained above.

Because of the difference in handling Coal before charging, the following differences are observed:

6.1 Coke Quality and Productivity

Since all the coal is compacted before charging the coal into the coke oven, stamp-charging can increase the charge density by 35%-40% and correspondingly the bulk density of coke produced by Stamp Charging i.e., nearly 30 % higher than achieved in a conventional top charging system. This could have increased the productivity also proportionally; but due to increased heating times required for full carbonisation (due to high compaction movement of the plastic layer becomes slow), the total increase in productivity of stamp charged coke is claimed to be only 10-15% higher than top charging but the coke is of high strength which may help in achieving a reduction in coke consumption in Blast Furnace. Besides, stamping, enhancement in coal charge bulk density can be obtained by following also:

1. adjusting the grain size distribution
2. optimizing moisture content
3. adding oil in the blend
4. partial briquetting

Other methods of enhancing bulk density can be applied to existing top-charged batteries that are already in operation. Coke production using the technology of partial briquetting of coal feed consists of briquetting (compacting) a part of the coal blend (usually 20–30%) and then mixing this part with the remaining part which has not been compacted. The compacting process is usually carried out using hydraulic presses. Increased coal charge bulk density due to the addition of briquettes results in the formation of cokes with lower porosity. In most cases, the mechanical strengths of cokes produced with briquettes were better.

6.2 Blending Of Coal/Use of Low-Quality Domestic Coal

In the top charging system, normally blending with non-coking coal or low-grade coal is avoided and thus only use of Hard Coking Coal (HCC) or Medium Coking Coal is practised. However, in the Stamp Charging system, the same has been successfully achieved by almost all and thus this has helped steel producers, particularly in India to minimize dependency on imported coal and maximizing the use of indigenous coals. It is observed that Tata Steel has achieved a blend of 80% of a local high volatile, high ash, poorly coking coal (West Bokaro) and 20% of HCC. This ability to produce higher CSR coke with lower quality coal means that the overall operating cost for stamp-charged coke is lower than for top-charged coke. Thus, in our context, the implementation of stamp charging can reduce imported HCC requirements by around 15-20%. Some producers have even blending petroleum coke to increase carbon and minimize operating costs e.g., JSW has used petroleum coke as well as non-coking coal up to 25% with Stamp Charging System.

High PCI rates in the blast furnace (~ 200 Kg/t of hot metal), to reduce specific coke consumption (kg/t hot metal), require significant coke quality improvements, particularly coke strength after reaction (CSR) above 62. Coke quality improvements have been achieved by a combination of coal blend optimization and technology improvements.

It will be desirable in the coke-making industry to use low-cost coal blends that produce a high-quality coke and ensure a safe operation of coke-oven batteries. One of the methods of improving the economics of coke production is to introduce cheaper components into coal blends, such as semi-soft coals, non-caking coals or other carbon-bearing materials (coke breeze, charcoal, biomass, plastic wastes, etc.). However, the proportion of these is likely to be limited. In general, an increase in the proportion of the aforementioned materials is always detrimental to coke quality, due to their deleterious effect on fluidity development, providing inferior (compared to hard coking coals) or non-fusible constituents. Collection of plastic waste is the major issue in its use in coal blend.

For optimizing the blend based on the availability of various types of coking coals for a particular coke plant in future, models for prediction of metallurgical coke quality parameters based on maceral composition and properties of coals shall be increasingly employed.

6.3 CO_2 Reduction and Heat Recovery

Presently, CO_2 emission in the Indian Iron & Steel industry is one of the highest and almost 40-50% more than the world average. While upgrading existing coke coven batteries, efforts shall have to be made to adopt coal dry quenching (CDQ) instead of wet quenching so that CO_2 reduction is achieved by utilizing sensible heat of the coke. This shall help in minimizing dependency upon external coal-based power supply as additional power can be generated by recovering heat which shall also reduce water consumption in the coke making. To an estimate, adopting CDQ will help in reducing nearly 100,000 T of CO_2 emission for every million tons of iron production. Besides CO_2 reduction, moisture in coke will also be reduced considerably which shall help in improving the performance of Blast Furnace.

6.4 Recovery Vs Non-Recovery Coke Ovens

Coke ovens are categorised as 'recovery' or 'non-recovery' depending upon the recovery of the by-products generated during the coking process. By-products include substances such as tar, sulphur, coke oven gas, crude benzol and ammonium sulphate, which can be utilized in other industries. Non-recovery, or heat recovery, ovens combust the by-products directly in situ and the waste heat can be recovered. This method provides more flexibility for coal selection and variable blends than conventional slot ovens. However, to fully benefit from the heat contained in the waste gases from a non-recovery oven, it is necessary to build steam boilers and electricity generating set with their own attached CAPEX and OPEX expense.

The flexibility associated with non-recovery coking means that the technology can make use of poor-quality coal. Typically, non-recovery operations can apply cheaper, low-quality coals to reliably produce a mid-CSR product but, through extended heating times or the use of better coal, are also able to produce a higher CSR coke product. These options are, however, more costly and extending heating times lowers the productivity of the operation. Non- recovery operations, however, can channel the hot waste gases and use them to drive electricity-generating equipment; the resulting electricity generated can then be either utilize to meet plant power requirements or additional power if any sold into the national grid. For smaller capacity plants, by-products may not be an attractive option but for higher capacity plants, still choice may be for by-product recovery plants.

Looking into the continuous increase in blast furnace sizes in expansion in India, it is expected that non-recovery coke ovens may gradually reduce, as there will be an increased requirement for high-CSR coke.

6.5 Improvement In Conventional Coke Ovens (Slot Type Ovens)

There has been gradual enlargement in chamber height to over 8 m and in oven volume to over 90 m3. Linked to an increase in the net volume of oven chambers, the oven width has also been expanded from the conventional 450 mm to typically 590 mm. Fewer production cycles per day per oven result in a major reduction in thermal and mechanical stresses. As a larger amount of coal is held in an oven, it significantly reduces the number of pushing per day and hence emissions per ton of coke produced. Besides, it also reduces space, machinery and handling operations per ton of coke produced.

Increased oven volume is also obtained by making the ovens taller, as the width is limited by heat transfer considerations. Higher oven chamber height maximizes the battery capacity. For example, the annual output of two 4.5 m tall top-charged batteries is equivalent to a 7 m tall top charged battery.

Fugitive emissions at coke oven batteries have been drastically reduced by improved sealing systems for oven doors and other oven closures. Oven doors pressure regulation system, for the individual control of oven chamber pressure, has been developed for virtually emission-free battery operation during coking and charging. Future batteries shall be equipped with this kind of system.

7. Emerging Coke-Making Technologies

7.1 Single-chamber-system Coking Reactors[5]

Single-chamber-system (SCS) coking reactors are large-volume coke ovens that are 45 to 85

Centimetres wide. Single-chamber reactors are separate, process-controlled units with rigid walls that can absorb high coking pressure. The single-chamber design allows much thinner heating walls than in other systems. This enhances heat transfer and combustion and allows for greater design flexibility in the plant. The load-bearing capacity of the single-chamber reactor walls means that a greater range of coal blends can be charged than in conventional coke ovens. The large-dimension oven in the SCS

[5] Emerging Energy-efficiency and Carbon Dioxide Emissions-reduction Technologies for the Iron and Steel Industry, Ali Hasanbeigi, Marlene Arens, Lynn Price; Ernest Orlando Lawrence Berkely National Laboratory, 2013

design reduces environmental emissions compared to those from multi-chamber reactors. SCS coke ovens are expected to take the place of current multi-chamber coke ovens whose walls have more limited flexibility. SCSs are 38- to 70-per cent more thermally efficient than other coke ovens. The SCS technology is currently under development.

The following benefits are reported for SCS coking reactors compared to conventional multi-chamber coking reactors.

1. Reduced coking time
2. Improved productivity
3. Increased coke strength
4. Reduced space requirement, which increases design flexibility of the plant
5. Ability to use a wider range of coal, particularly coals containing less volatile matter, thus yielding a higher coke output
6. Lower coke reactivity index values and higher coke strength after reaction values compared to those of conventional systems using similar coal
7. Lower capital investment costs than for conventional ovens, particularly from implementing lean oven chambers
8. Estimated capital investment of approximately $40 million for 100,000 t of BF coke per year: estimated cost savings of 10 $/t coke or higher
9. Utilization of coke oven gas as an energy carrier, a reducing agent for iron ore, and synthesis gas in chemical plants
10. Lower maintenance costs
11. Reduced CO2 emissions

However, technology has not become popular.

7.2 SCOPE21 Process[6]

A new coke producing process named the super coke oven for productivity and environmental enhancement toward the 21st century (SCOPE21), has been developed in Japan. Nippon Steel Corporation implemented the SCOPE21 process in constructing new coke ovens at Oita and Nagoya Works.

[6] SCOPE21 Cokemaking Process; Kazuya UEBO* Makoto MATSUURA, Yukihiro KUBOTA Masaki SASAKI, Hideki FUJIKAWA Hiroki NAKAI, Kazuhide DOI Toshihiko NOGUCHI, Kenji TANIZAWA; NIPPON STEEL TECHNICAL REPORT No. 123 March 2020

SCOPE21 technology aimed at coke producing method capable of significantly enhancing oven productivity, using coal resources more efficiently (increased use of non- or slightly caking coal), and improving energy efficiency and environmental protection. It is characterized mainly by coal pre-treatment methods where coal, before being charged into the coking chambers of the coke oven, is dried and rapidly heated through pre-treatment facilities. First, coal is blended and crushed as by the conventional process, and then it is fed to a fluidized-bed dryer to be pre-heated, dried, and classified into coarse and fine coal. While the coarse grains are rapidly heated to roughly 330 to 380°C through a pneumatic preheater, the fine grains are agglomerated in heat, then mixed with the heated coarse grains, and charged into coking chambers at high temperature. During these pre-treatment steps, especially the rapid preheating, the coal is modified, and its coking properties are improved. As a result, it is possible to produce coke of a sufficient strength even at markedly high blending ratios of non- or slightly caking coal, which is inadequate as the feedstock for the conventional coke making process. A simplified flow diagram of Scope-21 is shown in the figure below:

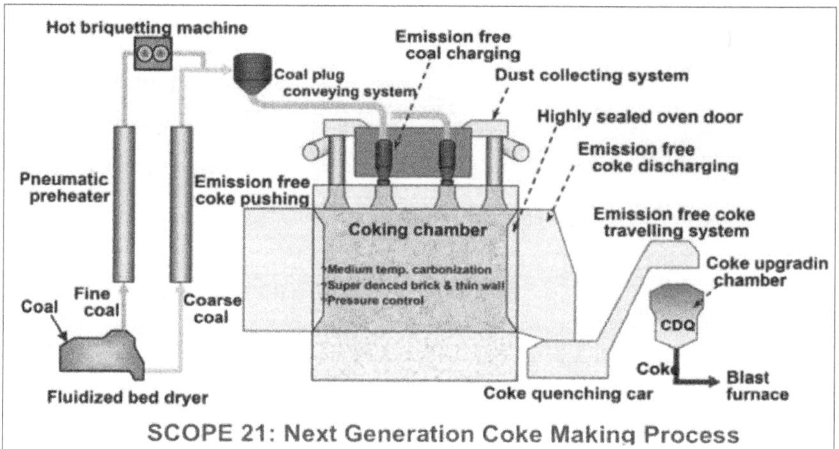

SCOPE 21: Next Generation Coke Making Process

Source: NEDA, 2008

For the SCOPE21 development, the target blending ratio of the non- or slightly caking coal was raised from 20% of the conventional wet coal charging process to 50%. Charging of hot and dry coal shortens the carbonizing time in coking chambers enhancing the plant productivity, and as a result, greatly decreases energy consumption. In addition, the agglomeration of the coal fine is advantageous for increasing the bulk density of charging coal, as is coal drying. The agglomeration serves also as a measure to prevent adverse effects of coal fines on oven operation. Owing to the above pre-treatment, SCOPE21 require a significantly shorter carbonizing time and a smaller number of coking chambers.

Though the commercial SCOPE21 plants at Nippon Steel's Oita and Nagoya Works reported stable and high-efficiency operation[7], the process has not seen wider adoption. High capital cost seems to be an inhibiting factor.

7.3 Conclusions

1. Coke is indispensable for iron and steel production through the blast furnace route. Coke making facilities, i.e., coke ovens, constitute an important part of integrated steelworks.
2. Several alternatives have been developed, and some new concepts are also emerging for making steel without coke. Initiatives like sustainable steel making, low carbon steel making etc. all envisage steelmaking without coke. But in the next decade, the BF-BOF route with coke ovens shall remain a major choice for the expansion of steelmaking capacity.
3. Coke plants which shall come up with the BFs in the coming future, shall mainly be with stamp charging and dry quenching and with more environmentally friendly facilities.
4. It can be expected that few new coke ovens shall be built in India also.
5. Thus, coke making shall continue, as blast furnaces are likely to continue till cheaper ways of production of cheap hydrogen and safe way of hydrogen transportation and storage are developed and commercialized, which may take 15-25 years.
6. Higher productivity and lower construction costs will be essential for newly built ovens.
7. To produce high-strength coke adequate for charging into blast furnaces, coal excellent in caking properties are needed, but since the resources of such good caking coal are limited, technologies that allow increased use of non- or slightly caking coal-like stamp charging will be sought to cut the coke production costs and use coal resources more effectively.

---

[7] SCOPE21 Cokemaking Process; Kazuya UEBO* Makoto MATSUURA, Yukihiro KUBOTA Masaki SASAKI, Hideki FUJIKAWA Hiroki NAKAI, Kazuhide DOI Toshihiko NOGUCHI, Kenji TANIZAWA; NIPPON STEEL TECHNICAL REPORT No. 123 March 2020

8. Fundamental studies are needed in coke structure and relationships between structure and quality. Models of prediction of metallurgical coke quality parameters based on maceral composition and properties of coals being carbonized will be essential to optimally utilize coking coals from several economically available sources.

About the Author

Dr Mukesh Kumar, Director of Steel Research and Technology Mission of India-SRTMI is a Metallurgist and got his PhD from IIT Delhi in Mechanical Engineering. He is having more than four decades of experience in Metallurgy, Oil & Gas, Technology Development, and Project Management in India and abroad. Has worked in senior management positions with organizations like Engineers India Limited (EIL), Vedanta Resources PLc, and also advised other organizations viz. Hindustan Copper Limited, Mishra Dhatu Nigam Ltd, KPMG India in the capacity of an independent consultant. Presently spearheading programs for the steel sector related to resource efficiency and circular economy, GHG reduction, productivity improvement, and development of innovative and disruptive technologies.

XII.2. The Concept of Coke Oven Doctors... Why and How to Implement It In India

Tarun Chatterjee

I would like to dedicate this Chapter to the late Dr Sergei Kaftan, the father of KOKSEKHIMSTANCE, Victor Nosov, Alexiv Royzen and my Japanese friend Nakamura of Nippon Steel & Sumitomo Metals. I had the privilege to work with all of them at different times and I consider them as my teachers who shared their knowledge that gives me the confidence to write this chapter. I also dedicate this chapter to all my Indian teachers during my service life in various plants.

1. Preamble

Way back in 1971 as a young engineer working at Bhilai Steel, I met octogenarian Dr Kaftan and he told me, "Coke Oven Battery is like a human being and always give shock and surprises. It can collapse any day without giving any prior notice. All it needs is good care like any human being. It is made of Silica Bricks of more than 500 shapes of around 15,000 MT with millions of mortar joints and protected by a strong protective anchorage. During the course of operation, thermal shocks are inevitable and added to that there are technological violations due to indiscipline which must be controlled to get the best service life of the Battery."

As I write this article I am in a bit of a philosophical mood and hence preferred to make few sketches with paper and pencil thereby putting my heart and soul into it instead of using computer-aided drawings or sketches.

2. What Should Be the Life Of A Coke Oven Battery?

In India, many believe that 20-25 years is a Good Service Life for a Coke Oven Battery.

Now, let us look at the life expectancy of the average Indian which has doubled from 34 in 1950 to 68 in 2020 and everyone knows how this happened.

Coming back to Coke Ovens let's ask the question of why the Indian Coke Oven Batteries cannot meet the Japanese Standard of 60-70 years of life and how to break the myth?

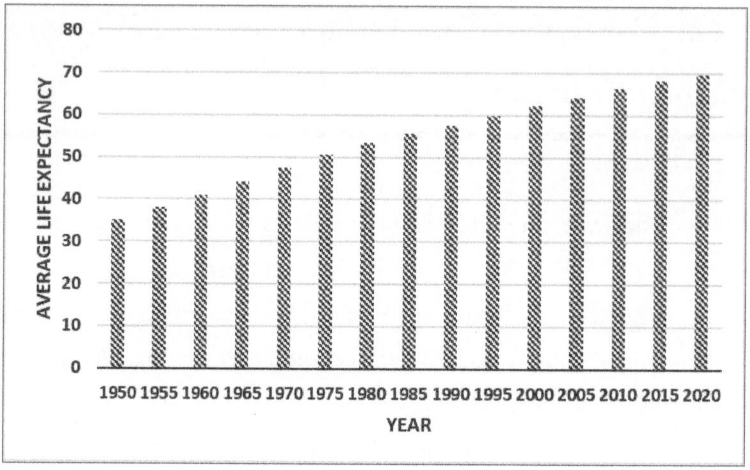

INDIAN AVERAGE LIFE EXPECTANCY 1950-2020

Chart by Ramkrishna Batabyal. Student. SNIT

Coke Ovens have always given sudden shocks and surprises to stakeholders, and it costs money (around 2000 million INR for a single 1 million tons for the battery alone) and repair or replacement takes a long time (about 18 to 24 months) causing serious problems for the integrated operation of the Steel Plants. The one and only solution is GOOD CARE from the day of commissioning.

3. What Affects the Life Of The Coke Oven Battery?

The lifeline of the Battery is the SILICA BRICK which is preferred due to its high thermal conductivity, strength and stability at elevated temperatures and other properties. But it has poor resistance for thermal shocks and other abuses. It undergoes allotropic changes on heating and is hence very sensitive. During the operation, the brickwork is subjected to un-avoidable thermal shocks and stress during the coking process and oven pushings.

The technological violations and indiscipline contribute a big way towards faster deteriorations which in medical terminology is called a **Cancer Attack** as shown pictorially in Figures 1 & 2 followed by its impact on the protective steel anchorage at Figure 3.

HOW THE CANCER DAMAGES THE SILICA BRICK WALL OF COKE OVEN BATTERYCANCER)

By Tarun Chateerejee July 2021

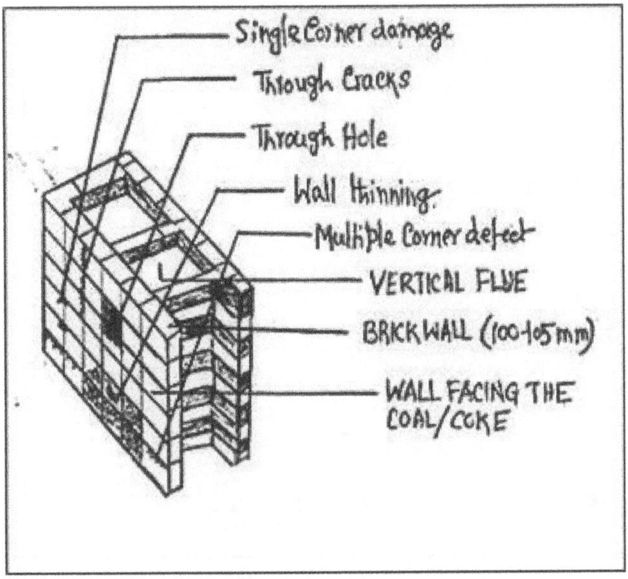

FIGURE 1 - TYPICAL DAMAGE (WALL CANCER)

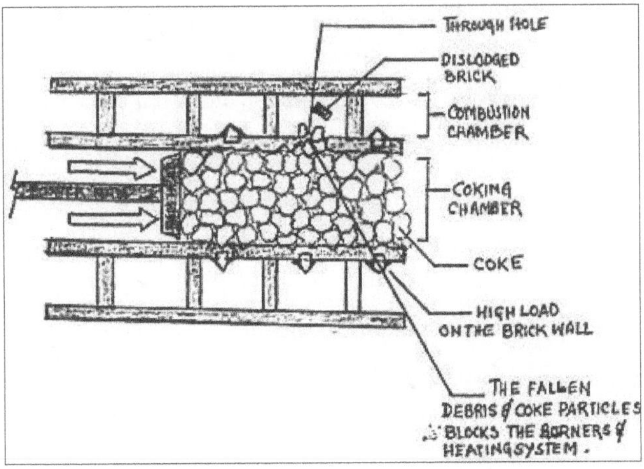

FIGURE 2 – HOW THE DAMAGE OCCURS

HOW THE CANCER AFFECTS ANCHORAGE

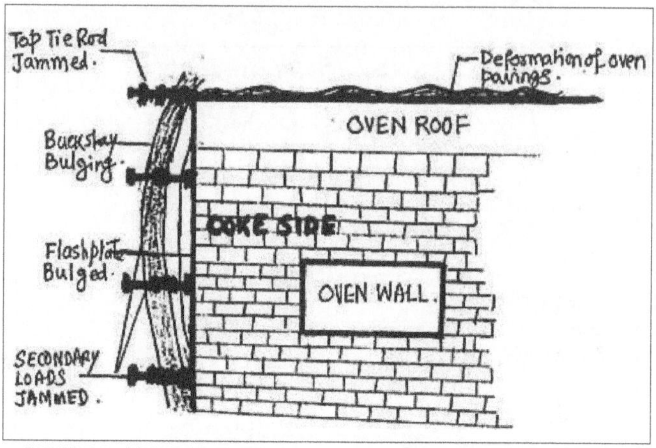

FIGURE 3 – TYPICAL VIEW OF A POORLY MAINTAINED BATTERY

Diagram By Tarun Chatterjee. July. 2021

4. What Are the Technological Violations?

Strict rules must be followed as per the documented principles and guidelines with respect to the following regimes of operation, and deviations should be considered to be Technological Violations that affect the service life of the battery

1. **BLEND REGIME:** The blend must be carefully chosen so that it does not exert excess swelling pressure (0.5 to 1 psi max for industrial value or 2 psi max as per moving wall pilot coke oven test). Ensure adequate coke shrinkage on full carbonization, Ash chemistry with low alkali content (which reduces spalling), ensure uniform moisture content 7-8% and lastly avoid frequent changes in the blend model.
2. **THERMAL REGIME:** Maintain vertical flue temperature as per set parameters twice in a shift. Maintain cross wall temperature as per the schedule and take corrective steps for any deviations. Under roof space temperature is very important as higher values will lead to carbon deposits and cracking of raw gas. The tar density and the content of high molecular weight resins (insoluble in Benzene/Toluene/Quinoline or Acetone) are also indicators of the thermal regime.
3. **HYDRAULIC REGIME:** The pressure and draft parameters for both heating gas circuit and raw gas as per set values at different locations.

4. **FLUE GAS REGIME and CROSS LEAKAGES:** Ensure full combustion of the heating gas from the flue gas analysis lab reports and regular inspection and corrective measures for the Cross-wall Leakages.
5. **PUSHING & CHARGING REGIME**: The time management for these two operations are essential and deviations beyond the set parameters are not allowed. Extra time of keeping the oven doors is strictly forbidden.
6. **ANCHORAGE REGIME:** The health survey of the flash plate, buckstays, the oven top tie rods (both transverse and longitudinal), the spring loads for the top and bottom tie rods as well as auxiliary loads as per set values and frequency of measurements.

Figures 4 and 5 depict the impact on Coke Oven Service Life due to the above violations.

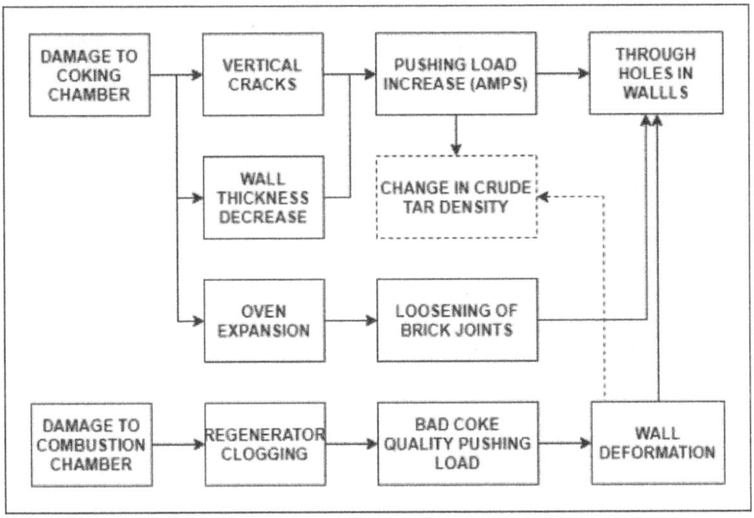

FIGURE 4 – SCHEMATIC FLOW OF OVEN DETERIORATION

(Source: Sakida, Sugiura, Nakamura, Noguchi, Awa, Nakashima, Kasai, Tsukamoto)

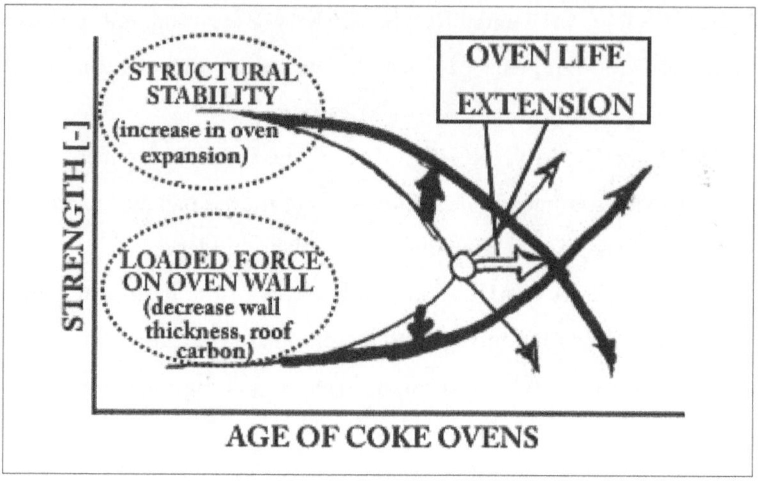

FIGURE 5 – CONCEPT OF COKE OVEN SERVICE LIFE

(Source: Sakida, Sugiura, Nakamura, Noguchi, Awa, Nakashima, Kasai, Tsukamoto)

5. What TOOLS are needed for the DIAGNOSTIC & CARE CENTRE

a) Portable Automated Flue Temperature and visual monitors (insulated, water-cooled with colour pyrometers and camera for video and still images from the Oven Top.

b) Load Cells on the Top tie rods for continuous strain measurements.

c) Liner CCD Camera for range as per oven dimensions and Laser Range Finder for Chamber Width mounted on special Service cars on the Pusher Tracks.

d) Data Processing Unit at the Diagnostic Control Centre

6. The Coke Oven Care Center for Continuous Monitoring Of Health (The Diagnosis Unit For Early Detection Of Cancer)

Thousands of data are generated round the clock which needs to be analysed for corrective actions. This data generation needs to be fully computerised without any human interface to avoid any possibilities of manipulations. Based on these data, a HEALTH CARD of the ovens and the Battery need to be prepared in the form of MIS and the monthly EXECUTIVE REPORT should be seen by the Principal Stakeholder/ Owner/ Investor.

7. The Role of Coke Oven Doctors & Care Devices

Though it may sound funny that the Japanese call the Repair Engineers, the Coke Oven Doctors, one has to understand the philosophy behind this.

The Special Service Cars should have the Repairing apparatus which based on the images obtained can reach the deformed/defective zone and carry out the repair work like flame gunning or ceramic welding. This machine is capable of taking the image and immediately carrying out the needed repair work within 4 minutes at any location inside the chamber without any human intervention. India has a large pool of highly experienced Engineers and Specialists and Repair Technologists and technicians. This resource needs to be nurtured and properly recognized.

8. From Concept to Business Model

All Concepts must be converted to a Business Model. India in 2021 has a large number of Metallurgical Coke Oven Batteries of mixed age group. New Coke Oven Projects are also taking shape at JSW, JSPL, JSSL, Arcellor Mittal etc. NMDC's Coke Plant is ready for commissioning and thus in the next 5 years, the batteries of the generation of less than 10 years will outnumber the older generation. This is a Big Market and opportunities are also Big.

India has core competence in IT-related industries and hence the business can be developed with indigenous know-how and bare minimum imported hardware components.

The Plant Owners and the Outsourced agencies can join hands for making the successful business.

This business will not only aim to enhance the service life of the coke ovens but also ensure better eco-friendly operation and funds for such projects could be supported by International Research Funds for Environment.

ANNEXURE

I. Glossary of Coal & Coke

1. Coal Classification and Rank

Coal is a sedimentary rock composed predominantly of solid organic material, with a greater or lesser proportion of mineral matter. Coal is derived from the accumulation of plant remains in a sedimentary basin and is altered to solid rock by heat and pressure applied during the basin's development. The process of gradual physical and chemical change takes place during the progressive development of coal from plant remains, through peat, lignite and bituminous coal to Anthracite. It is characterized essentially by decreasing moisture and volatile matter and increasing carbon content. The different coal varieties are products of different parent materials, different conditions are accumulation and different tectonic histories.

1.1 Peat

Peat is the first stage in the conversion of vegetable matter to coal. Bed moistures are high, often greater than 75%, and plant remains are visible.

1.2 Lignite

Lignite is low-rank coal containing high moisture. Generally, coal is considered to be lignite if it contains greater than 20% bed moisture (classification schemes for lignite are generally based on calorific value). Other characteristics of lignite are low reflectance, high volatile matter and higher oxygen and low carbon levels and often the presence of some woody structure. In general, the term is synonymous with brown coal.

1.3 Sub-Bituminous Coal

Sub-bituminous coal is the next highest coal in rank after lignite. Typical bed moisture levels are 10-20%, and calorific value also enters into the classification scheme.

1.4 Bituminous Coal

Bituminous coal is that coal which in rank is between sub-bituminous coal and semi-anthracite. Volatile matter on a d. a. f basis range from between 10% and 14%to 40% and over. Usually divided into three sub-groups low volatile, medium volatile and high volatile.

1.5 Coking Coal

Coal which because of its characteristics, is suitable for carbonising to produce blast furnace coke. Important properties required technically to produce good coke are good coking and caking properties (such as fluidity, dilatation, crucible swelling number etc.) allied with appropriate rank (indicated by reflectance values around 0. 9-1. 5%). Other properties which are important for commercial reasons are ash and moisture (should be as low as possible), and properties that are important because of the deleterious effects on pig iron are sulphur and phosphorus.

1.6 Anthracite

Anthracite is the highest rank coal and is characterized by low volatile matter (always less than 10%) and high carbon content. It has a semi-metallic lustre and is capable of burning without smoke. Semi Anthracite is coal midway between low volatile bituminous coal and anthracite.

1.7 Rank

Coal varies in composition and properties per the extent of alteration or degree of coalification of the original plant material from which they are derived. The concept of rank is used to indicate the stage of alteration obtained by particular coal, the greater the alteration, the higher the rank of the coal. Thus, lignites and sub-bituminous coals are of low rank, while semi-anthracites and anthracites are of high rank.

2. Terminologies Of Coal

COKING COAL: Coal that exhibits the behaviour of softening, caking agglomeration, subsequent swelling and the re-solidification with shrinkage, over certain temperature intervals, usually during slow heating. Bituminous coal is suitable for the production of coke. The bituminous coal exhibits plasticity on slow heating.

HARD COKING COAL: A coal that when coked alone produces a hard coke; it is normally a strongly coking low volatile or medium volatile bituminous coal.

SEMI HARD COKING COAL: A marketing term for coal that because of higher ash or poorer coking properties is marginally inferior to soft coking coal.

SEMI SOFT COKING COAL: A marketing term for coal that because of higher ash or poorer coking properties is marginally inferior to soft coking coal.

SOFT COKING COAL: A coal that when coked alone produces a relatively soft coke. It is normally high-volatile coking coal. It is often a valuable component of coking blends.

FAT COAL: Term used in several European countries for coking coals having medium volatile matter.

GAS COAL: Coal that is suitable for the production of gas.

HIGH-RANK COAL: More mature coal of the 'high-rank coal' class. It indicates semi-anthracites and anthracites. The lower limit of high-rank coal, according to the ECE3 classification system, has been proposed as 2. 0%mean random reflectance of vitrinite.

HIGHER RANK COAL: Geochemically mature Coal, as defined quantitatively in AS 2096(gross calorific value>MJ/Kg(daf)and 21 MJ/Kg(afm)). Includes the more mature sub-bituminous coals, bituminous coals, semi -anthracites and anthracites.

LONG FLAME COAL: Bituminous coal of high volatile matter.

LOW-GRADE COAL: Coal that has only limited uses owing to undesirable characteristics e. g high mineral matter content.

LOW-RANK COAL: Coal in the early stages of the coalification process.

LOWER RANK COAL: Coal that is geochemically immature as defined quantitatively in AS 2096 (gross calorific value less than 27MJ/Kg(daf)or 21 MJ/kg(afm)). This class includes the less mature sub-bituminous coals and all brown coals.

MEDIUM RANK COAL: Coal that is geochemically immature as defined quantitatively in AS 2096 (gross calorific value less than 27MJ/Kg(daf)or 21 MJ/kg(afm)). This class includes the less mature sub-bituminous coals and all brown coals.

NON-COKING COAL: A coal that does not agglomerate under slow heating conditions, e. g in a coke oven. In general, such coal has a very low crucible swelling number. Gray-King coke type and dilatation characteristics.

ORGANIC COAL: A hypothetical substance that includes only organic carbon, hydrogen, oxygen, nitrogen, and sulphur of coal.

PRIME COKING COAL: A coking coal that can be used alone or blended with poorer coking coals to produce coke suitable for metallurgical use.

RAW COAL: Coal that has received no preparation, other than possible screening or preliminary crushing.

RUN OF MINE COAL: Coal produced by mining operations, before the screening, crushing or preparation.

SOFT COAL: In classification, a term referring generally to lower rank coals. Coal is relatively friable and subject to degradation on handling. European term for coal lower rank than hard coal; coal having gross calorific value (moist mineral matter free) less than 24MJ/Kg.

STONY COAL: Highly carbonaceous sedimentary rock having dull lustre and uneven fracture.

THERMAL COAL: A marketing term for coals used for combustion purposes, e.g., for steam raising.

WASHED COAL: Coal is prepared for the market by preparation operations in addition to crushing and sizing.

3. Coking Properties

3.1 Gray -King Coke Type

A simple coking test. The finely crushed coal is heated slowly in a glass retort and the shape, texture and appearance of the coke residue are compared to standards to give a letter which is Gray -King type. Values range from A (no coking characteristics at all) to G, then G1 to G9 (superior cooking properties).

3.2 Crucible Swelling Number

One of the most common, simple caking tests. The finely crushed coal is heated rapidly in a crucible and the coke button obtained is compared with a series of standard profiles to give a number (CSN). Values range from 0 (no coking properties) to 9 then 9+ (superior coking properties). The results obtained in the test can be affected by oxidation, size consist and moisture of the sample and care should be taken in the interpretation of any results. In the US known as the free swelling index.

3.3 Petrography

MACERALS: **Macerals** are the microscopically recognizable individual organic constituents of coal. They are recognized based on their reflectance and morphology. A given maceral may differ significantly in composition and properties from one coal to another-- for some macerals, the variation depends mainly on the rank of the coal. There are three maceral groups, namely vitrinite, exinite and inertinite.

MACERALS ANALYSIS: Obtained by the microscopic examination of coal and is a volumetric distribution of macerals in a coal sample. Fine coal (not pulverized) is set into a small block of epoxy type material and one face is polished. A number of different points on polished face (usually at least 500) are examined by microscope and the macerals species observed are recorded. The maceral analysis is the result of these observations recorded. The maceral analysis is the result of these observations presented as percentages. The maceral analysis is an important parameter because vitrinite and exinite are more reactive in both combustion and coke making terms than the other species in coal.

REACTIVE: Reactives refers to the reactive macerals, i. e those which burn readily during combustion and those that become plastic during carbonization in the coke oven. Includes vitrinite and exinite maceral groups and a portion of the semi fusinite from the inertinite group.

VITRINITE: **Vitrinite** is the most common and most important maceral group in bituminous coals. Vitrinite is generally subdivided into vitrinite A, which has the normal reflectance value for the coal and vitrinite B which is of slightly lower reflectance.

Vitrinite is considered a reactive maceral, burning readily during combustion and becoming fluid during carbonization.

VITRINITE REFLECTANCE HISTOGRAM: A histogram presenting frequency of values versus reflectance values obtained during the reflectance determination. The histogram is probably the best means of assessing whether coal is a blend or not because single seam coal has a reflectance histogram that is symmetrical and covers generally only 2-3 vitrinoid types. A blend on the other hand will likely cover a greater number of vitrinoid types and present a non-symmetrical distribution. Also called a reflectogram.

VITRINOID TYPE: These are individual reflectance measurement groupings covering 0. 1%range e. g vitrinoid type, V6 includes all individual reflectance from 0. 60%-0. 69inclusive. V7-0. 70-0. 79%, V8=0. 80-0. 89%; V12=1. 20%-1. 29%etc. A vitrinite reflectance histogram (also called a reflectogram) can be drawn by plotting the number of individual measurements in each vitrinoid type versus reflectance.

EXINITE: The exinite group comprising spornite cutinite and resinite is the most reactive maceral group of coal and contains more hydrogen than vitrinite. It is derived from plant and cell secretions, cuticles and outer membranes of spores and pollen grains. and is normally the minor maceral component of coal.

INERTINITE: The inertinite maceral groups is generally considered not reactive, although some members do exhibit an amount of reactivity that is difficult to quantify. Micrinite, macronite, fusinite scleronite and ineretinite are generally considered to be un-reactive members of the inertinite group. semifusinite is partially reactive, but just how reactive is unclear but probably dependent on rank.

INERTS: Refers to those macerals which are not reactive, i. e. the inertinite group with the exception of a portion of the semi-fusinite.

REFLECTANCE: Obtained by the microscopic examination of macerals. Fine coal (not pulverized) is set into a small block of epoxy type material and one face is polished. Oil immersion of macerals at this polished face allows the measurement of using incident light. Vitrinite is by far the most common maceral for which reflectance is measured, although measurements are made for some inertinite species. Vitrinite is bi-reflective which means that with a polarizer, the reflectance varies as the microscope stage containing the block and vitrinite particles is rotated in the plane perpendicular to the incident light. Under these conditions, two maxima are observed at 180 degrees to each other. The mean maximum reflectance is thus the mean of a number of measurements of reflectance of separate vitrinite particles (usually at least 50) where the stage is rotated to give maximum readings. If the stage is rotated and the polarizer is not used, the result obtained is the mean random reflectance or average reflectance. Individual reflectance measurements fall above and below the mean and will generally cover one to three vitrinoid types in each direction. Vitrinite reflectance increases with coal rank and is a very commonly used and very significant parameter for assessing coals for coke making potential.

3.4 MAXIMUM FLUIDITY, DILATION & CONTRACTION

GIESELER PLASTOMETER: A coking test used very commonly in Japan and the USA which measures the fluidity characteristic of coal. Fine coal (not pulverized) is heated slowly and as it melts and passes through its plastic range, its fluidity is measured. Results are expressed as maximum fluidity in dial divisions per minute(ddpm). Characteristics temperatures and re-solidification temperature. The plastic range, which is the temperature range during which the coal is in its plastic state, is also important. All coking and caking tests are sensitive to oxidation but the Gieseler test is by far the most sensitive. There are two types of Gieseler plastometer, automatic and manual. The manual equipment gives results which are 50% to several times higher than that from the automatic equipment.

AUDIBERT-ARNU DILATOMETER: A caking test most commonly used in Europe that measures the expanding and contracting characteristics of coal. The finely crushed coal is compressed into a pencil, which is heated slowly and as the coal passes through its plastic range, initially gets shorter (contracts)and then gets longer(expands). Measurements taken are the maximum contraction and maximum dilation(expansion)both expressed as a percentage of the initial length, such that the maximum contraction and maximum dilation are also recorded. Results from this test are very sensitive to the oxidation of the coal being tested. The Ruhr dilatometer is similar to the Audibert-Arnu dilatometer, the major difference being in the compaction of the coal pencil. Ruhr dilatometers show less contraction and more dilation than Audibert-Arnu dilatometers.

ROGA INDEX: A coking test whereby finely crushed coal is mixed with anthracite and heated to produce a coke button, which is then submitted to a standardized drum test. The Riga Index is then calculated by formula from the amount minus 1mm material after the drum test. This index is used in the international classification as an alternative to the crucible swelling number, and can be compared as follows:

Crucible swelling number: ROGA Index

0 -½ 0-5

1- 2 5- 20

2. 5 - 4 20- 45

>4 <45

STRENGTH INDEX: An index calculated from maceral analysis and vitrinoid type reflectance data for coal which in very broad terms is related to the strength of individual granular structures in the coke resulting from that coal. Used in the prediction of ASTM stability factor.

COMPOSITION BALANCE INDEX: An index calculated from maceral analysis and vitrinoid type reflectance data of coal which is very broad terms is related to the strength of the intra-granular union in the coke resulting from that coal. Used in the prediction of ASTM stability factor.

3.5 Plastic Properties

PLASTIC LAYER: The 'fused layer 'between coal and semicoke in a mass of coking coal undergoing carbonization.

PLASTIC RANGE: The temperature range over which bituminous coal undergoing carbonization exhibits plastic properties. The Gieseler plastometer test is measured as the difference between the initial softening temperature and the solidification temperature.

PLASTICITY: The degree of softening experienced by some coals under the influence of heat. It can be measured by the Gieseler plastometer test.

4. Proximate & Ultimate Analysis

4.1 Proximate Analysis

The analysis of coal or coke in terms of moisture, ash, volatile, matter and (by difference) fixed carbon.

ASH: The inorganic residue after the incineration of coal to constant weight under standard conditions is less than the mineral matter because of the chemical changes occurring during incineration, with the most important differences being the loss of water of hydration, loss of carbon dioxide and loss of sulphurous gas from sulphides.

ASH-ANALYSIS: Ash is composed of very complex oxides and the ash analysis expresses this composition in terms of its component oxide. Generally, the ash consists mostly of silica and alumina. The presence of a large amount of the oxides of iron ($Fe2O3$), calcium (CaO), sodium ($Na2O$) Potassium ($K2O$) generally indicates ash with low ash fusion temperatures. The ash analysis differs from the composition of the minerals in the parent coal.

ASH-FUSION-PROPERTIES: The fusion properties of laboratory prepared coal ash are demonstrated by the heating of the ash in a mildly reducing or oxidizing atmosphere. The temperature range generally used is 900 degrees C up to 1600 degrees C. The temperature which can be recorded is initial deformation temperature, softening temperature, hemisphere temperature and fluid flow temperature. Almost invariably temperatures recorded under reducing atmosphere are lower are equal to those recorded under oxidizing atmosphere. Of the characteristic temperatures the initial deformation and flow generally the most difficult to reproduce.

VOLATILE MATTER: The loss in mass, less than due to moisture, when coal is heated under standard conditions and out of contact with air. This test is very empirical and results are very sensitive to operating conditions. Results obtained from different standard methods will not necessarily give the same result. The ASTM method generally gives higher result than other methods.

TOTAL MOISTURE: The moisture in the coal as sampled and removable under standard conditions.

BED MOISTURE: The moisture content of coal as it lies in the seam prior to mining. The moisture holding capacity is often said to be equivalent to the bed moisture. Equivalent to inherent moisture in ASTM. Also sometimes referred to as the in-situ moisture.

INHERENT MOISTURE: Often used to indicate air dried moisture, particularly in Australia. In ASTM equivalent to bed moisture or equilibrium moisture.

SURFACE MOISTURE: The difference between total moisture and moisture holding capacity. Called free moisture in.

FREE MOISTURE: The difference between total moisture and air-dried moisture. No equivalent in ASTM.

AIR DRIED MOISTURE: The moisture in the coal sample after achieving equilibrium with the laboratory atmosphere by exposure to it. Called residual moisture in South African standards. No equivalent term in ASTM, although air-dried moisture can refer to sample dried in an oven with warm air.

MOISTURE HOLDING CAPACITY: The moisture in the coal after achieving equilibrium with an atmosphere of 96% relative humidity and 30* C called equilibrium moisture in ASTM.

4.2 **Ultimate-Analysis**

The moisture in the coal is expressed in terms of carbon, hydrogen, nitrogen, sulphur and oxygen. The analysis refers to the carbonaceous material only and hence is often expressed on a dry ash-free basis or dry mineral matter free basis. Oxygen is generally estimated by difference although there are methods for its determination.

SULPHUR: Can be part of the carbonaceous material in coal or part of the minerals as sulphates or sulphides. forms sulphur dioxide during coal combustion which is a serious pollutant, most countries having regulations regarding emissions to the atmosphere. Undesirable in coking coal because it partly accumulates in the hot metal and requires desulphurization to remove if above about 0.02%.

FORMS OF SULPHUR: Sulphur in coal is considered to be present in three forms pyritic sulphur, sulphate sulphur and organic sulphur. Where the sulphur is pyritic or sulphatic it is part of the mineral matter and its content can be lowered by washing the coal. Organic sulphur is distributed through the carbonaceous part of the coal and cannot be washed out by conventional techniques.

Analytical methods are well known for pyritic and sulphate sulphur, but organic sulphur is determined by a difference from the total sulphur value. The presence of sulphate sulphur often indicates the coal has experienced some oxidation during its life. Pyritic sulphur is often blamed for spontaneous combustion problems.

PHOSPHORUS.: To be avoided in coking coal because it accumulates in the hot metal and gives undesirable properties to the resultant steel. Can create problems also during combustion by the formation of hard phosphate deposits inside boilers.

FIXED-CARBON: A component of the proximate analysis is calculated by difference. i. e. 100% less the sum of moisture, ash and volatile matter. Intended to vive and indication of char field. Not to be confused with carbon in the ultimate analysis.

NITROGEN: Part of the organic material is coal. Under certain combustion conditions, a portion will be emitted to the atmosphere as the polluting nitrogen oxides unless removed from flue gas.

MINERAL MATTER: Mineral matter comprises the inorganic components of coal. When coal is burnt the composition of the mineral matter is altered by the evolution of the water of hydration of the minerals (this is not determined during the total moisture determination) and other components such as carbon dioxide and sulphur oxides. The end product, the ash, thus has a different composition to original mineral matter. The mineral matter content of coal is difficult to determine, but there are well known equations to estimate it.

ALKALI MATTER: The most important ones in coal are sodium and potassium. They are undesirable in steaming coals because they can be given rise to fouling and slagging problems in boilers, although small additions can improve electrostatic precipitation performance. They are undesirable in coking coals because, they tend to increase coke reactivity in the blast furnace.

TRACE METALS: Also, often called heavy metals. Both terms are not strictly correct but generally refer to those elements in coal which are considered harmful to the environment and are present in very small quantities. Examples are mercury, arsenic, selenium, fluorine, cadmium

CHLORINE: An element in coal which if present in large enough quantities can cause corrosion problems in the boiler under poor operating and maintenance conditions. Less than 0. 2% chlorine is considered low whilst over 0. 5% is considered high. When sodium is associated with chlorine it is often suggested the chlorine can play a role in fouling during combustion.

4.3 Other Laboratory Measurements

AS RECEIVED BASIS: Analytical data calculated to the as received moisture content.

AIR-DRIED BASIS: Equivalent to as analysed basis when the analysis have been performed on an air dried sample.

DRY BASIS: Analytical data calculated to a condition of zero moisture.

DRY ASH FREE BASIS: Analytical data calculated to a condition of zero moisture and ash (i. e a first approximation to pure coal) to allow comparison of different coals. This is strictly a hypothetical basis because the ash is only being generated on the incineration of the coal but is used frequently because of convenience. Dry mineral matter free basis is more precise, but less easy to obtain.

DRY MINERAL MATTER FREE BASIS: Analytical data calculated to a condition of zero moisture and mineral matter and a better approximation of "pure coal". Is used for comparing different coals and estimating coal rank. Where a mineral matter has not been determined or is difficult to determine it can be estimated by calculation.

REPEATABILITY: Reproducibility (between laboratories) for an analytical procedure is the maximum acceptable difference between the mean of acceptable replicate determinations carried out at different times in the same laboratory on the same analysis sample by the same operator using the same apparatus. This is sometimes called within laboratory tolerance.

ABRASION INDEX: Indicates the abrasiveness of coal by monitoring the loss in weight of four metal blades that mechanically stir a sample of sized coal. The index is the number of milligrams of metal abraded from the metal blades per kg of coal used. Based on a test known as the Yancy, Geer and Price test.

DRUM TESTS: Used for examining the strength properties of coke. Sized coke is placed in the drum which has lifters on the inside. As the drum is rotated for a set time or number of revolutions the pieces of coke are lifted and then dropped to the bottom of the drum. The coke is then sized again, and the various drums indices are obtained from the amount of oversize or undersized material. Commonly used drum tests are ASTM, Micum, IRSID and JIS.

HARDGROVE GRINDABILITY INDEX: Indicates the relative grind ability or ease of pulverization of coal in comparison to coals chosen as standards. High values indicate coal is hard to pulverize and low values indicate coal is hard to pulverize. Hard grove grind ability index is rank dependent and increases as does rank, although anthracites have low Hard grove grind ability indices.

(FOR METHOD OF TESTS AS PER INDIAN STANDARDS REF. I. S 1350)

4.4 Common Terminologies for Metallurgical Coke

COKE: Coke is the agglomerated product of coal carbonization, generally at a temperature above 900 degrees C. Carbonization is the process of de-volatilization of finely crushed coal on heating in absence of air thereby producing carbon enriched solid product namely coke, liquid products namely tar, benzol etc and gaseous product namely coke oven gas.

CARBONIZATION: The carbonization of coal is carried out commercially for the production of coke for the Iron and Steel Industry's Blast Furnaces. The process involves heating coking coal in the absence of air to a temperature of around 1100degree C. this is carried out in a Coke Oven and the Coking time is around 14-20 hours. The gases and liquids that evolved during the process are used on sight for their energy contained or solid as plant products. In a process very little used today, town gas can also be prepared by carbonization, with the resulting coke being used as a domestic fuel.

MICUM OR COLD STRENGTH OF COKE: A standard tumbler test for determining coke strength. A 25 kg sample of plus20mm is tumbled in a specified drum for 100 revolutions and strength indices are calculated from a size analysis of the product. The M40 index is the percentage of coke plus 40 mm; The M10 index is the percentage of coke plus 40 mm; The M10 index is the percentage of coke minus 10 mm.

COKE REACTIVITY: A laboratory test designed to stimulate the loss of coke through reaction in the reducing atmosphere, as the coke makes its way down the blast furnace. Coke is heated up to 950 degrees Celsius in an inert atmosphere and held at that temperature in an atmosphere of carbon dioxide. The coke is cooled down under the inert atmosphere and the loss in weight expressed as a percentage is the reactivity.

COKE STRENGTH AFTER REACTION: A laboratory test designed to indicate the strength of coke after being exposed to the reducing atmosphere of the blast furnace. Coke after exposure to the higher temperature and carbon dioxide atmosphere of the coke reactivity test, is tested in a tumbler device to determine its strength.

POROSITY OF COKE: The volume fraction occupied by the pores in the pieces of coke. It is defined by the following expression:
1 - (relative density of lump sample) / (relative density of analysis sample

GRADES OF COKE:

1. BLAST FURNACE COKE: Coke specially prepared for use in blast furnaces; Usually strong coke having a narrow size range.
2. BLOCKY COKE: Approximately cubic pieces of coke.
3. FINGER COKE: Long narrow pieces of coke or coke that break into long pieces
4. FORMED COKE: Coke formed by the coking of agglomerates (usually briquettes) of coal, or partly carbonized coals or chars, to produce coke lumps of uniform size and shape determined by the size and shape of the original

agglomerates.

5. **GREEN COKE:** Coke that has not attained the required carbonization temperature in the coking process.
6. **HARD COKE:** Strong coke produced from low volatile or medium volatile coals at high temperatures.
7. **HIGH TEMPERATURE COKE:** Coke produced where the minimum temperature in the charge during carbonization exceeds 900 degrees C.
8. **LOW TEMPERATUR COKE:** Coke that has been formed by the influence of an igneous intrusion or other natural processes.
9. **NATURAL COKE:** Coke that has been formed by the influence of an igneous intrusion or other natural process.
10. **SIZED COKE:** Coke screened between specified particle size limit.
11. **SOFT COKE:** Low strength coke having a sponge like appearance.
12. **SPONGY COKE:** Low strength coke having a sponge like appearance.
13. **BREEZE COKE:** Small coke generally below 10mm in size.
14. **NUTS COKE:** Graded coke of size varied in local practice. The top size is generally of the. order. of. 25mm, and the lower size 10mm.
15. **BEEHIVE OVEN COKE:** A coke oven in which a mass of coal is carbonized by heat produced by combustion of the coal volatiles above the supper surface of the mass. Because volatile materials are burnt, no by products are obtained.
16. **BY PRODUCT OVEN COKE:** A coke oven in which a mass of coal is carbonized by heat conducted through the oven walls from adjacent flue chambers in which a battery of alternating oven chambers and flue chambers. Volatile materials are collected and recovered as gaseous and liquid by products.
17. **COKING PRESSURE:** The maximum pressure, expressed in kilopascals, exerted by the oven change on the walls of the oven during coking. It is recommended that the area in contact with the coal be reported along with the coking pressure. In reality, the area of coke in contact with the oven wall at the time of maximum pressure may be much less than the original contact area.

A measure of the rate at which a coal charge is heated in a by-product coke oven. It may be calculated in two ways, as follows:

The width of the oven is divided by the time between charging and pushing the oven, expressed in millimetres per hour.

The width of the oven is divided by the time between charging and the attainment of a specific centre of charge temperature (900 degrees C to 1000degree C)

COAL AND COKE ASH:

Analytical: The inorganic residue after the incineration of coal or coke to constant mass under standard conditions. In general, it differs in mass and composition from the original mineral matter.

Industrial: The residue of predominantly inorganic matter after the industrial combustion of coal or coke.

Note: In the case of lower rank coals, ash can differ considerably from the mass and composition of the mineral matter in the coal and can also vary conditions and sampling procedure

1. **ASH BOTTOM:** The ash deposited in the bottom of the furnace is distinct from the fly ash that is carried out with the products of combustion.
2. **ASH EXTRANEOUS:** Ash arising from that part of the mineral matter associated with but no inherent in coal.
3. **ASH FURNACE:** A mixture of ash, unburned fuel, riddling, and clinker resulting from the combustion of coal or coke, which collects in furnace bottoms as distinct from fly ash.
4. **ASH INHERENT:** Ash arising from finely dispersed mineral matter present in the original plant material or from which the solid mineral fuel was formed or from mineral matter incorporated intimately in the sold mineral fuel during the coalification process.
5. **ASH INSTANTANEOUS:** The maximum ash percentage of any particle in any given relative density fraction mass.

6. **ASH CURVE, CHARACTERISTIC:** Any curve, obtained from the results of a float and sink analysis, that relates cumulative ash percentage to cumulative floats (or sinks) mass percentages. The cumulative ash percentage is plotted on the abscissa (horizontal axis) with the cumulative floats or sink is mass percentage plotted on the ordinate (vertical axis). Standard was ability curves of this type are the cumulative floats curve, the cumulative sinks curve and the instantaneous (incremental) ash curve. Yield ash curves, for either total plant or individual separating device performance predictions, are also of this type.
7. **ASH FREE BASIS:** In expressing an analysis, a basis that refers to a hypothetical condition in which the coal or coke assumed to be free of ash.
8. A hypothetical condition in which the sample is assumed to be free of ash, with moisture determined by its moisture holding capacity, used in classification systems, e. g. AS 20096, especially for the calorific value.
9. The partial or complete melting of ash constituents, possibly leading to the formation of clinker or fused or partially fused furnace ash deposits.

II. Glossary of Shipping Terminologies

1. Shipping Abbreviations and Terms

ABBREVIATIONS	DEFINITIONS
Afloat	Cargoes onboard vessel en route to a destination
ARA	Antwerp / Rotterdam / Amsterdam
Arbitration	Out-of-Court settlement of points of controversy by one or more professional men appointed by the two parties.
Ballast	A term to describe a vessel sailing without cargo but taking water into her tanks to give stability.
Bends	Short for 'both ends'. Used when any loading and discharging term for a vessel is the same for both loading and discharging, e.g., 6000 mt per WWD SHINC BENDS
Bill of lading (B/L)	A receipt for goods shipped on board a vessel – signed by the master or his agent; subsequent possession of this document gives the ownership of the cargo described to the holder.
Bunkers	Refers to the fuel used by the vessel.
Capesize	A vessel about 120,000 dwt.
CFR	Cost and Freight. A C & F price includes the cost of the cargo and the freight.
CIF	Cost, Insurance and freight. A CIF price includes the cost of the cargo, its insurance and freight.
dwt	Deadweight tons. The weight which a vessel is capable of carrying by way of cargo, plus bunkers, stores and freshwater, when loading to the maximum permitted marks. This is measured in tons of 2,240 lbs.
Demurrage	An agreed daily amount of money payable to owners by charterers is a penalty if the vessel takes more than the laytime for loading or discharging.
Despatch	The opposite of demurrage., an amount payable by the owner to the charterer for loading or discharging a vessel more quickly than the laytime. Traditionally half the demurrage rate.
Draught	The depth to which a floating vessel is submerged. When a figure is quoted in a vessel's description this refers to her being fully loaded including stores and water.

ETA	Estimated time of arrival.
ETR	Estimated time of readiness.
FOB	Free onboard. A FOB price does not include insurance and freight.
Force Majeure	Circumstances beyond reasonable control. Not a term of art in English law.
Gearless	The vessel has little or no means aboard for loading and discharging cargo.
Handysize	A vessel of about 45,000 dwt
Knot	The measure of speed, being one nautical mile of 6,080 feet traversed in one hour.
Laid up	A vessel being moored through lack of employment usually without full crew and with engine sealed.
Laydays	A specific period in days during which the ship must arrive and be ready for loading.
Laytime	The time available for loading or discharging a vessel's cargo without incurring demurrage.
LOA	Length overall.
LWT	Lightweight tons. This is the weight of steel in a ship and is usually given when a ship is sold for scrap when the price is given in dollars per lightweight ton.
NOR	Notice of readiness. This is tendered by the vessel to the receiver's agent (usually during office hours) to indicate it is ready for discharge. The time of tender is the start of turn time.
OBO	Ore / Bulk / Oil Carriers.
Panamax	A vessel of about 65,000 dwt
Running Days	Days of 24 consecutive hours including holidays, Saturdays, Sundays unless excepted, i.e., a vessel is told she will be loaded in 4 running days, whether or not she is worked upon; an alternative is to say 4 WWD SHEX – four weather (permitting) days Sundays and holidays excepted.
SATSHEX	Saturdays, Sundays and holidays excluded.
SF	Stowage factor. Used to determine whether a vessel's holds will be filled before the vessel reaches her maximum draught. The SF for coal, depending on the type, may be between about 37 and 48 cubit feet per long ton. A vessel's maximum cargo is determined by diving

	the vessel's cubic capacity by the cargo's SF. It may be expressed as m3 /mt.
SHEX	Sundays and holidays excluded. This term usually follows a loading or discharging per day figure.
Timecharter	The charterer arranged for a fixed period. Payment is either per day or per deadweight ton per month, but it excludes voyage costs. The charterer has the use of the vessel with the shipowner supplying the crew and provisions.
Trimmed	The trimming of cargo is when it is levelled off after loading.
Trip - Timecharter	The charter arranged for a single voyage, often with destination loosely described, e.g., Spanish Mediterranean, but redelivery point specified, e.g., pass Gibraltar. Payment calculated as per time-charter.
IT	Turn time Quoted in hours, being the time between the tendering of notice of readiness (NOR) and laytime commencing.
ULCC	Ultra-large crude carrier. Generally, means a crude oil tanker of 350,000 dwt.
Voyage Charter	Charter arranged for a single voyage with agreed loading and discharging ports. Payment is per metric tons of cargo loaded and includes all voyage costs.
WWD	Weather working days. This is usually qualified by SHEX or SHINC and means the exclusion of time lost through bad weather.

2. Average Bulk Ship Sizes

SIZE (DWT)		FEET			METRES	
10,000	427	64.1	26.1	130	19.6	8.0
20,000	513	74.7	31.2	157	22.8	9.5
25,000	568	78.7	33.0	173	24.0	10.1
30,000	600	90.9	34.3	183	27.7	10.5
40,000	623	94.8	36.5	190	28.9	11.1
50,000	682	101.7	39.8	208	31.0	12.1
60,000	738	105.6	41.0	225	32.2	12.5
80,000	840	107.3	46.1	256	32.7	14.0
120,000	863	131.4	53.6	263	40.1	16.3
150,000	942	146.7	55.7	287	44.7	17.0
200,000	1014	167.3	61.2	309	51.0	18.7

3. Draught Surveys

3.1 Introduction

Draught surveys are made to establish the weight of bulk cargoes, such as coal, in ships. They cover the largest bulk carriers down to the small river barges. The net weight of the cargo is the corrected loaded displacement (from a draught survey after loading) less the corrected light displacement (from a draught survey prior to loading). The concept of a draught is to measure the weight of the vessel by estimating the volume of water that is displaced by the hull as it floats. It is therefore important that the attitude of the ship is considered, i.e., is it on the even keel. As well as the depth at which the ship is floating. It is important also that the density of the water that the ship is floating in should be considered.

3.2 Reading of Draught

The first step in draught survey is the measurement of the draught of the vessel which is read at the fore and aft end as well as midships on both port and starboard sides. These measurements are reduced to a single value by a method of averaging (not simple average) which takes into consideration any deformation of the ship. This can take the form of sagging when the vessel is bowed like a banana with ends further out of the water than the midships. Hogging, the other form of deformation is the opposite case.

The single value of a draught is referred to as the mean of means corrected for deformation.

3.3 `Correction of Displacement

The vessel's displacement table lists the draught versus displacement. The displacement of the vessel is thus obtained using the mean of means corrected for deformation. This displacement value assumes two circumstances – that the vessel is lying on an even keel and is lying in fresh water at a particular temperature, normally 4 degrees centigrade (at which the specific gravity of water is 1,000). Corrections are normally applied to cover these assumptions, and they are known as the trim correction and the density correction.

Other Titles by the Author

100 years (1914-2014) of Metallurgical Coke Making in India

(for more information, contact the Author)

The Durgapur Projects Limited User's Manual on COAL

(for more information, contact the Author)

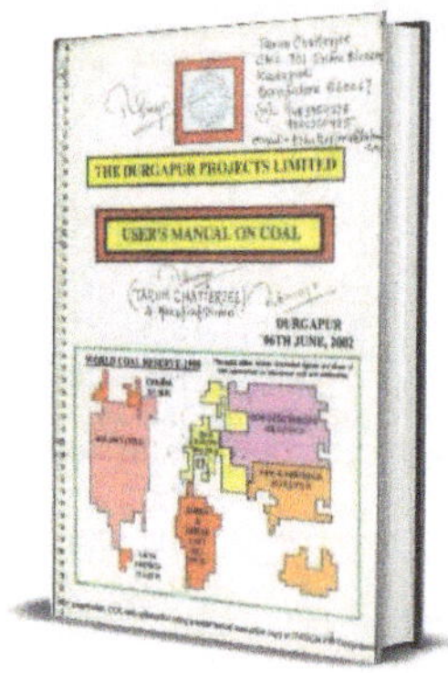

Gambling with Life

https://www.amazon.in/Gambling-Life-Tarun-Chatterjee/dp/1685093817/

https://www.flipkart.com/gambling-with-life/p/itm98c9224fbff07

STEEL TECH

STEEL TECH is a technical magazine published quarterly (January/April/July/October). This was started way back in 2006 under the leadership of Late Dr. Amit Chatterjee (as Chief Editor) a metallurgist of international repute who worked in Tata Steel in various position including MD of Tata Sponge & Mr. B P Sarkar, a metallurgist, ex-Tata Steel.
Currently Pritish K. Sen, ex-Tata Steel, the Chairman of MSME and Manufacturing Committee of Bengal Chamber of Commerce & Industries and Advisor and Interim In-charge of INSDAG is the Editor of STEEL TECH.

This magazine has been adjudged as the best technical magazine due to its rich technical content and quality and getup. The journal helps to remain abreast with the latest technologies in the iron & steel sector, global news on steel industry, trends to evolve suitable strategies for the decision makers. This has wide circulation both in domestic and overseas.

It has a strong Editorial Team and Advisors consisting of eminent personalities from Industry, Consultancy, Academia and Research Institutions. We also have a strong technical team with substantial experience of industry, marketing, external affairs, consultancy and research to provide professional services.
We also conduct international seminars on contemporary topics covering steelmaking technologies, raw materials, ferroalloys etc.

STEEL TECH

Pritish K. Sen
Sunny Apartments, 61 E, Manick Bandopadhyay Sarani
Kolkata – 700 040; INDIA
Email: infor@steeltech-India.com, psen2008@yahoo.co.in
Website www.steeltech-india.com
Contact: +91 9230639343/ 8777341510

TRUSTED NAME
IN STEEL & MINING INDUSTRY
META LOGIC
EXPLORE
MYMETALOGIC.COM
NEWS | INTERVIEWS | WEBINARS | REPORTS | RESEARCH
FOR STEEL & MINING VIDEOS
YOUTUBE.COM/METALOGICPMS
visit our website for more
METALOGICPMS.COM | MYMETALOGIC.COM
METALOGICPMS.COM | MYMETALOGIC.COM